EX·LIBRIS

Mary Jaffe

RELIGION IS NOT ABOUT GOD

RELIGION IS NOT ABOUT GOD

HOW SPIRITUAL TRADITIONS
NURTURE OUR BIOLOGICAL NATURE
AND WHAT TO EXPECT
WHEN THEY FAIL

LOYAL RUE

Rutgers University Press
New Brunswick, NJ and London

Library of Congress Cataloging-in-Publication Data
Rue, Loyal D.
Religion is not about God : how spiritual traditions nurture our biological nature
and what to expect when they fail / Loyal Rue.
p.   cm.
Includes bibliographical references (p.   ) and index.
ISBN 0-8135-3511-5 (hardcover : alk. paper)
1. Psychology, Religious.   2. Genetic psychology.   I. Title.
BL53.R82   2005
200′.1′9—dc22
2004007531

A British Cataloging-in-Publication record for this book is available
from the British Library

Manufactured in the United States of America

CONTENTS

PREFACE

I grew up in a home where the religious life was taken very seriously, and where the assumption that religion was all about God was never challenged. But somewhere along the line I began to see that behind the stories, hymns, and prayers there were serious human motives at work. Without these motives, I reasoned, there would be no religion. It was then a short step to the conclusion that human motives were ultimately both the source and the substance of the religious life. Religion was not really about God after all. Oddly enough, I have no clear recollection of when or how this fundamental change in perspective took place. It just occurred to me one day that I had been thinking this way for many years. From that day onward it was just a matter of time before this book would appear.

This book has been helped along by many conversation partners over the years. There is no way to list them all, but the ones that come most quickly to mind include the following: Thomas Berry, David Bishop, Michael Cavanaugh, Eric Chaisson, Terrence Deacon, Ursula Goodenough, John Grim, Marc Hauser, Philip Hefner, Jerome Kagan, Gordon Kaufman, Alan Macdonald, Bill Orme-Johnson, V. V. Raman, Philip Reitan, Michael Ruse, Conrad Røyksund, Kent Simmonds, Tex Sordahl, Terry Sparkes, Philemon Sturges, Lawrence Sullivan, Brian Swimme, Mary Evelyn Tucker, Edward O. Wilson, and Richard Wrangham.

In addition, the 1997–1998 fellows at Harvard's Center for the Study of World Religions were particularly helpful, as have been members of the Institute on Religion in an Age of Science. Periodic financial support from the John M. Templeton Foundation has been most helpful and greatly appreciated. Audra Wolfe, science editor at Rutgers University Press, deserves my gratitude for her expertise and support. A special thanks goes to Jennifer Tomscha for putting finishing touches on the manuscript and to Lyman Lyons for his able and meticulous copyediting. Finally, and yet again, I thank my family for their loving endurance and support.

Decorah, Iowa
Spring 2004

RELIGION IS NOT ABOUT GOD

INTRODUCTION:
RELIGION GENERALIZED
AND NATURALIZED

If religion is not about God, then what on earth *is* it about (for heaven's sake)? It is about *us*. It is about manipulating our brains so that we might think, feel, and act in ways that are good for us, both individually and collectively. Religious traditions work like the bow of a violin, playing upon the strings of human nature to produce harmonious relations between individuals and their social and physical environments. Religions have always been about this business of adaptation, and they will always remain so. This is not to say, however, that any particular religious tradition will remain adaptive. Religions sometimes outlive their adaptive utility and occasionally become positive threats to human survival.

It is one thing to make bold assertions like this and quite another to back them up. The purpose of this book is to do just that—to show how the ideas, images, symbols, and rituals of religious traditions have been designed to engage and to organize human neural systems for the sake of human survival, and then to examine the contemporary conditions that have compromised their adaptive utility. The three parts of the book attempt to pull together three distinct arguments.

The argument of part one is that a universal human nature exists and can be known by examining our evolutionary story, especially the evolution of behavior. The argument continues by showing how cognitive and emotional systems work together, and how they are conditioned by cultural influences. Part one ends with a transitional chapter on the nature of religion, arguing that religious traditions are best understood in terms of their powers to nurture our cognitive and emotional systems toward the achievement of personal and social well-being.

Part two seeks to validate this theory of religion with references to five of the world's great religious traditions. The argument advanced in part three is that traditional religious orientations are faced with a dual crisis of intellectual plausibility and moral relevance, and are therefore rendered incapable of nurturing human nature in ways that might prevent unsustainable patterns of human population and consumption from triggering a cascade of global environmental disasters. This argument finally gives way to speculations about the future of religion. An ambitious book, perhaps, but not a very complicated one. Each of the three parts opens with a brief introduction that maps out the territory and summarizes the argument. I encourage readers to page ahead to those sections for a more detailed orientation.

It is already clear that the key factor holding these three arguments together is the theory of religion presented at the end of part one. It may therefore be useful to consider what is formally at stake in this theory, and in the process to declare a few personal biases that readers will have to contend with. I will be proposing a general and naturalistic theory of religion.

By a *general* theory I mean one that tells us what religion is, where it comes from, and how it functions. General theories are premised on the belief that universal properties of structure and function can be found lurking behind the varying details of religious phenomena. The goal of a general theory is to show that all religious traditions may be seen as particular variations on a set of common themes.

By a *naturalistic* theory I mean one that reduces religious experiences and expressions to the status of natural events having natural causes. As such, a naturalistic theory of religion seeks to understand religious phenomena by using categories, concepts, principles, and methods compatible with the ones normally applied to non-religious domains of human behavior. Briefly stated, the central claims are: First, that it is possible to construct a satisfying general account of religion; and second, that this can be done without invoking supernatural principles of explanation.

DISCLAIMERS

I begin with three important disclaimers. First of all, this book is not meant to be hostile to the idea of God. I will not be arguing either for or against the existence of God. Perhaps there are gods, perhaps not. I will not pretend to know one way or the other. The question of God's existence simply doesn't come into the business of understanding religious phenomena. Both the existence of God and the non-existence of God are perfectly consistent with the claim that religion is essentially about fiddling on the strings of human nature. There is much to be said for the thesis that all theological formulations are equally and utterly dubious for the simple reason that God is inscrutable. The measure of a religious orientation is therefore not whether it gives an accurate account of divine reality, but whether it effectively manages human nature. It could be argued, of course, that religion would lack the power to manage human nature unless it is believed to offer truths about God. This may be the case, but even so, it is easy to see that *belief* is the thing, not the reality of any objects of belief. The religious question, then, is completely independent of the theological question.

Second, this book is not meant to be hostile to the religious life. Indeed, I hope the opposite message will come through clearly—that is, I regard religion generally to be a salutary thing. Religious phenomena are everywhere present in human life, and will undoubtedly remain so. As far as anyone can tell—and there is plenty of evidence to the point—there has never been a coherent human culture without a religious tradition. Religion comes naturally to human beings. It is a "given," an important universal feature of human affairs, God or not. This book should therefore not be seen as an attempt to undermine religious sensibilities. If anything, it hopes to kindle insights that will enable us to deepen them.

Finally, the thesis of this book is not original. The claim that religion is not about God has been advanced many times in the past, notably by the likes of Kant, Feuerbach, Marx, Durkheim, and Freud. Each of these authors believed, as I do, that regardless of what religion *says* it is about, it has to *do* fundamentally with meeting the challenges to a full life. Kant thought that religion was about achieving rational coherence in human experience; Feuerbach believed that religion was a covert way of coming to terms with self-alienation; Marx thought religion was about coping with the dehumanizing consequences of economic exploitation; Durkheim associated religion with a veneration of the social order; and Freud described religion as the projection of deep psychological dynamics. In each of these views the claim is that religion is about *us*, not about God. This book offers a fresh iteration of the thesis on the warrant that new insights into human nature have cleared a path toward a new theory of religion.

CAN RELIGION BE GENERALIZED?

The feasibility of general theories of religion is open to question, and there are good reasons to discourage the pursuit of such theories from the start. Indeed, the weight of informed scholarly opinion currently favors a moratorium on general theories. In his excellent treatment of these matters, Daniel Pals claims that "the course of the most recent discussions in the theory of religion has only deepened doubts and multiplied hesitation about all general formulations."[1] The sheer diversity of religious phenomena is itself discouraging. In the course of human history thousands of religious traditions have appeared, each with distinctive patterns of meaning. Some speak of thousands of gods, others speak of mere dozens, still others speak of only one, while some recognize no gods at all. Some religious traditions are rich, even baroque, in symbolic and ritual convention, while others are minimalist and informal. Some are militantly dogmatic, others tolerant. Some religious orientations are focused on community, while others center on the solitary individual.

The complexity of religious phenomena is no less daunting. It is difficult to identify any domain of human interest and activity where religious issues are not at stake. Politics, economics, personal morality, health, education, birth, death, sexuality, art, science—all of these, in some measure, affect and are affected by religion. Religion, then, is as large and complex as life itself. Many cultures, in fact, do not possess a word for religion, thereby lacking the means to distinguish religious phenomena from the rest of human experience and expression. Given the extremes of diversity and complexity associated with the religious life, one might reasonably doubt the prospects for a satisfying general theory. Theories broad enough to contain such extremes are likely to sacrifice substance and insight to vagueness, while theories that are specific in describing religious life will run the risk of neglecting or distorting relevant facts.

Each and every religious tradition is unique—unique in its cultural setting and historical development, unique in the set of challenges it has faced, and (perhaps most importantly) unique in the experiences and the constellations of meaning it has generated in the lives of individual men and women. Serious regard for the unique complexity of religious orientations has provoked a scholarly reaction against general theories in favor of a "particularist" approach to the phenomena of religion.[2] If I understand the particularist view correctly, it goes something like the following. To have a theory about something is to describe and explain what the thing is about, what it *means*. Thus, to have a theory of a particular religious tradition is to show what its various beliefs, values, rituals, and symbols mean to those individuals who practice the religion, within their own self-defining cultural context. Such a theory would attempt to capture the internal logic of the symbols and practices, to convey a sense of the "lived" tradition by somehow getting at what it feels like to be nurtured by it.

By contrast, to have a general theory is to show what religious phenomena mean, well, *in general,* apart from their particular setting. But the particularists insist that religious phenomena are always culture-specific, which implies that their meanings will be lost when they are transposed to an alien context of meaning (that of the philosopher or the social scientist, for example) where they are not self-defined. Particularists insist that each culture is sui generis, one-of-a-kind, irreducible to the generalities that pretend to make the meanings of one culture commensurate with the meanings of another. In the end, there is nothing comparable to being a Christian, or a Jew, or a Buddhist. The essence of each tradition is inexorably linked to particular defining events. These events, processed by communal intercourse, emerge in self-contained patterns of meaning and a unique rationality by which the entire range of human experience may be interpreted. One cannot apprehend the meanings of a tradition from the outside, and one cannot be simultaneously inside a plurality of traditions. Broad theoretical objectivity is therefore out of the question—a religious tradition *just is* its subjective meanings. One might, perhaps, venture a coherent interpretation of a particular religious tradition, but it is futile to generalize across the boundaries of incommensurate meanings. Thus the moratorium on general theories of religion.

Having already declared my intention to ignore this moratorium, it is only fair that I justify general theorizing in light of particularist objections. I stress that this can be done without in any way disparaging the genuine contributions of the particularist approach. In other words, one may applaud the positive program of the particularists without accepting their negative attitude toward general theories. Indeed, the particularist approach makes a substantial contribution to the generalist approach by providing both heuristic and corrective insights. No one would argue that botanists and zoologists, who study the unique adaptations of particular species, contribute to a bias against general theories of evolution. On the contrary, evolutionary theorists are dependent in many ways on detailed research emphasizing the uniqueness of species. Likewise, particularist interpretations that emphasize the culture-specific nature of religious experiences and expressions are important for formulating general theories about these experiences and expressions. To the extent that we are

better informed about the culture-specific meanings of religious traditions, we are better able to say something gainful concerning what
religion, in general, is about.

The particularist bias against general theories of the nature and
function of religion is tied to the claim that the essential and self-
defined meanings of a tradition are lost when one assumes a perspective that transcends cultures. And this claim is in turn tied to certain
assumptions about the nature of meaning. To escape the bias against
general theories one needs only to show that particularist assumptions about meaning do not constrain the general theorist. On this
point it is relevant to see that meaning is open to analysis at different
levels of generality.

On one level it makes sense to speak of *subjective meaning,* referring to various mental states of an individual, such as beliefs, desires, hopes, fears, regrets, intentions, and the like. These are all meaning states; that is, they are *about* something or another. For example,
Molly may hope that she will win the lottery, or she may believe that
her car is in the garage. The meaning of her hope is about the reality
and the outcome of an event (the lottery), and the meaning of her belief is about the reality and the location of an object (her car). To *be*
Molly is in an important sense to experience her meaning states. In
some measure, Molly's meaning states will be absolutely unique—
that is, it is probable that no one else is capable of thinking about her
car (or her brother, or her future, and so on.) in exactly the same ways
as she thinks of them. Or consider Molly's religious piety. It is likely
that no one else has religious beliefs and experiences that are precisely identical to hers. Molly's religious orientation—her peculiar
constellation of meaning states about God, the creation, her responsibilities, her destiny—is hers alone. If Molly is, say, a Christian, then
her Christianity is like no one else's.

The particularist view does not deny the importance of subjective
meaning states, but it wants to make a claim for an authentic level of
meaning that transcends individual subjectivity. Molly's Christianity
may be unique in some measure, but not completely so. There are important overlaps of meaning that Molly shares with other Christians,
and these overlaps define a coherent body of *conventional meanings.*
Conventional meanings lie outside the realm of subjective states.
They are in the public domain, making it possible to identify an ob-
jectively real thing called the Christian cultural tradition. A cultural
tradition is the sum total of its conventional meanings, meanings that
come to have an objective reality through a continuous historical process of social interaction and negotiation. A cultural tradition is both
a repository of subjective expressions and a reservoir of symbols for

nurturing subjective experiences. But its symbolic meanings are *not* subjective: They are independent artifacts belonging collectively and exclusively to those whose subjective meanings are drawn from them, and among whom these meanings are commensurable. As objective artifacts, the conventional meanings of a cultural tradition are open to the inquiry of particularists, whose purpose it is to construct coherent interpretations of symbols and their functions in the lives of those who share them.

The real question is whether theory can venture beyond these self-contained conventions. Particularists say no. It may be tempting to compare, say, Christians and Hindus on the basis of similarities in their patterns of worship, but because these patterns are uniquely constructed in different cultures by distinctive meanings and experiences, they are not comparable things. Coincidental, perhaps, but not commensurate. There exist no useful points of contact between the conventional meanings of Christian and Hindu cultures, which means that all attempts at generalized comparisons are off. The urge toward a general theory must therefore be resisted for the reason that religion, in general, does not exist.

If this were all there is to the business of meaning, then we might readily accept the particularist bias against general theories of religion. But there is more to be said. It seems clear that the particularist view is committed to the doctrine that everything relevant and interesting about subjective and conventional meanings is determined at the level of cultural dynamics. Take Molly's Christian piety, for example. It is demonstrably true that Molly would not have the subjective meanings she does were it not for the ministrations of the particular cultural tradition in which these meanings were nurtured. And further, it is demonstrably true that Christian culture would not have the conventional meanings it does were it not for any number of Mollys who, through the centuries, achieved significant overlaps of subjective meanings. Thus we are left with the picture of an on-going dialectic between individual meanings and the larger cultural context that both creates and is created by them. But this picture is arbitrarily narrow and circular, for it leaves out everything entailed by the fact that subjective and conventional meanings are ultimately constructed by human brains.

The various meanings that play into different aspects of human life have their origins in the dynamics resulting in brains organized to *have* meanings. A good share of these dynamics is cultural, as the particularist will be quick to point out, but the cultural matrix comes far short of telling the whole story. No less important, surely, are the dynamics of biological evolution that have assembled and organized the functions of the human brain over millions of years. Neither Molly nor Christian culture would have the meanings they do were it not for the ministrations of natural selection. The human brain sculpted by the evolutionary process has a complex modular organization. It is composed of distinct, yet interactive, functional units, each one adapted to perform specific operations. There are modules for perception, for memory, for emotions, for language, and much else—each with its own evolutionary subplot, each with its appointed task, but all working in a cooperative hierarchy to serve the interests of the organism. This modular brain is the biological substrate for all meanings: for all individuals, for all cultures, for all times and circumstances.

In a profound sense these various modules of the human brain participate in the construction of meanings. It seems arbitrary to deny it. They are *about* something, as surely as subjective and conventional meanings are about something. It is very difficult to specify just how the algorithms—the deep meanings—embedded in neural systems become involved in the formulation of explicit beliefs, values, intentions, and the like. The "aboutness" of these deep meanings does not enter into conscious experience in the way the aboutness of subjective meaning does. Like the operating system in a computer, these deep meanings provide the general rules and defaults that enable and constrain the explicit meanings that eventually emerge into consciousness. I will refer to these as *adaptive meanings*, for they have been appointed by natural selection to direct on-site constructions of meaning having more specific aboutness, enabling individuals to think, feel, and act in ways that will be adaptive to local circumstances.

The obvious example is language. All normal humans are equipped with neural modules preparing them to learn whatever particular language their culture presents to them. Thus, if you place a Chinese infant in a French family, she will acquire the French language as surely as French children will. I believe it is appropriate to say that the adaptive meanings of the brain's language modules participate in the subjective meaning of every explicit linguistic formulation.

Furthermore, I believe it should be possible to specify the adaptive meanings that contribute to the formulation of religious beliefs and values. If so, then a general theory of religion is feasible. It is true that religion in general does not exist, but the same is true of language in general, and this has not precluded the construction of insightful general theories about the nature, the origins, and the functions of language. A general theory of religion should focus its attention on the adaptive meanings inherent in certain modular systems of the brain. Such a theory should be able to tell us what these meanings are about at the highest level of generality. Our best resource for such a general theory will be the discipline of evolutionary psychology. Evolutionary psychologists have been very helpful in describing the rules and strategies inherent in particular adaptations—for example, perceptual systems, emotional systems, and language systems. These descriptions are relevant to a general theory of religion because they contribute to the larger (and perhaps more speculative) task of understanding the general purposes that are characteristic of human nature.

At the most general level of all, we can say with confidence that the ultimate goal of human beings accords with that of all other life forms—that is, to maximize reproductive fitness. But every species has its own distinctive strategies for doing this. If we can identify these broad strategies for our species, then we will have in hand the general purposes of human nature. To know this is to know what our day-to-day projects and their meanings (including religious ones) are really about.

What, then, are the general strategies that may be said to be both universal and exclusive to human nature? This question will occupy much of the discussion ahead, but for the moment I offer the following summary. The general strategy of our species is to achieve personal wholeness and social coherence—that is, to develop healthy and robust personalities while at the same time constructing harmonious and cooperative social groups. To the extent that we succeed in these vital projects, we enhance our prospects for reproductive fitness. For other species the strategies will be slightly or vastly different, but for humans the name of the game is personality and sociality.

How, then, do we achieve these ends? We achieve them in large measure by formulating explicit on-site meanings about how things are in the world around us, and which things matter for advancing our individual and collective interests. That is, we construct and maintain shared worldviews composed of cosmological and moral elements. These shared traditions of meaning tell us who we are, where we come from, and how we should live. They give us an orientation in nature, society, and history, and thus provide us with resources to negotiate our way through the many challenges to a full life. The precise contents of cosmological and moral ideas will vary somewhat with the accidents of historical experience—which tells us that there are many particular ways to pursue the general goals of personality and sociality. Nevertheless, the general goals are universally and exclusively human.

Here we have the basic ingredients for a general theory of cultural and religious traditions. If the theory is correct, then we may say that the neural operating systems of our species (that is, our human nature) have prepared us to construct integrated narratives about how things are and which things matter, and that these shared meanings may be judged as more or less adaptive to the extent that they are conducive to the achievement of personal wholeness and social coherence. This leaves us in the odd position of asserting that while subjective and conventional religious meanings may be about God, religion in general is not—it is, rather, about influencing neural modules for the sake of personal wholeness and social coherence. An adequate general theory of religion should be able to bring substance to these claims (part one of the book). And further, it should be able to show us how conventional religious meanings have been honed to promote personal wholeness and social coherence (part two).

Developing a general theory of religion requires attention to questions at the level of the universal, such as why religious traditions are found in every human culture and only in human culture; why these traditions invariably develop narrative integrations of cosmology and morality; why they all address themselves to matters of personal therapy and social policy; why they harp on so many of the same vices and virtues; and so on. These formal characteristics, shared by religious traditions everywhere, are not accidental. They indicate species traits, transcending the contingencies of historical existence, and they constitute very real facts deserving of theoretical inquiry. These facts cannot be explained apart from a theory of human nature grounded in the disciplines of biology, anthropology, evolutionary psychology, and the neural sciences. These provide the primary conceptual resources for a general theory of religion. A general theory of religion is therefore necessary if we are to understand a substantial set of relevant facts. Yet a general theory—even the most sophisticated of them—could never predict the appearance of Judaism or Jainism.

I suspect that most of the tensions between particularist and generalist theories about religion arise from the fact that different minds find themselves attracted to different sets of facts. Consider: Some facts tell us about inherited similarities and differences, while other facts tell us about acquired similarities and differences. Inherited similarities include all the morphological and behavioral traits that determine the uniqueness of a given species, what we might call the "nature" of the species. Inherited differences provide the individual variations upon which natural selection acts. Without genetic variations of this sort there would be no biological evolution to speak of. Acquired similarities are determined by common patterns of learned adaptations to the environment. These patterns culminate in distinctive cultural traditions. Acquired differences result from unique experiences, providing the individual variations upon which cultural selection acts. Without such variations cultures would lack their most significant resource for innovative change. These four categories of facts generate most of the interesting questions about human beings, including interesting questions about religious behavior.

I may be completely wrong about this, but I have the impression that many people who think and write about religion find it difficult to accept that there may be interesting and important determining factors in each of these four domains. It is natural to suppose that all the real action plays out in one's own arena of curiosity. The ethnographer might very well concede that many traits are inherited, but then insist that none of these are relevant to insights about what religions are or how they work. And the generalist might agree that particular traditions have unique characteristics, but that these amount to insignificant details. Jealousies like this actually exist, and they create misunderstanding and defensive posturing, which is unfortunate because it tends to undermine the breadth of cooperation necessary for understanding the complexities of religion.

In contrast, we might begin to envision an atmosphere in which different levels of theoretical interest are mutually supportive. The complete student of religion will be curious about all sorts of facts, and will welcome resources from every direction. For example, a biographer of St. Teresa should be well grounded in the sciences of human nature. Would it be relevant to an understanding of St. Teresa's spiritual life if it were known that she suffered from epilepsy? Of course it would. It is equally important that general theorists of religion be well grounded in studies of particular traditions. Would it be relevant to a general theory if it could be shown that Buddhists are capable of emotional experiences having no equivalents in Christianity, and vice versa? Certainly. And would one's understanding of particular traditions be enhanced if a general theory could show what it is about human nature that makes such radical cultural differences possible? For sure.

I do not mean to obscure the difficulties. There will be much left to argue about. Disputation is, after all, the fuel of serious inquiry. In any event, I remain convinced that general theories of religion are beyond the merely possible; they are essential to the enterprise. This conviction is one of the premises of the book, and its justification is one of the promises.

CAN RELIGION BE NATURALIZED?

It should be evident that the quest for a general theory of religion is not without controversy. Even more controversial is a second major thesis of this book: that religion can be naturalized. The naturalistic thesis asserts that the experiences and expressions constituting the religious life can be seen to result exclusively from natural causes. This does not imply that religious phenomena can be completely explained—few events in nature can be—but only that the extent of our understanding is contingent on our efforts to reduce these phenomena to the terms of underlying natural processes.

Naturalism is a variant of metaphysical monism, the philosophical stance declaring that all meaningful distinctions pertain to observed or reasoned facts within a self-contained and continuous order of being. For the naturalist, the order of being is the order of nature—the natural is real and the real is natural. If, therefore, we have reason to believe that some entity or event is real, then we have precisely those reasons for believing it to be natural.

Naturalism may be characterized by its rather strict application of "Ockham's razor," or the principle of parsimony. William of Ockham, the most influential Western philosopher of the fourteenth century, was famous for his intellectual crusade to eliminate pseudo-explanatory categories. Ockham's rule of elimination—his razor—says: "What can be done with fewer is done in vain by more." The point of Ockham's razor is to achieve economy of explanation by shaving away concepts, principles, and categories that are not essential to the subject matter to be explained, and are not established by rigorous methods of observation and reasoning. Thus, naturalists oppose explanations that unnecessarily assume a transcendent order of entities and events having causal influence in the order of nature. Why posit two orders of being where one is sufficient?

It is precisely on this point of sufficiency of explanation that non-naturalists have pressed the case for supernatural accounts of religion. Naturalists, the argument goes, tend to over-economize in their explanations to the point of distorting and ignoring facts. So argued Mircea Eliade, an eloquent and prolific opponent of reductionism.[3] Eliade believed that religion could be generalized, but not naturalized—generalized because all particular forms of religion derive from human encounters with the *sacred,* but not naturalized because the sacred does not derive from the order of nature. The sacred, a realm of absolute transcendent reality, cannot be apprehended by natural categories. Nor can the religious life be apprehended apart from the sacred. Natural processes—those that are biological, psychological, and economic, for example—may have certain limited effects on religious phenomena, but these are peripheral. Ultimately and essentially, religion is *theogenic,* that is, its facts can be understood only by assuming the causal influence of a supernatural reality.

There are several other variations on the anti-reductionist thesis. Some of these maintain that there are patterns of religious meaning inherent in human history, which radically transcend the dynamics of nature. Others argue that religion cannot be naturalized because it is essentially about a moral order, and that moral values can be neither derived from nor explained by natural facts. In order to make sense of moral behavior, therefore, we are forced to transcend the vocabulary of naturalism. A similar argument says that religion cannot be naturalized because it necessarily involves something immaterial, such as a life force or spiritual awareness, neither of which can be sufficiently described in natural terms. Another alternative is to concede that certain forms of religion *can* be naturalized (idolatry, fertility cults, and false prophets, for example), but that "true" religion is divinely inspired and cannot be traced to natural causes. And finally there are default arguments, insisting that the many failures of previous attempts to naturalize religion (such as Freud's psychogenic theory or Durkheim's sociogenic theory) warrant the conclusion that the job simply can't be done.

These arguments all share in the view that religious experiences and expressions cannot be reduced to the status of natural phenomena. Even after the naturalists have taken their best theoretical shots, there remains something of decisive significance that eludes our understanding. Thus, in order to have an adequate theory of religion one must admit to the necessity of supernatural categories of explanation.

In the face of such arguments the naturalist is left with a single option: to produce. The only sure way to make the case for naturalizing religion is to actually do it, that is, to give an account of the origins and functions of religion that renders theogenic alternatives vapid and unnecessary. It may be observed that the history of inquiry is on the side of naturalism. Earthquakes, floods, astral displays, birth defects, diseases, and a good many other phenomena have been effectively naturalized to the undisputed satisfaction of all. And for the past century we have been steadily acquiring the theoretical resources for naturalizing human behavior. Recent progress in behavioral genetics, neuroscience, and evolutionary theory provides additional resources for extending the naturalistic program into the more sensitive areas of art, literature, morality, and religion.

It should not be assumed that all naturalists agree on a common worldview. Far from it. There have been many widely divergent variations on the naturalistic theme, including the syncretism of Confucius, the materialism of the Atomists, the substantialism of Aristotle, and the idealism of Hegel. Naturalists will agree that the natural order is ultimate, but beyond this point there is much room for dispute about the nature of nature. This book casts its lot with a version of naturalism I will call *consilient scientific materialism*.

MATERIALISM

A materialist worldview claims that all natural facts can be construed, in some minimal sense at least, in terms of the organization of matter. This should not be taken to suggest that all natural facts are "nothing but" physical facts, but only that whatever is or happens in nature is contingent on a substrate of material reality.

The picture of nature presented to us by contemporary science reveals a cosmic evolutionary process unfolding in a complex hierarchy of interlocking systems and subsystems that govern the organization of matter. As one follows the arrow of time, more complex systems emerge to organize matter, bringing new entities, properties, and relations into the order of nature. At the lowest level are subatomic particles, which are organized into higher-level complex systems called atoms. Atoms have diverse properties, enabling their organization into various molecular systems. Molecules are systematically organized to form a variety of complex structures, including rocks, minerals, and planets. Molecules may also be organized into living systems, composed of cells, tissues, organs, and so on. Living organisms are systematically organized into populations, communities, and ecosystems. The most highly organized material systems are found in human beings. Here, nerve cells are organized into various functional systems, which may be integrated into coherent personalities. Persons then interact in complex patterns to form social groups and cultural systems.

Some materialists maintain that all of this complexity in the organization of matter may be reduced to the dynamics of atomic and molecular systems. This is the strong reductionist thesis, which claims that it is in principle feasible to give a full account of higher order systems in terms of physics and chemistry. This view, now very much in the minority, is a good example of taking Ockham's razor too far. A more satisfying picture recognizes that while more complex systems are contingent on physical and chemical substrates, they nevertheless involve emergent principles of organization and manifest genuinely novel properties which cannot be fully described by the principles and properties known to the physical sciences. That is to say, not all natural facts are physical facts. There also exist biological facts, psychological facts, and cultural facts.

This book will assume a view of nature in which all natural facts may be resolved into four general categories or levels of material organization: physical, biological, psychological, and cultural. The distinctions between these four levels of nature may be seen to derive from different modes of information. Physical facts describe the behavior of matter insofar as it is organized by information inherent in physical systems. Biological facts describe the behavior of matter insofar as it is organized by information preserved in the genetic code. Psychological facts describe the behavior of matter insofar as it is organized by information stored in neural systems. And cultural facts describe the behavior of matter insofar as it is organized by information embodied in symbols. That's it! If something is a fact, then it is in principle reducible to these four modes of information, or perhaps a constellation of them, in the case of complex facts, which most are.

SCIENTIFIC

The various disciplines of science represent attempts to organize our knowledge of physical, biological, psychological, and cultural facts. If we were to construct the academic curriculum afresh, in conformity with this view of nature, then we would do well to establish four major faculties. But, alas, we are left to contend with the disciplines that developed haphazardly over time. Still, there is a rough correlation between the existing disciplines, subdisciplines, and interdisciplines and the four categories of natural facts.

At the level of physical facts are the corresponding disciplines of physics, chemistry, astronomy, geology, astrophysics, and cosmology. Bridging the gap between physical systems and biological systems are the interdisciplinary domains of biophysics and biochemistry. Corresponding to the level of biological facts we find cell biology, genetics, physiology, anatomy, zoology, botany, and ecology. The neurosciences, together with evolutionary psychology, bridge the gap between biological and psychological systems. At the level of psychological facts are the various subdisciplines of cognitive psychology, developmental psychology, personality theory, and so on. Social psychology connects the levels of psychological and cultural facts. Attending to the organization of cultural or symbolic facts is an unruly hoard of disciplines, including sociology, anthropology, political science, economics, history, linguistics, mathematics, philosophy, and the various "critical" disciplines focused on literature, the arts, and religion. The major differences between these intellectual domains have to do with methods, and with lingering attachments to traditional agendas. What they share in common is their (selective) interest in the artifacts of human symbolic abilities.

Taken together, these many disciplines constitute science, the collaborative enterprise of systematically organizing our knowledge of the natural order. For the scientific materialist, all plausible explanations of natural phenomena will find their place among these disciplines.

CONSILIENT

Scientific materialism claims that all natural facts involve the organization of matter, and that the empirically grounded and self-correcting disciplines of science are to be considered normative in all our attempts to explain natural facts. A *consilient* scientific materialism goes a step further to advance a thesis about the unity of science.[4] Edward O. Wilson has rescued the concept of consilience from historical obscurity to characterize the ultimate prize of inquiry: a coherent, unified meshwork of ideas that renders intelligible the full scope of human experience.

The sciences, as we have them, tend to be fragmented into separate domains of inquiry, but such fragmentation is both artificial and unsatisfying. If nature is itself a unified meshwork of interlocking causal events, as the naturalist believes, then we should expect that existing gaps between the sciences will be significantly narrowed by further inquiry. Wilson demonstrates that the unification of knowledge is already remarkably complete among the natural sciences. Consilience of theoretical explanation from physics to chemistry to biology and well into the nascent field of neuroscience has already been achieved. What remains is to explore ways of thinking that might extend the consilience program to include the social sciences and the humanistic sciences.

Thus we are returned to the original question: Can religion be naturalized? To naturalize religion is to provide explanations for religious phenomena that are consilient with our scientific explanations for other natural facts. The focus of this inquiry will be on human nature. For a consilient theory of religion to succeed it will have to show, first, what human nature is and how it emerges in the process of evolution. And second, it must show how religious experiences and expressions emerge from the dynamics of human nature. To the extent that such a consilient theory succeeds, we shall have before us a satisfying naturalistic understanding of religious phenomena.

Suppose a consilient theory *does* succeed. What then? What can be said about the power of religion under the conditions of understanding it? Does an understanding of religion preclude religious understanding? Does the attempt to naturalize religion also effectively neutralize it? These are delicate questions that will be held off until part three. I say "delicate" because the religious life is the sanctuary of existential meaning, where attempts to naturalize may be experienced as hostile acts intended to destroy the meaning of life, to undermine faith and hope, to steal away the treasured grail. I will not attempt to deny the reality or the gravity of the experience. I have endured it myself. Nevertheless, the urge toward consilient explanation is strong. All I can promise from my own experience is that any existential losses incurred by naturalizing religious meanings may be fully compensated by an acquired sense for the mystery and sanctity of nature itself.

PART
I

ON
HUMAN
NATURE

No picture of human nature could ever be more than a caricature. We are too complex, too diverse, and ultimately too mysterious to be completely summed up. The account of our nature offered in part one, therefore, will be partial and tentative. It will attempt to sketch out the broad perspective of cosmic evolution, which is now the essential framework of the human story, and will then focus on the salient features that mark us as a unique species.

Chapter 1 will trace the evolutionary narrative from the first moment of creation to the emergence of life. The picture that emerges is one of increasingly complex systems for organizing the behavior of matter: physical systems, chemical systems, and biological systems.

Chapter 2 will continue the theme of emergent complexity as it traces the evolution of behavior from relatively simple molecular systems to the higher orders of complex neural systems. Each of these emergent systems—from simple reflexes to the capacity for symbol use—brings new patterns of diversity and complexity into the astonishing narrative of material organization. Here the argument will be that a distinctive human nature emerges in the inherited capacities that guide us in the construction of personality and sociality.

Chapter 3 will concentrate on those aspects of human nature that are open to manipulation by religious traditions. Especially important to this discussion will be the dynamic interactions between emotional, cognitive, and symbolic systems.

Chapter 4 makes a transition from a focus on human nature to a focus on spiritual traditions. Here the point will be to describe, in a general way, what religion is. The discussion will concentrate on the structure, the origins, and the functions of religious traditions.

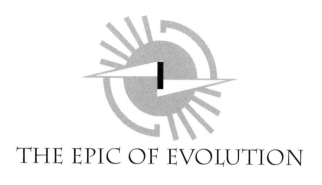

THE EPIC OF EVOLUTION

At the Seventh First Annual award ceremony for the Ig Nobel Prizes, Eric Shulman got up and recited the following two-minute history of the universe:

Quantum fluctuation. Inflation. Expansion. Strong nuclear inter-action. Particle-antiparticle annihilation. Deuterium and helium production. Density perturbations. Recombination. Blackbody radiation. Local contraction. Cluster formation. Reionization? Violent relaxation. Virialization. Biased galaxy formation? Tur-bulent fragmentation. Contraction. Ionization. Compression. Opaque hydrogen. Massive star formation. Deuterium ignition. Hydrogen fusion. Hydrogen depletion. Core contraction. Enve-lope expansion. Helium fusion. Carbon, oxygen, silicon fusion. Iron production. Implosion. Supernova explosion. Metals injec-tion. Star formation. Supernova explosions. Star formation. Con-densation. Planetesimal accretion. Planetary differentiation. Crust solidification. Volatile gas expulsion. Water condensation. Water dissociation. Ozone production. Ultraviolet absorption. Photosynthetic unicellular organisms. Oxidation. Mutation. Nat-ural selection and evolution. Respiration. Cell differentiation. Sexual reproduction. Fossilization. Land exploration. Dinosaur extinction. Mammal expansion. Homo sapiens manifestation. Animal domestication. Food surplus production. Civilization! Innovation. Exploration. Religion. Warring nations. Empire cre-ation and destruction. Exploration. Colonization. Taxation with-out representation. Revolution. Constitution. Election. Expan-sion. Industrialization. Rebellion. Emancipation Proclamation. Invention. Mass production. Urbanization. Immigration. World conflagration. League of Nations. Suffrage expansion. Depression. World conflagration. Fission explosions. United Nations. Space exploration. Assassinations. Lunar excursions. Resignation. Computerization. World Trade Organization. Internet expansion. Compositions. Extrapolation? [1]

The Ig Nobel event itself was full of spoofery, but Shulman spoke only partly in jest. The serious side of him was calling to our attention the fact that it is now possible to construct a science-based narrative of cosmic evolution. During the past fifty years or so, scientists have come to see the big picture of creation with a clarity and coherence that has never before been possible in human history.

The epic of evolution is the sprawling interdisciplinary narrative of evolutionary events that brought our universe from its ultimate origin in the big bang to its present state of astonishing diversity and organization. In the course of these epic events matter was distilled out of radiant energy, segregated into galaxies, collapsed into stars, fused into atoms, swirled into planets, spliced into molecules, captured into cells, mutated into species, compromised into ecosystems, provoked into thought, and cajoled into cultures. All of this (and much more) is what matter has done as systems upon systems of material organization have emerged over fifteen billion years of creative natural history.

Several versions of this narrative of cosmic evolution have appeared in recent years to crowd the popular science shelves of neighborhood bookstores, each one unique in emphasis, but all sharing the same central theme: The universe is a single thing, one long continuous event of expansion, diversity, and complexity. Figure 1 is an attempt to help us visualize the epic of cosmic evolution.

The epic of evolution is fundamentally a story about the creation of matter from energy, followed by the successive emergence of novel systems for organizing the behavior of matter. Matter is no longer thought of in the old-fashioned way, as inert stuff that gets pushed into organized patterns by independent forces. Matter is inherently excited and loaded with potential for self-organization, and as it becomes organized it manifests new properties and dynamics, creating ever more complex conditions to elicit new patterns of organization.

The narrative begins with the big bang, roughly twelve to fifteen billion years ago. In the beginning, the universe—everything that would ever exist—was wrapped in a state of dimensionless existence so odd that it far exceeds the powers of human imagination to grasp. We create the illusion of understanding by saying that before the big bang, the universe was infinitely dense and infinitely hot. This uneventful singularity of space-time-energy suddenly destabilized and

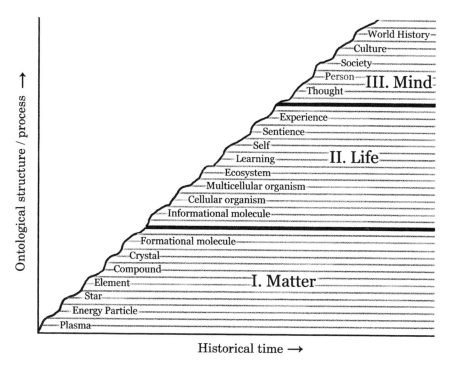

Fig. 1. A model of cosmic evolution (adapted from Holmes Rolston, *Science and Religion*, New York: Random House, 1987).

events began to unfold. The events during the first instants after the big bang were indescribably chaotic because conditions were too hot for any organization to occur. But after a few microseconds of rapid expansion and cooling, energy began to decay into particles and it was possible for free floating quarks to join together to form neutrons and protons, which would later combine to form the nuclei of atoms. But for 300,000 years the universe remained too hot for nuclei to capture electrons into the systematic organization we call the atom. When conditions were sufficiently cool, copious amounts of atoms appeared in the forms of hydrogen and helium.

For the first billion years or so, the organization of matter was limited to the microcosmic assembly of simple atomic systems. But as clouds of hydrogen and helium expanded, the universe entered into a phase of macrocosmic organization. Because the density of the expanding cloud of gas was not uniform, it grew lumpy and fragmented into more than a hundred billion galaxies, each one moving away from the others. Eventually lumps within these galaxies gave birth to stars, as many as a hundred billion per galaxy. Stars are giant infernal systems, held in relative equilibrium by the opposing forces of gravity and radiation. The internal regions of stars can generate heat sufficient for the nuclear reactions that produce more complex atomic systems such as carbon, nitrogen, oxygen, and other elements that populate the periodic table. When large stars explode as supernovas, they produce the heavier elements (from iron on up) and splatter these newly minted atoms into space.

For the first several eons after the big bang the only discipline that would have made sense in our universe was physics. It was all physics. But when supernovas started exploding, many types of atoms were introduced, each type bringing into play new properties of matter and new potentials for further systematic organization. Chemistry begins!

As a galaxy becomes strewn with the debris of stellar explosions, conditions begin to favor the construction of many new stars. Second and third generation stars form in the usual way—gravity concentrates matter until friction heats it to the nuclear flash point. Then follows an equilibrium of forces: inward gravity in balance with outward radiation. This is precisely how our sun was formed nearly five billion years ago. In the vicinity of our sun, and bound into orbits by its gravitational force, were swirling bands of matter, rich in the elements that would make up our solar system. Planet Earth started out as a clump of matter in a swirling band of debris, and as it swept around the sun its own gravitational field collected additional particles until it had vacuumed up most of the matter in its orbit. As the earth grew larger, it became hotter, eventually reaching the point where most of it melted down, allowing for a lot of shifting and sorting of the earth's materials. Heavy molten iron gravitated to the center, forcing lighter materials toward the surface.

For its first 800 million years the earth remained very hot and was under constant bombardment by radiation and meteor showers. By 4 billion years ago the planet had cooled sufficiently for a rocky crust to form. Radiation from the sun, condensation of water vapor, radioactive decay, and periodic outgassing from the still molten interior produced an abundance of water and atmospheric gases. Much of the earth's crusty surface was broken up and recycled in this turbulence, leaving a few large fragments surrounded by water. These remnants (today's continents) continue to shift about on the earth's surface as their tectonic foundations adjust to motion in the fluid core.

The biosphere is the region of the earth's surface that supports life. More precisely, the biosphere is a highly complex, geologically based biochemical system that developed out of interactions between land (lithosphere), water (hydrosphere), and air (atmosphere). It was at the intersection of these major components of the young planet that chemical organization flourished to bring forth living creatures.

Nothing in this narrative so far sounds even remotely relevant to an understanding of how religious traditions interact with human nature. But in fact the deep history of the cosmos is profoundly important for understanding human nature. We human beings are highly complex organisms, owing the lives we have to the emergence of hierarchies of natural systems. We are ultimately the manifestations of many interlocking systems—atomic, molecular, biochemical, anatomical, ecological—apart from which human existence is incomprehensible. How these natural systems came to be is essential to the story of how we have come to be. At each point along the trajectory of cosmic evolution we are able to gather new insights to add to our sketch of human nature. From the earliest epochs of creation we learned that human existence is grounded in star stuff—we are made up, exclusively, of elemental atomic systems that were forged in stars and then released into space by supernova explosions. And from later epochs we catch another insight: Human beings are geological formations! Granted, the sketch isn't yet sharp enough to distinguish the human from, say, a mountain, but these insights—that we are star-born and earth-formed—are no less important to self-understanding than the many features that we recognize to be exclusively human.

For most of cosmic history it was all physics and chemistry. Biology, so far as we can tell, did not begin to make sense in our universe until approximately four billion years ago, when living creatures emerged spontaneously from the chemical soup of Planet Earth. Several scenarios have been hypothesized to account for this mysterious event, but in all likelihood we shall never know the full story. Among the few things that can be said with confidence is that life emerged from a chemical quagmire of molecular evolution wherein molecules competed with one another for the attention of free atoms. As fragile molecules decomposed, their atoms would be free to join in the composition of larger, more stable molecular units. The big winners in this process were those selected for having special properties, such as catalysis or replication. Life emerged when complex chemical systems managed to integrate the delicate functions of energy transformation, information processing, and self-replication.

A living organism is a self-regulating and self-replicating biochemical system that can manage to keep its rate of chemical composition comfortably ahead of its rate of chemical decomposition. Organisms failing in this task, for whatever reasons, are either dead or soon to be so. Living systems meet the challenges to chemical composition by transforming energy and processing information—that is, if an organism has the right combination of information and energy, then it might continue to metabolize and to reproduce. It would follow that any biochemical system coming into the world already endowed with a supply of energy and information would be odds-on to keep itself going. The exploitation of this principle of inheritance is responsible for the origin and continuation of life.

The diversity of life forms, meanwhile, can be traced to a pair of negative facts: First, the chemical mechanisms of heredity are not invulnerable to error, and second, the environments in which the first cells emerged were not homogeneous. The first of these facts accounts for genetic variation, and the second accounts for natural selection. If copy errors (mutations) occur in the process of replicating the genome, then new information is carried forth resulting in novel traits. Any new trait, structural or behavioral, may be either helpful or harmful to an organism as it negotiates a living from its natural environment. Random variations in traits are thus constantly being tested for their suitability in a changing environment. Given enough of this chancy sort of thing and the inevitable happens: There will emerge many different ways to be a living thing.

We are one of these "ways," along with all the other species that have come to populate the planet. Like the others, humans have evolved with a unique combination of traits for doing what every life form must do—that is, to endure and to reproduce. Whatever excellent qualities a species may come to possess will mean nothing in the end if they are not consistent with reproductive fitness, the wherewithal to endure and to reproduce sustainedly. This is what really counts in the game of life. So when we come, eventually, to consider the fancywork of human nature—our intelligence, our affections, our moral sense—it will be important to recall that these qualities are purchased on the account of traits that contribute to our reproductive fitness. This principle of reproductive fitness, therefore, is a superordinate principle that applies to all living systems.

My purpose in this part of the book is to say something gainful about human nature. By insisting that the study of human nature be undertaken within an evolutionary framework, I have followed a principle expressed by Alfred North Whitehead: What a thing *is*, is shown by *how it comes to be*. To this principle we may add a second: What a *living* thing is, is shown by *how it carries on*. At this point in the narrative we can say no more about human nature than might be said about any other living thing—that is, *we are by nature star-born, earth-formed, fitness-maximizing biochemical systems*.

More than a billion such systems have emerged during the epic of evolution, of which perhaps as many as thirty million remain extant. The complete story of how these species came to be, and how they manage to carry on, would be hopelessly long and involved, and mostly beside the point. In the next chapter, therefore, the focus will be narrowed to a single aspect of the epic having special relevance for our sketch of human nature: the evolution of behavior. The salient features of human nature have little to do with our appearance. When it comes to physical attributes, human beings are—let's face it—pretty unremarkable. What makes our species stand apart from all the others is the astonishing range and complexity of our behavior. To say that no other species comes close would be a gross understatement. To understand why will be the point of the next two chapters.

THE EVOLUTION OF BEHAVIOR

This much is obvious: All living things behave. Bacteria, algae, ants, and birds behave. There's no dispute about that. The story gets interesting only when we ask *why* they behave as they do. As we have seen, evolutionary biology has a short and simple answer to the Why question: Organisms behave in ways designed by natural selection to maximize their reproductive fitness. That is, all living things have it in their nature to propagate their own genes, and those of their close relatives, indefinitely into the future. Exceptions to this rule are doomed to extinction.

Another fairly obvious point is that all behavior is mediated by heritable systems. We commit a fundamental mistake if we assume that behavior per se is heritable. Nobody inherits behavior. What we inherit are genes that code for proteins that build the tissues of mechanisms for organizing behavior. So the evolution of behavior really comes down to a story about the evolution of *mediators* of behavior. The general thrust of the story is this: Over time, there has been a gradual process of systematic development in which behaviors have become mediated by ever more complicated mechanisms, enabling ever more complicated and variable interactions between organisms and their environments. Human nature, we may say, is embedded in the logic of these behavior mediation systems.

MOLECULAR SYSTEMS

The first (and still by far the most numerous) organisms to populate the planet were one-celled bacteria. Bacteria may not do a whole lot by some measures of achievement, but in fact they accomplish everything they need to: They find and ingest nutrients, they break down and build up molecules, they transform energy, they expunge wastes, and they reproduce. What more is there? Bacteria, and other one-celled creatures, perform all these behaviors by means of straightforward biochemical reactions. At the level of the cell all behavior is mediated by molecular systems.

As we enter the domain of molecular systems we discover a world dominated by the shapes of proteins. Most organisms are mostly protein—that is, proteins account for more than half of the dry weight of most living cells. Nearly everything a cell does—its construction and support, its mobility, its uptake and discharge of materials, its consumption of energy, its reproduction—all of these activities are made possible by the properties of proteins and molecules that interact with them. The properties of these protein molecules are determined by the unique shapes into which they twist, turn, and fold themselves after they are constructed according to genetic instructions.

There are thousands of distinct proteins in a typical cell, each with its own three-dimensional structure "keyed" to specific interactions with other proteins or non-protein molecules. For instance, all cells have receptor proteins on their membranes that serve as binding sites for very specific molecular substances. Once an external molecule snuggles into a receptor, additional events are triggered that depend on the nature of the stimulus. In some cases the receptor will change its shape to form a channel through which the molecule enters the cell. In other cases the receptor may change its shape to signal a cascade of internal events resulting in motion, secretion, or some other behavior. At the level of the cell, behavior is all about lock-and-key biochemistry.

Here follow two important points. First, this molecular level of mediation never disappears from any full account of behavior. In fact, we can say that all behavior—whether it's an alga swimming after a sunbeam or an ambassador negotiating a treaty—ultimately comes down to lock-and-key molecular systems. Whatever else may be said of an organism, if its biochemistry stops, its behavior stops. And the second point is that there are many higher-order mediation systems that have evolved to orchestrate the biochemical locks and keys. The evolution of behavior is a story about these emergent systems.

Higher-order systems for mediating behavior first came into play about 700 million years ago, with the advent of multicellular organisms. The great adaptive achievements during more than two billion years of unicellular evolution resulted in a broad diversity of metabolic pathways. Multicellular species took full advantage of these inventions, but then took the path of evolution in the direction of morphological diversity. Just how it happened that individual cells could be stripped of their autonomy and organized into cooperative relations is still a mystery. What is not mysterious is that multicellularity conferred enormous advantages in terms of specialized strategies for carrying on: for example, more sophisticated predation, energy storage, mobility, reproduction. If you have an abundance of cells, then you can dispatch different cell lines to carry out specialized functions, thus to negotiate forbidden niches and gain access to new sources of energy. But multicellularity does not come easy. For cells to behave coherently as a single entity, there must be higher-order mechanisms for the coordination and control of subsystems.

The first multicellular animals were probably loosely connected colonies of identical one-celled organisms. The emergence of true multicellular creatures would require increased interdependence among colonial cells, together with a gradual division of specialized functions carried out by groups of cells. The secret to success in these cooperative ventures would be the innovation of higher-order systems for transporting nutrients and information throughout the collective body. Perhaps the first of these higher-order regulatory systems to evolve were vascular networks for the direct transport of nutrients and chemical information. But another, much faster, type of higher-order system involved the nerve cell (neuron), specialized to carry information-rich patterns of electrical impulses to distant sites where chemical reactions are instigated. Neural systems are assemblies of neurons operating as functional units. Their job, ultimately, is to orchestrate the behavior of organisms in ways that allow them to carry on.

NEURAL SYSTEMS

The concept of neural systems (or neural modules) will play an important role in the discussions that lie ahead, so it will be useful to elaborate a bit before continuing the evolutionary narrative. As mentioned earlier, the great adaptive achievement of both unicellular and multicellular life had to do with diversifying structures to handle specific challenges posed by the environment. Limbs, shells, coloration, internal organs, sense organs, and musculature are just a few of the many adaptive modules that have been blueprinted into the genomes of various species. By the same Darwinian logic many neural modules have evolved as well, each one selected for its power to mediate adaptive behavior. If the body evolved by adding modules for particular adaptive functions, then why not the brain?

A century ago it was widely believed that connections among nerve cells formed a continuous mass throughout the brain, yielding a single organ of general intelligence, like the old village telephone party line where everybody gets to listen in on the flow of information. We now know this view is mistaken.[1] Instead, the new picture is more like a fluctuating hierarchy of many simultaneous conference calls, with many informational themes going on at once. There is no general, all-purpose intelligence, but rather a large number of coordinated neural assemblies that are specialized for carrying out particular information-processing tasks. The brain is a federation of these functionally specialized units.

There are on the order of 100 billion neurons in the human brain, each one an excitable calculating device with the potential for thousands of connections with other neurons. When a brain neuron receives information from neighboring cells in the form of patterned signals, it adds and subtracts electrical charges to determine whether and how to respond. Although each neuron is discrete, they normally fire only in cooperative arrangements with other neurons. These cooperative arrangements—these computational neural modules—are capable of serial and parallel connections with other neural modules, enabling behaviors ranging in sophistication all the way from a blink to a wink. All behaviors may therefore be resolved into layers of subfunctions, all the way down to the lowly neuron.

The origins of these information-processing networks is a subject of intense interest among neuroscientists. Is their architecture and development wired in by heredity, or are they assembled through learning? This issue has flummoxed researchers for generations. But lately there has been a growing consensus that the "either-or" formulation is misguided. It is more accurate to say that all neural modules are influenced by genetic constraints *and* by environmental input. Even the most "hard-wired" of elementary reflex systems are open to slight modulation by experience. And even those "software" modules that require experience and learning have an undeniable genetic component. The old nature-nurture controversy has therefore been redrawn. The debate is no longer about the relative influences of heredity and environment, but rather about whether inherited systems for learning are few and general, or many and specific. The best evidence to date strongly favors the latter.[2]

Neural modules are not always what the term leads one to expect—that is, they are not merely like add-on computer chips that upgrade the system. Some neural modules are densely packed gobs of tissue having all the structural integrity of an appendix or a liver. But others are, as Steven Pinker describes them, "more like roadkill, sprawling messily over the bulges and crevasses of the brain."[3] Neural modules appear to be flexible when it comes to spatial requirements—just as some modern corporations are localized in a suite of offices while others are laced out in cyberspace. But this is not the point. The important thing about neural modules is that they are functionally unified structures, tenured in by natural selection to carry out specialized information-processing tasks for the purpose of mediating adaptive behaviors. And it is in the designs of these genetically prescribed systems that we shall find further substance for our sketch of human nature.

Returning now to the evolutionary narrative, we find a story about the emergence of various types of neural modules. The general drift is that the sequence in which animals faced environmental challenges was a determining factor in the evolution of neural systems. New environmental circumstances selected for abilities to process new forms of information. Thus we see a diversification and specialization of neural modules, coupled with emergent systems for coordinating their functions.

REFLEX SYSTEMS

The first animals to possess genuine nervous systems had something in common with your toaster. That is, they were probably equipped with grid-like patterns of neurons evenly distributed between layers of body tissue. Like the filaments in a toaster, these nerve nets fired en masse when stimulated, engaging the entire organism in a unified reflexive response. Still found in animals like jellyfish and corals, simple nerve nets are genetically hard-wired mechanisms producing automatic and invariant responses to stimuli. Like your toaster, they deliver the same behavior every time you throw the switch. Nerve nets earned their genetic tenure by satisfying the same standard that nature would apply to all subsequent neural modules: the ability to produce adaptive behavior. For simple nerve nets the behavior was motion. These mechanisms were just the thing for organisms whose lives might be saved by making quick getaways from dangerous situations. When a threatening contact triggers a convulsive response, then the organism can often propel itself to safety.

Whenever a biological strategy proves its utility in one domain, there is a good chance it will find new applications in others. Nowhere is this principle of conservatism more evident than in the evolution of behavior. If reflexes are useful for one type of motion, then why not for others? If useful for escape, then why not for capturing food, finding a mate, or communicating with other members of the species? But consider the problems inherent in a plurality of hard-wired neural modules. Suppose an organism is simultaneously stimulated in ways that provoke incompatible responses. What then? Obviously, any organism having a plurality of these systems had better be equipped to coordinate their functions—that is, to inhibit some while enabling others. Different reflex modules, serving a variety of specific behavioral objectives, had to be integrated by centralized programs for "deciding" which modules would fire, and when. The simplest central nervous systems (for example, the brains of flatworms) are basically clearing houses for coordinating a plurality of reflex systems. Complicated brains, like ours, process orders of magnitude more information, but the basic logic of "input $\longrightarrow$ decision $\longrightarrow$ output" remains the same.

Like the molecular systems they modulate, reflex systems never drop out of a full account of behavior—they stay in the picture as fundamental units of reaction, usually incorporated as components of more global and broadly integrated mediation systems. Indeed, none of the so-called higher neural functions would be conceivable without the continued support of localized reflex systems. The most familiar human reflexes are those responsible for swallowing, gagging, coughing, sneezing, eye blinking, knee jerking, and the like. But many others are critically involved in mediating functions in circulatory, respiratory, intestinal, endocrine, and sensorimotor systems.

PERCEPTUAL SYSTEMS

Reflex systems are terrific, and we could never carry on without them. But if that's all you've got to orchestrate your biochemistry, then your life won't be very interesting. New principles of neural organization were brought into play by evolutionary arms races. As multicellular animals diversified their niches, and as new relationships of competition and cooperation developed, selective pressures favored more sophisticated sensory and motor abilities. Arms races between predators and prey rewarded sensory designs for detecting the whereabouts of other organisms. Any breakthroughs in processing information about motion, distance, color, sound, or smell would have carrying-on value and would become fixed in the genetic code. Adaptations on the sensory side would tend to be evenly matched by new motor traits as well, for new sensory powers would have limited value without the ability to make a catch or to avoid one. So the sensory enhancements were balanced by motor enhancements favoring greater speed, accuracy, strength, or endurance. But certainly neither of these paths of evolution could go forward without the logistical supports of increasingly sophisticated inter-neural systems. Input patterns from upstream sense organs would have to be assembled and assessed before output information could be constructed and released to downstream motor systems.

With the world out there as complicated and dangerous as it is, one might reasonably suppose that the more information we have about it, the better. Generally speaking this is true, and at a first glance omniscience looks pretty appealing. But up close it becomes evident that an omniscient being would be not only freaky, but downright disadvantaged. It turns out that having *all* the information about the world is worse than having a small amount of the *right* information—that is, information relevant to an organism's interests. Who needs to see infrared and ultraviolet radiation when all the important stuff is limited to the visible spectrum? Information processing is energy intensive, time consuming, and takes up space, all of which suggests that it pays to be highly selective about how much and what kinds of information to get involved with.

As animals moved into more complicated ecological niches, the needs for information increased. Powerful selective forces would have favored perceptual systems that could do more with less—that is, process a lot of information without growing too large or becoming too slow or inefficient. Simple elaborations on the reflex system would reach the point of diminishing returns fairly soon. Reflexes are too linear, too stimulus-bound, too all-or-nothing to answer the challenge of doing more with less. A better way would be to come up with systems that could play the averages. Consider the polling industry. Opinion pollsters do not withhold their judgments until every last household has been contacted by phone. Instead, they get the gist of things and then make an educated guess. By limiting the information to a representative sample, a staff of a dozen callers can accomplish what might otherwise require thousands. They do more with less.

Highly evolved perceptual systems operate along similar lines. A fraction of the information on offer from the world around is sampled by selective sensory receptors and sent along to processing centers in the brain where the information is amplified and interpreted to produce a coherent perception. In other words, the inter-neural systems of the brain—between upstream senses and downstream effectors—act as information enhancers, integrating memory with active sense data preprocessed in separate parts of the brain. In visual perception, for example, the size, color, and motion of an object are each processed in distinct neural modules before their final assembly into a mental image. The actual amount of visual information coming into the brain is far too sparse to determine anything close to a coherent mental image—a few bits on color, a few bits on shape, a few more on angle, motion, illumination. Nothing but fragmentary clues, really, all arriving via independent neural pathways, leaving the interactive modules of the brain to cobble together a plausible guess at what's out there.

Perceptual systems vary quite a bit from species to species, ultimately because they have faced different environmental challenges in their evolutionary histories. Some animals have highly refined auditory systems, but a poorly developed sense of smell. Others have a high degree of visual acuity, but cannot hear well. Further, some animal species lack specialized neural modules needed to integrate sensory modes. Lizards, for example, can both see and hear, but since they cannot associate the two modes, they have no sense that what they see and what they hear may be the same object. The ability to integrate these sensory modes requires yet another specialized neural module.

Perception, it may be said, is largely guesswork, but the guesses are not left entirely to chance. Far from it. They are calculated guesses, worked out by perceptual modules that have, through countless generations of improving odds, acquired genetically biased rules for seeing the features of the world—more-or-less, approximately—as they really are.

PHYSIOLOGICAL DRIVE SYSTEMS

As we have seen, the challenge of carrying on is to find strategies for keeping one's rate of chemical composition safely ahead of the rate of decomposition. Relevant perceptions about the external world are essential for this task, but equally important is information about the internal states of the body, especially for animals living in fickle environments. Consider the special challenges encountered when organisms left the sea for life upon the land. Take dehydration—not exactly an occupational hazard for occupants of marine habitats, but for land animals it remains a constant threat. Water is not always immediately available upon land, and even when it is, factors like wind, extreme heat, and direct sunlight can quickly evaporate an organism's body fluids. Furthermore, marine environments are abundant with reliable food supplies, but many land animals are forced to endure long periods without nourishment. And finally, the body temperature of most marine animals conforms to the relatively stable temperature of the water around them; by contrast, land animals are exposed to daily and seasonal, and often extreme, changes in air temperature.

Despite the possibility of environmental fluctuations in food, water, and temperature, the internal chemistry of all animals has strict requirements for visceral constancy. When cells deep in the body need fluids, nutrients, or heat (more or less), they mean *now*, else they suffer tissue damage and decompose. Thus there were many selective pressures for the evolution of homeostatic (steady state) mechanisms enabling organisms to monitor internal physiological conditions and to compensate for disruptive changes in the environment. Homeostatic systems operate like the thermostat on your furnace. Internal sensors detect minute changes in physical conditions. When this happens, a neural signal is sent to a control center in the brain (the hypothalamus), where a physiological response is initiated.

The trick is to make an appropriate response before damage is done to body tissues. When the response is effective, the internal chemistry is restored to equilibrium. When a body is stressed by heat or cold, the hypothalamus may order up a reflexive response like shivering, sweating, panting, changing metabolic rate, constricting or dilating blood vessels. These reflex systems help to maintain stable body temperature by conserving, generating, or dissipating heat. Similar reflex programs regulate certain aspects of feeding, drinking, and mating behaviors. Reflexive responses are machinelike, stimulus-bound, invariant behaviors—what we would call involuntary. Like all behaviors, though, they represent the brain's answer to the question: What would help, in this situation, to carry on?

One can easily see how natural selection might wire in an extensive repertoire of specific answers in the form of neural modules for reflexive behaviors. And indeed it has. But let's not forget the principle of doing more with less—that is, the adaptive advantage of making the most effective use of neurons under the constraints of space, time, and energy. There is no way that a reasonably sized brain could contain a stock of reflex systems sufficiently varied to handle all the specific challenges to carrying on in a terrestrial environment. Therefore, if brains were ever to enlarge the repertoire of behavioral answers to environmental questions (which they obviously have done), they would have to transcend the reflex strategy by introducing more flexible systems.

A simple analogy to the retail paint industry might help to clarify. Guess who makes more money: store A, which satisfies demand by stocking a wide range of specific colors, or store B, which stocks a single color (white), but has the dyes and formulas on hand for matching virtually every color? Store B obviously does more with less. Neural modules for physiological drive systems transcend the powers of reflex systems in roughly the same way—that is, a generic substrate (in this case, the pleasure principle) acquires the power to organize a wide range of ad hoc behaviors in response to the many specific demands of the environment.

By the pleasure principle I mean the psychological dynamics that produce sustained goal-directed, reward-seeking, hankering behaviors—the sort you find in hungry rats prowling through a maze hunting for a satisfying morsel of food. Try as you may, you will never get a lizard to run a maze, because lizards can't sustain an interest. It's no exaggeration to say that lizards don't *want* anything. They just sit there, literal numbskulls, waiting for something to come along and trigger the next reflex, and not caring one way or another if it never does. Lizards are short on longings; they can't be moved by deficit experiential states to explore potentials for producing satisfied experiential states. To be so moved requires something more than reflexes.

Physiological drive systems evolved as new strategies for mediating behaviors. They are higher-order homeostatic systems featuring emergent properties of subjective experience—thirst, hunger, fatigue, stress, pain, cravings of various qualities. When an animal is aroused to thirst, it will engage in exploratory behaviors directed toward circumstances where it can drink, whereupon it will experience pleasure. The desire is thereby satisfied and the system is restored to equilibrium.

Visceral reflexes and physiological drives have much in common. They are both inter-neural modular systems that mediate behavior by the logic of homeostasis, and they both function to regulate the internal chemistry of organisms. The control centers for each are located within the nuclei of the hypothalamus. But there are major differences between them. Drives are sustained over much longer time intervals, and they are multiphasic—that is, involving potentially long series of discrete behaviors. Drives are also much more integrative, involving feedback routines from sensorimotor and memory systems. And drive systems are open to alternative trajectories to the same target, an openness that enables the inhibition of some behavioral options in favor of others. This last feature captures the essence of a drive system: What happens during the longer interval of a driven response is an on-going process of evaluation, where each successive option for behavior is judged for its utility to bring about a satisfying and pleasurable experience.

The evolution of physiological drive systems provided solutions to many of the special challenges faced by terrestrial animals. They are effective means for getting organisms to take proactive measures against the slings and arrows of changing environmental fortunes. The hunger factor, for example, is a good device for getting animals to fatten up as a hedge against tissue-damaging famines, just as thirst helps to prevent dehydration. And further, drives (unlike reflexes) are interruptible processes, freeing organisms to take emergency calls. Finally, looking ahead, the evolution of drive systems laid the neural groundwork for subsequent modules that would continue the trend of shifting the burden of adaptation from inherited traits to acquired abilities. In particular, the reinforcing power of the pleasure principle prepared the way for more sophisticated learning and memory modules, and the evaluative component set the stage for the emergence of emotional systems.

LEARNING/MEMORY SYSTEMS

The principle of doing more with less has been a consistent theme throughout the evolution of behavior, as new environmental challenges continued to select for efficient information-processing mechanisms. As we have seen, one strategy for this is to operate by the logic of statistical sampling, as perceptual systems do. Another strategy is neural plasticity. Here the idea is to design a range of flexible neural mechanisms that can be modified as organisms interact with particular environmental features. Plasticity of neural systems was especially adaptive for terrestrial animals whose greater mobility could potentially bring them into more varied ecological settings. Acquired behavioral traits are no less adaptive, and far more efficient, than inherited traits. Under conditions of sudden variations in environment, therefore, it made good evolutionary sense to develop heritable systems for adaptation by learning.

Learning and memory are two aspects of the same process. Memories are the artifacts of learning, such that as learning takes place, memory is acquired. But where do memories reside? Memories may exist in many systems throughout the brain, basically wherever experience has resulted in new connections between neurons, or wherever new firing thresholds have been established to make neural activity more or less probable or potent. Learning is the experiential promotion of changes in neural structures, and memories are the behaviorally relevant effects of these changes. The functional unit of memory is the "memory trace," or the *engram,* a term invented a century ago by Richard Semon to denote "the enduring change in the nervous system that conserves the effects of experience across time."[4]

Experimental psychologists have turned up lots of fundamental information about the different types of learning and memory. So much, in fact, that the field can be confusing. We find distinctions between short-term, long-term, and iconic memory; associative and non-associative forms of memory; habituation and sensitization; and then we have declarative and procedural memory, semantic and episodic memory, and classical and operant conditioning.[5] The point here is not to present a full account of learning and memory theory, but merely to give some sense of the diversity and extent to which plasticity of neural systems enters into the mediation of behavior. Virtually all of the so-called higher functions rely on neural systems that have been modulated by learning. Object recognition, concept formation, emotional experience, anticipation, planning, problem-solving, comparing and contrasting, decision-making, language use—none of these abilities would be possible apart from neural systems that generate memory objects as the raw data of mental activity.

The plasticity of neural systems amounts to a promise of freedom from the rigid constraints of genetic determinism. Consider the humble bacteria: They are condemned to behave pretty much as their genes declare. But in many organisms the genes have relinquished their tyranny by building behavior mediation systems that utilize the coding of engrams in tandem with the genetic code. The emergence of extra-genetic information systems was no trivial event in the grand scheme of things.

In the introduction I indicated that the view of nature informing this book makes distinctions between four basic types of natural facts: physical facts, biological facts, psychological facts, and cultural facts. According to this view, the first ten billion years of cosmic history amounted to a lifeless drama of physical facts. Biological facts livened up the story when the behavior of matter became in part organized by information encoded in genes, the units of inheritance. I am now suggesting that psychological facts first emerged in the epic of evolution when the behavior of matter came under the organizing influence of information encoded in engrams, the units of memory. While genes preserve changes wrought by natural selection across generations of species, engrams preserve the changes effected by learning throughout the lifetime of an individual.

The adaptive value of mechanisms for learning and memory has been profound. If you have engrams for where food and water are located, or where danger lurks, then your chances of carrying on are greatly enhanced. Adaptation by genetic change is always a game of chance, whereas adaptation by learning and memory eventually leads to a game of choice. The engram is therefore more than a unit of memory, it is a unit of freedom as well. The plasticity of the engram strategy proved to be so efficient and so effective for mediating behavior that one is left to wonder why the idea was not carried to further extremes than it has been. Why not go beyond mere plasticity to create ultimate learning machines: Creatures so free and variable in their behavior, so fluid in their nature that they may be said to possess no determinate nature at all, apart from their powers of self-determination?

Indeed, many social scientists and existentialist philosophers have insisted that human beings just are such creatures. Philosophers may say what they like, of course, but no one has yet shown how a creature of indeterminate nature could possibly exist. To survive by engrams alone would require a brain the size of Chicago and a curriculum of learning that would last for centuries. And even then, such a being would spend most of its time and energy on the aimless construction of engrams that would be totally irrelevant to its survival interests. The natural world has no tolerance for such monstrosities.

The so-called tabula rasa that figures into much social science and philosophy is a ridiculous fiction, not something natural selection would come up with. A far better option was to coordinate the engram strategy with genetic information. Organisms are advantaged by learning only when they learn the right stuff—that is, when engrams are relevant to the design of adaptive behaviors. Learning must therefore be selective, systematically constrained by innate biases, for otherwise it would be far too random and inefficient. Thus, for example, we are born with fixed biases for attending to some types of information while ignoring other types. Some learning biases may themselves be learned, but most of them are genetically endowed. And another thing: The advantages of learning often depend upon learning the right stuff in the right order. That is, some abilities are absolutely prerequisite to others. Developmental psychologists have identified several "universal competencies," providing clear evidence of genetically prepared learning mechanisms.[6] It is also important that behavior mediation systems be equipped with genetically controlled defaults, to assure that *some* behavior is forthcoming in the event that relevant learning fails to occur.

The phenomenon of *working memory* provides an excellent example of the coordination of genes and engrams. Consider that we have many different and relatively independent neural modules whose structural changes embody specific engrams of different types—some for colors, others for shapes, smells, sounds, facts, events, concepts, words, and so on. By what means and by what rules are these engrams activated so they can play their important part in the mediation of behavior? Should we expect to find a "super module" somewhere—some higher order neural mechanism serving as an executive memory retrieval system? The habit of thinking in evolutionary terms seems to encourage this idea. As organisms developed greater capacities for learning and memory, it seems reasonable that there would be selective pressures favoring an emergent system that could enhance the adaptive value of these plastic systems by retrieving and integrating their information in an orderly fashion. An ascendant view among neuroscientists is that such a system did in fact emerge in the form of working memory. Working memory is an attentional control system that holds several (but not many) pieces of information in a temporary mental "workspace," allowing the organism to perform various integrative functions on them. Working memory creates a mental arena where the challenges of the moment can be addressed by the influences of the past, both hereditary and learned. It generates memory objects relevant for determining what is momentarily at stake for the organism and how it should respond. In addition, and very importantly, *the assessment process appears to be genetically biased in favor of retrieving information that has been recently or repeatedly stored, or has been marked by unusually strong rewards or punishments.*[7]

The point to be emphasized for our purpose is that memory systems work as well as they do only because they are constrained by genetic programs. To be more precise: Engrams function as integral parts of computational systems, which are governed by algorithms encoded by the genes. Genetically endowed rules bias and constrain the operations of neural systems in many ways. Without them, mental development would be haphazard and memory systems would be chaotic. These rules are directly relevant for our sketch of human nature; indeed, they are at the core of human nature, for they account for many of the universal characteristics that we find in human development and behavior.[8]

EMOTIONAL SYSTEMS

The nature of humanity is shown by how the species has come to be, and how it continues to carry on. Any adequate understanding of human nature must take into account the evolution of systems for mediating behavior. Thus the narrative of this chapter started with molecular systems and then moved onward to reflex systems, perceptual systems, drive systems, learning/memory systems, and now emotional systems. The trouble with this linear approach is that it risks creating the wrong impression about biological evolution. It is tempting to view the emergence of higher-order mediation systems as coming aboard one by one, like a series of discrete computer upgrades. But such a view would be—except in the most superficial sense—highly misleading. A more fitting analogy might be the bizarre process of constructing a house with materials delivered randomly from a junkyard, where the architectural plans for the house change from day to day based on what the builders were able to accomplish with the previous delivery. The idea is that components of new systems are often on hand in some form before it becomes clear how they might be co-opted for adaptive responses to new environmental challenges.

The emotional life will play a central role in the following chapters, so it is especially important to come to terms about the evolution of emotional systems.[9] But here our troubles begin. Constructing an evolutionary account of emotional systems is notoriously difficult for various reasons. For one thing, it is almost impossible to get hold of reliable information about the anlagen, or precursors, of emotional systems. The most relevant information about emotional anlagen would be found in the brains of our common ancestors with the reptiles. But, alas, there are no extant species that closely resemble the common ancestor, which means that researchers are left to compare mammalian brains to those of modern reptiles. Looking to modern reptiles for the emotional anlagen is speculative, however, since we have no idea how far modern reptiles have diverged from ancestral lines.[10]

Still, the results of such research are highly suggestive. It has been shown, for example, that the amygdala, a structure in the ancient limbic area of the brain, is responsible for controlling defensive behaviors in reptiles. It has also been demonstrated that the amygdala is "the heart and soul of the fear system" in rats, cats, dogs, rabbits, monkeys, and humans.[11] This is not to suggest that modern reptiles are capable of experiencing emotions like fear, but there is a plausible evolutionary hypothesis here: That neural structures in the limbic region of the brain gradually took on new functions, leading to the development of emotional systems in mammals (and to some extent in birds).

Another inherent difficulty in constructing the evolutionary story is that emotional systems are almost certainly products of a convoluted process of coevolution, involving both affective and cognitive pieces in the puzzle. The limbic structures active in human emotional responses are massively connected by neural pathways to many sites throughout the brain. We know from clinical observations that lesions in the pathways between the limbic area and cognitive systems in the prefrontal cortex severely impair emotional responses.[12] And we know from developmental studies that emotional competencies have definite cognitive prerequisites.[13] It appears, then, that emotional modules are widely diffused throughout the brain—indeed, throughout the entire autonomic nervous system. These are highly complex systems with many component parts—hormonal, perceptual, mnemonic, cognitive—all working together in functional unity.

Emotional systems exemplify what cognitive scientists call parallel distributive processors—that is, higher-order mechanisms that perform computations "by the intricate system of connections within which a large number of extremely simple processing units are joined."[14] To construct a precise account of the evolution of these complexities would be a Herculean task—the story of just one system would involve evolutionary subplots to narrate the intricate coevolution of the system's many component parts.

These difficulties indicate that we will probably have to wait for some time before any measure of consensus is achieved regarding the evolution of emotions. To make matters worse, very little agreement exists among emotion researchers even on the basics—what emotions are, how many there are, how they work, and how they interact with other neural systems. Progress on these fundamental issues has been confounded by experimental constraints and insufficient technologies, as well as by the sheer complexity of the subject matter. So for the moment, at least, speculation reigns.

A plausible evolutionary scenario is that emotional systems emerged and diversified with the elaboration of neural connections between limbic structures and various engram-enriched sites throughout the cerebral cortex. Key mutations resulting in more complex neurocircuitry and more diverse neurochemicals would have made it possible to stimulate existing limbic structures in completely new patterns. And thus emerged heritable systems capable of making subtle cognitive assessments of what is at stake (harm or benefit, threat or opportunity) for an organism in its ongoing encounters with the external world.

The first emotional traits to distinguish themselves were probably simple systems for anticipating pain and pleasure. Thus one might view fear (or disgust) and desire (or longing) as the primitive anlagen to further elaborations of dissociative and affiliative emotional responses. The adaptive value of fear and disgust would be in motivating animals to avoid dangerous predators and harmful substances, while the adaptive value of desire and longing would be to generate the conjugal and maternal bonding typical of mammals. But additional emotional systems would follow, as selective pressures in the environment continued to favor genetically induced variations in neurochemistry and neurocircuitry. Thus, for example, subtle computational subroutines might modulate the fear response to produce emotions such as awe, respect, or humility. And modulations of desire and longing might produce affection, sympathy, or gratitude. In time, a multiplex of emotional mediation systems would emerge, each with its own inherent rules to govern the homeostatic events of arousal and restoration.

But what were the selective factors that might account for the evolution of a multiplex of emotional systems? Bearing in mind that emotional traits are primarily mammalian, we may rephrase the question as follows: What was it about the ecological circumstances faced by early mammals that favored the evolution of emotions? Nothing special, really, except that the planet was transformed sixty-five million years ago by the impact of a huge asteroid, an event that extinguished the great reptiles and accelerated the evolution of mammalian species. Suddenly, new ecological niches were up for grabs, setting off a process of intense competition among the early mammals.

The first mammals were tree-dwelling creatures, subject to strict limits on the number of offspring that could be raised effectively. Consider as well the central fact of mammalian life—nursing—and one gets a picture of females precariously transporting their dependent young from limb to limb, highly vulnerable to predation. Such conditions would have placed a high premium on social cooperation, both for defensive and nurturing purposes. Mammals were forced by circumstance to exploit the paradoxical principle that the best strategies for individual competition often involve a measure of social cooperation.

Mammals are not the only social animals, as a trip to any beehive or anthill will show. But social interactions among insects are governed entirely by reflex systems, whereas mammals owe their sociality to the patterns of competition and cooperation that emerge when individuals are motivated by the positive and negative effects of homeostatic emotional systems. In other words, emotional systems were selected for their utility in biasing individuals toward social behaviors.

Emotional traits did not develop uniformly across all species of mammals. Those leaving the trees early on, to pursue niches upon the ground, came to rely on various physical adaptations as well (size, speed, strength, endurance, weaponry, armor, and so on) and were, in proportion, less dependent on the benefits of social skills. But among the remaining tree dwellers (that is, the remaining primates), the evolutionary premium on sociality would continue to increase, leading to more refined emotional traits having the power to generate and regulate ever more complex social systems.

The evolution of mechanisms that bias individuals toward social behaviors remained a mystery to evolutionary theorists for nearly a century following Darwin's death. The problem was this: How could genes promoting a sacrifice of self-interest possibly evolve? Wouldn't such genes compromise the reproductive success of those who carried them? Natural selection, it was thought, should favor genes for selfish behaviors and eliminate genes for sacrificial behaviors. But then came the theory of *inclusive fitness,* which showed that genes prompting sacrifice would be selected if the sacrificial behaviors tended to enhance the survival and reproductive chances of close kin, who carry the same genes. Genes for helping kin, the theory goes, would be preserved in the kin helped.

For example, say a monkey is motivated by affection to sacrifice her life in the act of saving her child. Her own genes (including those responsible for affection) would be lost, but her child's genes (including those responsible for affection) would be spared for another chance in the reproductive game. By contrast, a parent who is emotionally indifferent toward her children would be too selfish to risk her life, leaving the children (and their ungenerous genes) to perish. This shows us how genes for affiliative emotions (such as affection, sympathy, gratitude) might be selected in environments where social solidarity and cooperation are beneficial. The principle at work here is known as *kin selection,* which specifies that individuals will inherit traits (in this case, emotional systems) predisposing them to behave altruistically toward kin, in proportion to their relatedness.

Once affiliative emotions became fixed to regulate interactions among kin, there was a good chance that they might be extended, conditionally, beyond the family circle, thus to enhance the prospects for broader social cooperation. The principle of *reciprocal altruism* describes these conditions. This principle says that individuals will tend to sacrifice for (or cooperate with) non-kin in proportion to favors already received, or when there is a good chance that favors given will be returned. Surely, however, there had to be limits beyond which sacrificial behaviors would have become maladaptive. That is, there had to be braking mechanisms to prevent individuals from becoming emotional doormats for selfish opportunists. Thus, to mitigate the excesses of self-giving emotions, we find the self-preserving forces of fear, resentment, and anger standing ready to deliver resistance and retaliation.

The primate brain is, almost literally, a social artifact. Like all brains, it evolved as an adaptation for carrying on in its environment. But given that the environmental challenges facing primates favored complex social behavior, the primate brain became specialized for negotiating a social world. Such brains were shaped by a coevolutionary race between emotional and cognitive traits. Clues to the coevolutionary process are embedded in the twin functions of emotional systems: to motivate and to communicate.

The motivational function is already familiar: An individual aroused by an emotional system is held in a deficit state until it performs some variation on a genetically prescribed behavior (attack, retreat, approach, assist). When the action is performed satisfactorily, the system returns to the equilibrium state. The communication function arises from the fact that as the emotional process unfolds, it manifests socially relevant information that can be picked up by conspecifics. That is, emotions are expressed in a variety of sounds, facial gestures, eye movements, body language, breathing patterns, bristling hair, and so on.

Cognitive abilities for decoding these expressions would have coevolved step-by-step with the emotions themselves, for the obvious reason: Correctly assessing the emotional states of others enables an individual to make reliable predictions about their behavior, and thus to adjust one's own course of action with advantage. Adjustments of this sort are very likely to involve an emotional factor themselves, since individuals often find much at stake in the impending behavior of others. The emotional expression of one individual might trigger an emotional arousal in another, and another, setting off a ripple effect with unpredictable consequences. It is easy to see how the coevolution of emotions together with the cognitive abilities to appraise their significance might lead to a complicated social fabric. Picture this:

A is motivated by affection to bestow a favor on B.
B is consequently motivated by gratitude to repay A, perhaps by aggressing against C, an opponent of A.
C becomes fearful of B and thus wary of D, a known ally of B.
C thus avoids D's overture to an alliance.
D, sensing rejection, becomes visibly saddened.
E is moved to sympathy by D's visage and offers comfort.
Etc., etc., etc. . . .

It takes no more than a few basic emotional traits to construct a kind of social mobile, with individuals bobbing, twirling, and swinging to-and-fro on their strings, moved and counter-balanced by the twin functions of motivation and communication. In the course of primate evolution the emotional anlagen (fear, disgust, desire) diversified to include a repertoire of basic "social emotions," such as affection, sympathy, gratitude, humility, resentment, and anger. These provide the dynamics of reciprocity underlying the formation of primate mating patterns, familial groups, cooperative alliances, and social hierarchies (and religious traditions, as we shall see). The influence of these social emotions appears to stand in proportion to the complexity of a species' social system.[15] The more emotions there are to generate and regulate interactions, the more complex the social mobile is bound to get.

It would take increasingly discriminating brains to negotiate an increasingly complex social environment. Individuals failing to read emotional signals properly would tend to be excluded from the reproductive game, leaving the breeding pool better stocked with socially talented brains. But see how the plot continues to thicken. Brains sophisticated enough to appraise the significance of subtle emotional cues can also be expected to come up with various unruly insights, like this one: "Hey, it is possible to fool these emotional systems, thereby to manipulate the behavior of others to my advantage." Once the wrinkle of deceit factored into the social mobile, matters became even more challenging. Now if you want to negotiate the social environment successfully, you must be capable of even more acute appraisals of what is at stake in your encounters. Now you must determine accurately whether or not you're being put on.

The potential for manipulating emotional systems accelerated a cognitive arms race between deceivers and detectors. The worst deceivers were selected out by good detective skills, leaving only the best deceivers to weed out the poor detectors.[16] And so on. The process spiraled to produce brains capable of discerning between appearance and reality on the basis of minute signifiers, and brains artful enough to falsify signals deliberately—that is, brains in possession of important cognitive prerequisites for symbolic communication.

What can be said of such brains? First, we may say that they must possess considerable capacity for learning. Negotiating the primate social environment required engrams for recognizing particular individuals, and for recalling details of past encounters with them. If X abused me or cheated me last week, then my appraisal of what is at stake in my next encounter with X should take this into account. Perhaps my appraisal will arouse me to fear, in which case the system will prompt me to follow a defensive strategy—to run away or to recruit an ally. If these strategies look unpromising then I may try to appease X with a gift of food. Alternatively, my appraisal may result in anger, which would prompt a strategy of retaliation. In any event, my appraisal of the situation will depend on the emotional colorings of my memories of X. But if my memories fail me then I am likely to handle the encounter poorly.

Relevant as memories are to the appraisal process, anticipating the future is equally important. If I am to behave adaptively, I must be able to construct a running account of how things are going forward in the encounter, which strategies look most promising, and how matters are likely to turn out under alternative options. In other words, my brain must construct an on-going narrative of the encounter, a kind of flow chart where each option for behavior branches off to describe an alternative outcome. This ability requires more than engrams—it requires also the operating theater of working memory, where emotionally marked mental objects from the past can be integrated with sensory images of the present to project alternative possibilities for the future. This sort of mental gymnastics requires a blizzard of looping subroutines to be carried out along neural pathways connecting many parts of the brain. As we shall see, it is very probably this narrative process that underlies conscious experience.[17]

Antonio Damasio has argued that the operations of working memory are essentially evaluative—that is, the narrative thought process is driven by fundamental values inherent in biological regulation.[18] These implicit values account for the fact that organisms can selectively process vast amounts of information in a focused, coherent fashion. They describe the goals—the *teloi*—the adaptive meanings that organize and energize the psychological process. At the level of ultimate value we must say that all living things have it in their nature to carry on, to survive, and to reproduce. But in the strategies of a species' mediation systems are inscribed various instrumental values—indeed, a hierarchy of teloi—that function as the means to reproductive success. There are teloi inherent in molecular systems, in reflex systems, perceptual systems, drive systems, memory systems, and emotional systems. The nature of a species is defined by the teloi embedded in these mediators.

All mammals, and especially primates, are emotional creatures, which means that the teloi of their emotional systems are to a large extent responsible for organizing their experience and behavior. To fulfill the teloi of these systems is in some important measure to fulfill one's nature. Thus we may say, for example, that an angry chimp seeks to fulfill its nature by executing a retaliatory attack on some offender. But this is a small part of a chimp's nature; anger is only one of a suite of emotional systems which, taken together, say something about the defining teloi of primate nature. More broadly considered, primate life is organized by inherent algorithms that promote both selfish and sacrificial behaviors. The teloi scripted into primate emotional systems tell us that here are creatures that often value the integrity of relationships as much as they value self-preservation. Indeed, the primate brain itself was assembled in this dialectic between self and society.

The next chapter will focus in greater detail on the operation of emotional systems. Our task here has been merely to sketch out the story of their evolutionary development. In the process we have turned up new material to be added to our sketch of human nature. Humans are, after all, primates, which means that we, too, are emotional creatures, destined by natural history to work out our salvation on the strings of a social mobile. The distinctive nature of humanity must be sought in new strategies for doing this.

COGNITIVE SYSTEMS

Aristotle wisely observed that all humans are motivated by the desire to know. He would have been even wiser had he extended the principle to other species as well. That is, all living things have it in their interest to internalize some measure of information regarding states of affairs in the external world, whether it be something very simple, like the angle of the sun, or something complicated, like the causes of a political revolution. For neural animals these matters regarding internal representations of external affairs fall into the complex domain of cognitive science.

One of the most presumptuous book titles in the history of publishing has to be Steven Pinker's *How the Mind Works*. But to Pinker's credit, he is always the first to crack jokes about the hubristic title, and further to his credit is that the book makes a darned good stab at the subject. But perhaps there is less presumption than polemic in the title. Pinker is delivering the final, if not posthumous, deathblow to behaviorism. During the hegemony of behaviorist psychology there reigned a dogmatic prohibition against all theories of mind. Speculations about what is really going on in the black box where stimuli become responses were considered unscientific. Science should not traffic in concepts about unobservable mental events and their underlying mechanisms.

The ascent of cognitive science has given new license to theories about how the mind works, and Pinker's book is one of several calling attention to the fruits of some important post-behaviorist insights, among which are: (1) that the brain is not a general purpose organ of intelligence, but rather a collection of computational modules; (2) that these task-specific modules were assembled and coordinated by the dynamics of evolution; and (3) that scientific inquiry about how they work is possible and legitimate.

One might suppose that the ultimate goal of these inquiries is to produce a definitive map of the modular brain: what its many modules are, where they are located, how they process information, how their functions are arranged hierarchically, how they overlap one another, and so on. If this is the goal, then we are not there yet, but cognitive scientists are making significant headway (so to speak) in understanding how the mind works.

We are still on the lookout for material to add to our sketch of human nature, and in this section our interest is focused on the mechanisms allowing us to internalize information about the world around us. At one level we can say that genes themselves represent an indirect kind of theory about the external world. Genes do not offer up specific information about how things really are in the world, but they do make a wager about them. The genes say, "We're betting that the world is such that if you build the cognitive systems inherent in our blueprints, then you'll have a very good chance of surviving." This wager points to the only decent argument we have to support our philosophical confidence that human sense and reason are reliable measures of reality. It seems to make good evolutionary sense: The genomes that wagered poorly have been ushered off the racetrack. But even so, we should be humbled philosophically by the fact that lots of other winning species remain on the track with us, and there is good reason to believe that many of their ways of internalizing external affairs are substantially different from ours.

Does this imply that human cognitive systems are really no better as measures of reality than those of, say, dung beetles or woodchucks? If so, then there is probably something faulty about the racetrack analogy. Perhaps it would be more fitting to say that different species wager their genomes in different races. A woodchuck's understanding of the world may appear impoverished when compared with the human's, but it is well suited to the niche occupied by woodchucks. The real test of cognitive superiority comes whenever the human project is carried into the habitats of other species. And when that happens it turns out that humans create losers by the score. We have a mass extinction going forward, in part because humans are flatout better able to internalize external affairs than any other species.

It stands to reason that the woodchuck race would favor increased cognitive abilities, too. Why, then, aren't woodchucks smarter? The answer has to do with the trajectory of competition. In the woodchuck niche, having a slight edge in cognitive abilities was presumably less important than having other traits, such as strong forelegs and sharp teeth. If so, then selection for thinking traits would be relaxed relative to selection toward digging and chewing traits. And this would make all the difference.

What can be said about the primate race is that the trajectory kept veering in the direction of cognitive abilities—that is, the abilities to take in a widening swath of information, coupled with abilities to process the information in new ways. We are as smart as we are because the odds on cognition wagers kept increasing, with the result that we (not dung beetles or woodchucks) have come to occupy a niche favoring neural modules capable of generating grand theories about the world. The most interesting challenge, relative to a sketch of human nature, is whether we can find a satisfying way to think about these modules.

MENTAL OBJECTS

The mind works by performing a wide range of operations on a wide range of mental objects, thus to address a wide range of adaptive problems. At the formal level this looks like a fairly simple process to describe. All we need to know is what mental objects are, how we come to have them, and what sorts of operations we subject them to. What makes this difficult is that everything the mind does is a function of neural matter—that is, the wires and juices of the brain. How these can produce mindworks is one of the ultimate questions of human self-understanding, and it would not be surprising to discover that it exceeds our powers to answer it. A favorite quip among neuroscientists warns that if the brain were simple, then we would be too simple to understand it.

According to Pierre Changeux, mental objects include percepts, memory images, and concepts, all of which may be featured in working memory.[19] *Percepts* arise when sensory information is assembled and interpreted by the brain. The relation between sensation and perception may be illustrated by the familiar "face-vase" problem. Subjects are shown an ambiguous drawing that may be perceived either as a pair of interfacing human profiles or as a single vase. The visual information transmitted from the eye to the brain is exactly the same for each subject, yet the information may be resolved differently. Two subjects may therefore have the same sensations but vastly different perceptions.

Once formed, percepts may become fixed in memory by formation of an engram that can be reactivated in working memory. However, the process of remembering a percept is not a simple matter, precisely because perceptual engrams are complex entities. It happens that various sensory data (color, shape, and motion, for example) are processed and stored in different parts of the brain. It is therefore possible that engrams might be only partially reactivated in memory events. And further, memory recall is always affected in subtle ways by the mechanisms of retrieval.

In other words, a percept is stored in memory as fragments of an original whole, and the act of remembering the original event is actually a process of selectively reactivating bits and pieces of engrams and using them to construct a new kind of mental object, the *memory image*. Memory images are not percepts, nor are they exact reproductions of original percepts. We are closer to the reality if we say that memory images are themselves original compositions, unique patterns emerging from the computational mixture of engram fragments and present information.[20]

Concepts are original compositions, too, but of a different order. This is not to say that concepts are not encoded by neurons, which they certainly are. It is rather to say that conceptual mental objects represent various features that cut across particular percepts and memory images. Say little Johnny has no conception of "teacher"— all he knows is that Ms. Jones is there in the classroom each day directing a host of activities. But as Johnny hears reports from playmates in other classrooms he eventually works it out that there is a role, composed of similar functions, which may be performed by various individuals in a variety of settings. Suddenly, Ms. Jones joins a general category populated by many persons who may perform the role. The concept "teacher" excludes all the particular fragments making up the memory images of Ms. Jones *except* those that compare with other teachers. It is as though Johnny has selected certain fragments from his Ms. Jones engram to allow their variable attribution to a range of other individuals. Now that he is equipped with a newly emergent mental object (the concept of teacher), Johnny is predisposed by a set of expectations to behave adaptively in future encounters with other teachers.

Concepts are engrams, too. The constituent neurons of the complex engram for teacher are no doubt laced out over several parts of the brain, but they may be selectively activated by retrieval mechanisms and brought forward to participate in the construction of new mental objects. Fragments of the Ms. Jones engram will participate in composing the concept of teacher, and fragments of the teacher engram may link up with fragments of the priest and policeman engrams to form a concept of authority. Nobody knows for sure how percepts, memory images, and concepts are interrelated at the neural level, but the presumption is that there is some sharing of neural matter that can be modulated to produce variable states and patterns of activity.[21]

Here is the picture: Fragments of percepts become involved in the composition of memory images; image fragments contribute to the construction of other images; images also play a role in the construction of concepts; concept fragments may be activated in the neural assembly of new concepts; and concept fragments can also contribute to the construction of images and even to the construction of percepts. Thus, existing mental objects of each type may be relevant for the brain's generation of new mental objects of each type. The power of the brain to compose new engrams from the constituents of pre-existing engrams accounts for our singular ability to construct elaborate and sophisticated theories about how things are in the world of external reality.

It is by virtue of this power that a select group of species manages (more or less) to transcend the "out of sight, out of mind" condition that is characteristic of most animal species. As the brain splices together various image fragments, the mind can behold objects and events that are no longer available to our senses. And beyond that, the mind can behold inferred realities that could *never* present themselves to the senses. For example, I can imagine Socrates presenting a multimedia seminar on artificial intelligence, and I can imagine three-headed tartan-clad dragons with an addiction to romance novels.

In the course of a few years this engram-generating power of the brain builds up an enormous supply of mental objects that interact to produce a complicated internal encyclopedia representing a manifold of external affairs. There can be no question that there is enormous adaptive potential in this astonishing ability to assemble an internal world of thought which presumes to be an adequate stand-in for the features of an external world of reality.

MENTAL OPERATORS

Our questions now turn from mental objects to the operations we perform on them. That there are many such information-processing operations is evident from a moment's reflection. What is not so evident is how many there are, how we come to have them, and by what rules they are governed. Much is already known about how the mind receives upstream sensory information, and much is known about the neural execution of downstream motor systems. Most of the mystery has to do with the rules governing inter-neural cognitive systems, precisely where the definitive aspects of human nature reside.

Any model of mental operations must take into account the functional requirements of the brain. All good brains must have rules enabling them to: (1) encode information representing the external world, (2) encode information representing the organism's vital interests, and (3) bring these forms of information together in the process of devising appropriate behaviors. These functional requirements suggest a tripartite model for thinking about the types of rules that govern the operations we perform on mental objects. We may hypothesize, therefore, that the brain's mindwork is achieved by three types of modules, which I will call reality operators (corresponding to facts), valence operators (corresponding to values), and executive operators (corresponding to outcomes).

By *reality operators* I mean rules and mechanisms that govern neural systems in processing information about how things are (or at least appear to be) in the external world. That is, information about the kinds of things that exist, what properties they possess, what sorts of events take place, and how things, properties, and events are related.

Valence operators are rules and mechanisms that guide neural systems in mapping information about an organism's bioregulatory systems—the biological value system—onto mental objects generated by the reality operators. By virtue of this process various shades of relevance and value are attributed to a value-free world of objects, events, properties, and relations.

Executive operators are rules and mechanisms for bringing value-laden mental objects to bear on the process of generating and assessing options for behavior in response to whatever challenges and opportunities life sets before us.

The most basic types of reality operators are modules governing the initial buildup of mental objects. Here I have in mind innate rules and mechanisms enabling us to form categories and concepts for objects, events, properties, and relations based on the most salient elements in perceptual experience. All humans come into the world prepared by these operators to render their experience intelligible by formulating categories. Thus it happens that humans everywhere form categories based on color, form, pattern, texture, movement, size, speed, frequency, sequence, sound, proximity, duration, function, derivation, assembly, decay, and many other salient characteristics of perceptual experience.

Such categories provide the rudiments for constructing reliable knowledge about the external world. Nobody comes into the world with ready-made categories, but we all arrive with a common set of rules and mechanisms for creating them. The point about these innate modules is not that they enable individuals to bypass learning, but rather that they prepare individuals for specific learning tasks—

that is, those that have been most common or crucial in the evolutionary history of the species.

Steven Pinker believes there is enough specific information in the algorithms of innate modules to assure that humans everywhere will be competent intuitive scientists:

> People in all societies have words for abstract conceptions, have foresight beyond simple necessities, and combine, compare, and reason on general subjects that do not immediately appeal to their senses. And people everywhere put these abilities to good use in outwitting the defenses of the local flora and fauna. We will soon see that all people, right from the cradle, engage in a *kind* of scientific thinking. We are intuitive physicists, biologists, engineers, psychologists, and mathematicians.[22]

The concept of intuitive science is provocative. One might surmise that most animals—mammals, at least—inherit the means for constructing their own species-typical intuitive science. How the world looks to a woodchuck is not how it looks to a chimp. A human intuitive physics amounts to a wide range of easily acquired knowledge about the kinds of objects that exist in the world and how they will behave under given conditions. Some things bounce or bend, others break. Some things float, others sink. Add formless water to formless dirt and formable mud results. When a big thing breaks apart there is left a plurality of smaller things. Push something and it will move, bigger things require more force. Pinch a berry and it will squirt, pinch a stone and it won't. Tap something lightly and it makes a sound, strike it hard and it makes a louder sound. Wind can move some objects but not others. Some things are alive and able to initiate their own movement, others things need help to move. Everybody everywhere picks up on these salient features and constructs similar mental objects about them. All that is required is minimal exposure, and the modules for intuitive math and physics take it from there.

The same goes for intuitive biology. Everybody everywhere picks up on the salient similarities and differences among flora and fauna, and constructs mental objects accordingly. There are no cultures that catalogue tree-dwelling mammals with birds, or ground-nesting birds with rabbits. And further, when monkeys go from limb to limb, people everywhere call it a leap, but when birds go, they fly. Every human population develops a keen sensitivity to the cycles of life, including birth, growth, decline, and death. In addition to knowledge of our own species, we are biased to learn about patterns in the life cycles of many different species of plants and animals. Plant and animal materials (such as skins, bones, roots, leaves, bark) are exploited in every culture for use in medicine, clothing, shelters, tools, weapons, recreation, and adornment. If there were no innate modules for organizing biological knowledge, then we could reasonably expect this knowledge to be far more arbitrary and variable than it actually is. The decidedly non-arbitrary character of folk biology tells us that natural selection has biased the brains of all humans to jump to the same intuitive conclusions.

I have been arguing that human nature includes various reality operators—innate neural mechanisms that bias our processing of information about salient features in the external world. These operators are sufficient to produce a suite of universal intuitive theories about the world. But reality operators alone are not sufficient for good brains. It is important to recognize that salience is not the same as relevance. The salience of a fact may trigger us to pick up on it, but it carries no information about why the fact may be important. Intelligibility is one thing, significance is quite another. That is, the meaning (relevance, value) of external facts is not objective, not something "out there" to pick up on.

This discrepancy between facts and values has been formulated by philosophers as the so-called naturalistic fallacy. This doctrine says that facts in themselves are value-neutral, which means that however many facts you line up, you can never get them to imply anything about the value of something. One does not detect value as a salient feature of any object, event, property, or relation for the simple reason that there are no values out there to be detected. This does not mean, however, that values may not be assigned to facts, which is something we do all the time, and without hesitation. Thus we automatically experience snakes as dangerous, and thunderstorms as fearsome. Our tendency to do so is not a function of reality operators, but rather a function of innate valence operators interacting with reality operators in the construction of mental objects. Valence operators are the rules and mechanisms that evaluate the significance of external facts relative to the biological teloi of our species.

The point here is that humans come into the world equipped with a biological value system, or what in broad terms might be called a species-wide intuitive morality. In fact, every species may be said to possess an intuitive morality, by which I simply mean that each species may be described as having various goods and evils relative to its own biological teloi. The values of human intuitive morality are embedded in the goal-directed workings of our basic drives and emotional systems. These valence operators have been biased by natural selection to evaluate certain patterns of incoming information as positive (those conducive to biological goals), and others as negative (those detrimental to biological goals).

The interactions between reality operators and valence operators occur in the neural pathways looping back and forth between the cortex and the limbic area. In these interactions information about the value-neutral external world is integrated with value-laden information about internal body states. In this process the naturalistic fallacy is transgressed with a vengeance, as facts become overlaid with values. Interactions between reality operators and valence operators are mutual and continuous, which means that our affective states can bias our perceptions of reality, and our perceptions of reality can modulate our affective states. Indeed, it is this on-going process of interaction that results in our experiencing a world infused with meaning.

Intuitive science and intuitive morality are integrated in what I will call an intuitive worldview. By this term I mean a coherent set of mental objects that build up over time under the guidance of innate operators for determining how things are in the external world and which things matter for achieving our biological teloi. For the moment I am ignoring the important fact that no one has ever encountered the human intuitive worldview in its pristine form. But this does not rule out the assertion that humans are genetically endowed with a universal species-typical default mentality that interacts with cultural influences in various ways. I will return to this issue shortly, but first it is necessary to say something about the executive modules which enable us to make good use of our intuitive ideas about how things are and which things matter.

WORKING MEMORY

Mental objects do not enter into the behavior mediation process at random. The engrams that constitute our intuitive mentality can be counted on to slumber right where they are until they are called upon to send impulses to a centralized system where the final mediation is carried out. Many philosophers and psychologists react strongly against any reference to an executive center because it conjures up the image of a ghostly homunculus closeted in the neural machinery, there to observe everything going on in the mind so that it alone can freely decide what to do. There is no doubt that the postulate of a well-informed, wide-awake transcendental tyrant can be misleading. For one thing, it restores the view that the mind is a general intelligence device. Nevertheless, it is obvious that highly organized, goal-directed behavior happens, and if it happens there must be a coherent centralized process behind it. Working memory is this process.

Steven Pinker suggests that we envision working memory as yet another neural module that operates according to algorithms for processing information:

> The society of mind is a wonderful metaphor . . . But the theory can be taken too far if it outlaws any system in the brain charged with giving the reins or the floor to one of the agents at a time. The agents of the brain might very well be organized hierarchically into nested subroutines with a set of master decision rules, a computational demon or agent or good-kind-of homunculus, sitting at the top of the chain of command. It would not be a ghost in the machine, just another set of if-then rules or a neural network that shunts control to the loudest, fastest, or strongest agent one level down.[23]

These are the questions: What is working memory? Where is it located in the brain? How does it do its mindwork? We may think of working memory as a highly integrated system of neural modules whose global function is to orchestrate the influence of subsidiary systems for mediating behavior. It juggles mental objects and winnows their competing impulses so that the organism is able to construct coherent and satisfying responses to challenges and opportunities at hand in the environment.

Neurobiologists now have a pretty good picture of where the principal components of working memory are located in the brain. Joseph LeDoux suggests that "the lateral prefrontal cortex is involved in the executive or general-purpose aspects of working memory."[24] This part of the brain is believed to exist only in primates, and is much larger in humans than in our simian cousins. Another area of the frontal lobe, the anterior cingulate cortex, is also implicated in working memory. These regions of the brain are anatomically interconnected, so that "it is tempting to think of the general-purpose aspects of working memory as involving neurons in the lateral prefrontal and anterior cingulate regions working together."[25] If you had to design an executive center in the brain—a place where attention and intention happen—you would make sure it was well connected to sensory, mnemonic, visceral, and motor systems. These structures in the frontal lobe fit the specifications perfectly.

The mindwork of the executive operators is probably much more complicated than present models suggest, but generally it involves the assembly and manipulation of mental objects, receiving input from various systems throughout the brain.[26] LeDoux's functional model (figure 2) places working memory at the juncture between top-down and bottom-up procedures. Information from engrams in long-term

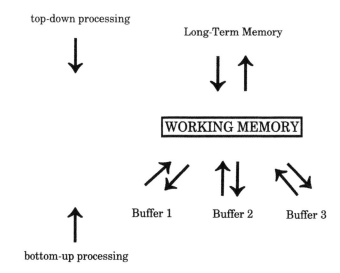

Fig. 2. A model for working memory (adapted from Joseph LeDoux, *The Emotional Brain*, New York: Simon and Schuster, 1996).

memory is assembled in working memory along with input from several independent buffers. Buffers are short-term memory stores for various information-processing systems. Each sensory mode has one or more temporary buffers that enable perception to occur. For example, visual buffers make it possible to connect early phases of an event with later phases, so that the event is experienced as a continuous unity. There are also language buffers enabling us to keep in mind the first part of a sentence until we hear the last part. Without language buffers we could not apprehend the unified meaning of a sentence.

Working memory is the virtual workspace where several mental objects can be juggled simultaneously as cognitive operations are performed. Considerable research has suggested that the capacity of working memory is about seven items (plus or minus two), although some theories allow as many as twenty cognitive units.[27] The idea is that the working memory can attend to only a limited number of items while cognition proceeds. When we try to add more, we displace other items from the workspace. What gets into working memory is determined by *rules that are biased for retrieving recent, recurrent, and heavily valenced information.*

The operations of working memory include reasoning, problem solving, language comprehension, memory searching, decision making, sequencing of priorities, and construction of narratives. The challenge for the executive operators is to assemble a sufficient amount of information in the attentional workspace, and to hold the items in an active state long enough to get these cognitive tasks completed. Above all, the working memory has to produce coherent goal-directed responses—that is, it must create the neural conditions for orderly and adaptive behaviors. This is an extremely difficult thing to pull off because the executive operators have to contend with the simultaneous demands of a plurality of homeostatic motivational systems. Humans have many goals to pursue, but they cannot all be achieved at once. The working memory must therefore accomplish what politicians can only promise—that is, to satisfy a multitude of special interests with efficiency and fairness, and for the common good. Orderly outcomes are accomplished in a blizzard of neural computations, countless and continuous "if-then" operations performed on the items maintained in working memory.

Many neuroscientists now believe that working memory is the theater of consciousness, the seat of the soul, where mental objects are generated by lightning-fast inferences and finally merged into intentional designs. It might well be that consciousness is a function of the capacity of working memory. That is, perhaps the threshold to consciousness is crossed only when the working memory manages to maintain and manipulate a sufficient complexity of mental objects. What we normally mean by consciousness is the awareness of a subjective perspective on the world, the familiar subject-object encounter. We may envision this experience of subjectivity as arising with an individual's ability to imagine and assess alternative narratives about the future. In order to do this an individual would first have to construct mental objects representing different possible future outcomes based on different hypothetical behaviors (*if* this action, *then* that consequence). Then the individual would have to compare these outcomes and appraise them relative to the valences representing different biological teloi. As the voting process moves forward, an intentional subject emerges. More on this in the next chapter.

SECONDARY OPERATORS

I have argued that human beings are genetically endowed with behavior mediation systems that bias the construction of a species-typical worldview (that is, intuitive science and morality). There is a universal human nature predisposing all humans toward a common mentality. Why, then, has no one ever observed this universal worldview? Why, when we move from culture to culture, do we encounter fundamentally different ideas about how things are and which things matter?

Trained observers may detect a few subtle differences in behavior between two groups of chimpanzees, but for the most part one chimp society carries on pretty much like the next. We may speak with confidence about a common mentality among chimps. But when it comes to humans we have to work hard to identify elements of an intuitive worldview. If humans are so profoundly alike, then why are they so radically different? There must be features in our shared nature that allow for the emergence of diverse ideas and values. That is, we must have in our nature the resources for overriding the biases of our innate reality operators and valence operators.

I will hypothesize that the working memory capacity of human beings makes possible the ad hoc construction of *secondary operators* that compete with our primary operators in ways that inevitably and differentially override our intuitive worldview. If this is the case, then we can reconcile an affirmation of cultural relativism with the claim that there exists a universal human nature.

The reason why humans everywhere do not think entirely alike is that many of the rules by which we construct, manipulate, and evaluate mental objects have been acquired by extra-genetic means—that is, by individual invention or cultural convention. Bird categories may be universal and inevitable given both our genetic makeup and the ubiquity of birds, but what about the highly contrived category defining all red objects smaller than a house that one encounters within ten minutes of leaving the dentist's office? Nothing universal or inevitable about this one. Nor is there anything intuitive about the periodic table of elements, the second law of thermodynamics, a commitment to democracy, or embarrassment about nudity. The variability of mental objects is almost without limit, which is compelling evidence for the influence of culturally determined rules for human cognition and valuation.

Thus we may distinguish between primary and secondary operators. Primary operators are the algorithms we inherit, scripted in the genes to bias our construction and manipulation of mental objects. By contrast, secondary operators are the algorithms we invent—they are themselves mental objects that guide us in the ad hoc construction and manipulation of other mental objects. The interesting question about secondary operators asks how they arise to compete with the primary operators.

First the secondary reality operators. The argument here is that an increase in the capacity of working memory would establish the conditions under which extra-genetic rules for category and concept formation might emerge. If the capacity for working memory were to increase even slightly, then more items—and more complex inferences—could be attended to in the workspace. For example, instead of "if A, then X," one might get "if A and B, then not X." In other words, when the complexity of inference in working memory increases, then so too would the potential for detecting novelty or discrepancy in the environment.

Individuals with more working memory capacity might pick up on non-salient features that cannot be assimilated by the rough and ready categories and concepts of intuitive science. Under some conditions intuitive categories will prove unsatisfying, and the individual will remain in a state of cognitive arousal. Still curious, the individual will be motivated to run deeper memory searches, scanning for fragments that might contribute to more satisfying and more sophisticated inferences. Individuals will now be able to reason more indirectly, and to posit the objective existence of unseen realities that would be inconceivable to a more impoverished working memory. This process of suppressing the primary operator in favor of something more promising will not always succeed, but when it does, it creates a new kind of rule for subsequent mental operation. When the mind cannot readily make sense of novelty or discrepancy, it takes license to construct ad hoc rules enabling it to do so. The urge to understand, like pressurized blood in a clogged artery, makes new channels.

By virtue of secondary reality operators the categories and concepts of intuitive science may be subject to refinements, or they may continue to be overridden and virtually displaced by new habits of mind. And further, it is likely that these new habits of mind will contribute to subsequent detections of novelty or discrepancy, with the result that individuals gradually increase their reliance on secondary operators. Thus it may happen over the course of a lifetime that two individuals of the same species, who share a common intuitive mentality, will come to organize their experience of the external world in radically different ways.

Nevertheless, the primary operators remain lurking in the background, always insinuating their wisdom into our thoughts and problem-solving strategies. After all, they represent the default mentality of our species. This is important to bear in mind for the following reason: If there are no means by which secondary operators can be objectified and transmitted to other minds—if there is no symbolic language—then whatever cognitive diversity is achieved by individuals will remain subjective, and will be lost when the individuals die. When old dogs die their tricks die with them, and each new generation defaults back to the provisions of intuitive science. Such was presumably the way things were among humans before the invention of language, prior to which there was very little cultural diversity.

If cultural variations regarding matters of fact arise from the influence of secondary reality operators, then one might reason that variations in values can be traced to secondary valence operators—and that these, too, are contingent on the uniquely human capacity for working memory. Or so I shall argue.

Primary valence operators are, to repeat, the very rules and mechanisms that govern our physiological drives and emotional systems. They represent the deep biological goals and values of the species. We may say that these homeostatic mechanisms are components in a comprehensive system of self-maintenance. That is, together they monitor internal states and register what the organism needs in order to maintain itself. They evaluate the potentials in any situation relative to biological goals and motivate the organism to respond adaptively. This is the gist of what I have called a universal intuitive morality.

But a moment's reflection reveals that humans pursue many and diverse goals that often transcend, and even contradict, the teloi of our intuitive morality. Such pursuits reveal the work of secondary valence operators—that is, extra-genetic rules that regulate an emergent value system superimposed upon the biological self-maintenance system. Basically, the idea is that an increase in working memory capacity will result in an enhanced sense of selfhood and a new process of self-monitoring, which in turn will dramatically modulate the operations of our emotional systems. This new process creates the conditions for values to proliferate and diversify.

SELF-ESTEEM

Prior to the second year of life children classify other persons as highly complicated objects. But at about eighteen months the child begins to appreciate that persons are more than mere objects, they are active agents—that is, unique centers for initiating behavior. Eventually the child sees that the agent properties it has learned to attribute to other persons are also possibilities for itself. In other words, the child applies the person-as-agent concept reflexively, thus to perceive the self as an actor, the subjective center of social narratives.

This enhanced sense of self-understanding amounts to a fresh orientation toward the social world, resulting in an intense interest in social roles. As children assimilate the logic of social roles they learn to apply the standards of performance implicit in those roles. And once the standards are internalized the child will apply them reflexively in a narrative process of self-monitoring. The point of self-monitoring is to observe the outcome of one's own behavior, and to judge the outcomes (good or bad) by the standards of performance. In the course of countless particular performance assessments, an individual is led to make inferences of a general nature about the worth of the self. Consistently poor performances will contribute to a sense of negative self-esteem, while good outcomes will count as evidence favoring a sense of positive self-esteem.

Enhanced working memory in humans resulted in an entirely new homeostatic system, which motivates individuals to maintain a positive image of self-esteem as judged by socially induced standards of performance. Once this self-monitoring system is established, it functions as a powerful organizer of an individual's experience and activity, influencing nearly all aspects of information processing, including attention, perception, memory, concept formation, and all manner of judgments concerning which things matter. It is no exaggeration to say that of all our behavior mediation systems, self-esteem is the most dominant. As Howard Kaplan notes, "The self-esteem motive is universally and characteristically . . . a dominant motive in the individual's motivational system."[28]

The potential of the self-esteem motive to dominate an individual's behavior is one of its most distinctive features. But just as important is the plasticity of the system. Like the futuristic automobile that can burn anything as fuel, a person's self-esteem may be linked to almost any goal or outcome. Achieving a sense of self-esteem is always contingent on performing well against internalized standards, but there are few constraints on what these standards might be. Furthermore, there is nothing permanent about self-esteem linkages. The triumphs that made you feel good about yourself as a child are probably meaningless to you now.

This openness of the self-esteem system has high adaptive potential because it allows for changes in attitudes (and thus behavior) in response to changing environmental circumstances. In fact, the self-esteem system may be characterized as the principal resource for adaptive social change. Individuals will tend to modify their self-esteem linkages in response to social signals of approval or disapproval. For example, if you get praise for recycling your newspapers, and scorn for not recycling, then you will eventually link your self-esteem to recycling. The power of social signaling is truly awesome. Just consider what people are motivated to do in the name of self-esteem: We risk our lives in battle for the adulation of heroism, we pierce, paint or starve our bodies to meet social standards of beauty, we endure surgery for attention-getting breasts, and we steal, deceive, and murder to acquire status symbols.

The variability of the self-esteem system is a mixed blessing. On one hand it provides a mechanism for social groups to adapt their behavior to changing circumstances. But on the other hand it opens the door to more diversity of values than any orderly society can tolerate. Conformity to a common set of standards is essential to unity of purpose, so if members of a social group do not share at least a core of self-esteem linkages, then social cooperation will begin to falter. Thus in every culture there will exist tension between conformity and innovation, between continuity and change.

Self-esteem linkages are precisely what I mean by secondary valence operators. In one manner of speaking these operators are merely mental objects—derived rules, not inherent—that exert a top-down influence on emotional systems. That is, they impact the emotional modules in such ways that these systems take on new algorithmic properties for valencing the world. Under the influence of these operators, objects, events, properties, and relations are assigned positive or negative valences according to whether and how they might enhance or harm a person's self-esteem.

To illustrate, picture two men playing golf. Both men understand the rules of the game thoroughly, and both turn in extremely bad scores. The first of the two men is indifferent about his performance, but the second is utterly devastated. The difference is that the indifferent golfer has not internalized the standards of golfing excellence in the same way as the devastated golfer. Never mind why—the point is that one golfer has a self-esteem linkage to high performance and the other does not, even though they comprehend the standards equally well.

Neurobiologically, the difference has to do with patterns of connectivity between the cognitive mental objects (in this example, the golfing standards) and the structures of the limbic system. We noted earlier that emotional systems emerged on the scene when it became possible to stimulate limbic structures in brand-new ways. A variation on this phenomenon appears to have been made possible by increased capacity for working memory. The cognitive mental objects constituting the enhanced sense of self and the self-monitoring process have reconfigured cortical-limbic loops in ways that make it possible to stimulate the limbic areas in still newer ways. Thus the conditions are set for more refined and discriminating emotional traits. New emotions—secondary emotions—emerge to play a central role in the mediation of behavior. Secondary emotions (including guilt, pride, jealousy, envy, regret, compassion, and others) are beyond the experiential reach of primates lacking the human capacity for working memory.

One of the implications of this view is that it appears possible (perhaps even likely) that new ways of managing the self-esteem system can create the conditions for unprecedented emotional qualities and genuinely novel motives that might vary in subtle but important ways across time and cultural circumstances. The next chapter will focus more closely on the workings of the emotional systems, but for the moment we are left with the point that the self-esteem system makes it possible to transcend our intuitive value system in diverse ways. To this point, however, must be added the reminder that secondary values operate by extra-genetic rules. If there are no symbolic means to objectify and transmit these traits across generations, then we will default back to our intuitive morality.

SYMBOLIC SYSTEMS

No traits are more often cited as emblematic of human uniqueness than language and culture. The human monopoly on these traits has been challenged by successful attempts to teach the rudiments of language to our closest cousins, the chimpanzees. The irony, however, is that this research has widened the gap as much as narrowed it. A few chimps have managed to acquire enough linguistic skill to convince most experts that they have the neural prerequisites to qualify as symbol users. But the success of these experiments has been so minimal in proportion to the intensive labor expended on them that one is left more to marvel at normal children than precocious chimps. There is a big difference between a chimp that can acquire a tentative grasp on a modicum of language after a Herculean effort, and a child who can sop it up whole with ease. The difference is that humans have an instinct for language. Our brains have been designed by evolutionary processes to be invaded by and dominated by a symbol system.

The general claim that there is an evolutionary story behind the human language instinct is no longer controversial. Nevertheless, there is little prospect for consensus about the details of how and when language abilities emerged. The dating estimates range all the way from nearly two million years ago to a mere forty thousand years ago.[29] The first direct evidence we have for the use of symbols dates to about forty thousand years ago, when cave paintings and carvings began to appear. For those who like to pin their speculations to the archaeological record, this has been a convenient date for the origin of language.

This date has the advantage of being consistent with a general and sudden diversification and sophistication of artifacts. The archaeological record shows very little evidence of variation or inventiveness prior to this time. But then suddenly it reveals an array of harpoons, darts, bows and arrows, fishhooks, needles, ropes, lamps, artworks, and musical instruments—in a word, progress. The inference is that language was the key unlocking the door to creativity and cultural innovation.

The later dating also places the origin of language into proximity with the extinction of Neanderthals. Neanderthals and modern humans coexisted for nearly one hundred thousand years, but then suddenly Neanderthals drop off the map. Why? The reasoning is that the moment modern humans struck up a conversation (forty thousand years ago) they became too formidable for Neanderthals to compete with. Language did them in.

There are several major problems with this account, but it is difficult to dismiss it out of hand. It strains the imagination to believe that humans could be in command of language for hundreds of thousands of years and not exploit its full potential for cultural innovation. But if they did exploit it, then where is the evidence? Opponents of the late dating for language look for evidence outside the archaeological record, that is, in biology. The trouble with the sudden emergence of language forty thousand years ago is that this date does not correlate with any relevant evolutionary events. The evolutionary events relevant to language had all occurred well before two hundred thousand years ago, when anatomically modern humans appeared. Could humans have possessed all the necessary gear for language but not used it for nearly two hundred millennia?

Terrence Deacon dates the origin of symbolic communication at nearly two million years ago.[30] He believes there is only one principal selective factor that can account for the evolution of the language instinct—language itself! Two million years ago *Homo erectus* stumbled upon symbolic communication, but only with great difficulty, perhaps on the order of difficulty experienced by contemporary symbol-using chimps. These early symbol systems would have been very primitive by modern standards, but they would nevertheless confer adaptive advantages on those individuals who could use them.

The adaptive advantage was sufficient to establish a coevolutionary process involving language and the brain. More complex language elements created selective pressure for brains that were more symbolically competent—preparing the way for still more complex language features, then more competent brains, and so on. For more than a million years this self-perpetuating process sculpted neural systems that made language both more complex and easier to use. There were language-relevant changes occurring in other anatomical structures as well, especially in the face and vocal tract. Thus it was that a humble ape-like creature was upgraded to become the symbolic species, a species that has been shaped by language to use language.

Deacon believes that the principal challenge for any evolutionary account of language is to show how the operation of symbolic reference first came about. Once there existed a rudimentary ability for symbolic reference, then primitive symbol systems could emerge to establish the language-brain coevolutionary process. But to get the knack of symbolic reference our ancestors had to pass through a bottleneck to a new mode of thinking. Deacon explains this process in terms of three modes of reference: iconic, indexical, and symbolic. These modes of reference amount to different levels of meaning and interpretation.

An icon is a sign that bears a resemblance to its referent object. A photograph, for example, is an icon, and so is a whistle when it mimics a bird song. A receiver interprets a sign iconically if the sign evokes a mental image of the referent object. If you mimic a dog barking, then you are using a sign iconically, and if it puts me in mind of a dog, then I am interpreting your sign iconically.

An index is a sign generated not by resemblance, but by some sort of physical or temporal association between a sign and its referent object. A thermometer is an index for the temperature of water, a bad smell is an index for the presence of a skunk, an alarm call may indicate the proximity of a predator, and so on. When you utter "snake" you are using a sound as an index, and if it puts me in mind of a snake then I am interpreting your sign indexically.

Being able to send and receive information by using icons and indices can be quite useful. In fact, these modes of reference are used routinely in animal communication systems. Eyespots on the wings of a moth are icons for the eyes of larger animals. These deceptive signs can be interpreted iconically by preying birds who, when they see them, will abort an attack. Alarm calls are indices, commonly used by many species. Vervet monkeys, who are especially adept at indexical signaling, employ specific calls (words, if you like) for different predators, including "eagle," "snake," and "leopard."[31] Vervets able to deploy and interpret these indices obviously had more descendants than those who were unable to use them.

Iconic and indexical signs are not yet symbols, but they are fundamental to language—they are the primitive elements on which the symbolic mode of reference depends. Our hominid ancestors probably made extensive use of icons and indices. Indeed, one might suppose that *Homo erectus* had a pretty impressive vocabulary of signs and gestures to work with. Such a vocabulary would enable individuals to communicate efficiently about the elements of their intuitive worldview. Even before the appearance of symbolic reference these forms of communication would function as selective forces favoring adaptive changes in neural circuitry and vocal apparatus.

The language bottleneck involves a passage from indexical reference to symbolic reference. Symbols represent a higher-order mode of reference. Icons and indices are generated by associations between signs and objects, where the linkage is made by resemblance (icons) or by contiguity in space or time (indices). Symbols, however, are generated by associations between different indexical signs, where the linkage is made by social convention, or agreement. The difference is between sign-object associations and sign-sign associations.

What makes the apprehension of a symbol so difficult is that the receiver keeps wanting to interpret it as an index. For example, a coat of arms can be interpreted as a symbol, but it might alternatively be interpreted as an index. If Pavlov had displayed his family's coat of arms each time he fed his dogs, they would have formed an indexical association. To Pavlov's dogs the coat of arms would *indicate* food, but to Pavlov himself the sign would *symbolize* a more complicated higher-order reference.

In order to make a symbolic interpretation, an individual must be able to hold several indices in working memory and then suppress the bias to make indexical interpretations while shifting the attention to alternative associations one step higher in the hierarchy of signs. This is a difficult gestalt task, and it requires the sort of prefrontal brainpower that is present only minimally in chimps. This is why it takes such an effort for chimps to use symbols: They keep defaulting to indexical interpretations.

Comprehending signs in the mode of symbolic reference is a difficult operation, as experiments with chimps attest. But for modern chimps—who have a symbol system waiting for them in the lab—the battle is half won from the start. The bottleneck was much tighter when our apish ancestors had to work from scratch. Consider this speculative scenario about how things might have gone. We begin with an individual S, who has been ruminating for days over what he perceives to be a discrepancy in some detail of the intuitive worldview. For some reason things are not cohering, so he keeps running mental objects through if-then operations. Frustration mounts. Then suddenly he has a flash of insight, a genuinely original thought. S has managed to combine mental objects in a counterintuitive way to construct a secondary reality operator. He can now generate new inferences and more satisfying meanings. In an effort to confirm his insight, S approaches R and attempts to express himself. He cannot find the words. He tries stringing together familiar sounds in new ways. He gestures wildly, but still no glimmer of recognition in R, who keeps trying to interpret the signs indexically, intuitively. Again and again S tries, but R fails to get it. S indicates "no, that's not it," and tries a different tack. More failure, more negative feedback, more frustration. Eventually R succeeds in suppressing her indexical bias and catches the higher-order symbolic reference. Joy all around.

Something genuinely novel emerges in the event of symbolic communication: A new system for preserving and processing information is established outside the body. A truly symbolic and nonintuitive language introduces intersubjective properties of meaning that transcend the psychological properties of sign-object associations. The elements of an intuitive worldview can be shared without the use of symbols (we are all wired to jump to roughly the same intuitive conclusions), but apart from a symbolic medium, secondary insights would remain wholly subjective and bound for eventual obscurity. Symbolic reference captures the information generated by secondary operators, but unless this information is both expressed by a sender and decoded by a receiver, it will dissolve into its engram substrate.

The new mode of symbolic reference transcends the subjective networks of neural assemblies to locate meaning in an objective network of social interactions. Symbolic reference presupposes indexical and iconic associations but is clearly not reducible to them—no more than the metaphorical use of "cat" can be contained in what is known about cats.

The final determinants of symbols, then, are located more in social dynamics than in individual brains. Symbols may be initiated subjectively, but they are socially received and refined into objective conventions, therewith to be fed back to new users. The rules and vocabulary making up a language system emerge and reside in a distributive network of users. In other words, symbol systems are both extra-genetic and extra-somatic: They exist outside the body, in the objective social domain, a sort of commons where their elements are negotiated and modified in relation to other elements. It is true that symbols also reside in the neural systems of the brain, but this is only incidentally the case. We might think of the brain and language by analogy to your PC and the Internet. Your PC is designed to download material from the Net, and it may even contribute in some ways to the Net, but in a final sense the Internet has real properties of its own, independent of its users. In the same way, our brains are designed instinctively to browse, download, and use language, but we do not have neural modules that could generate a symbol system on their own. A private language is an oxymoron.

How else could language and the brain possibly evolve, but by the logic of coevolution? In coevolutionary relations between parasites and hosts, both parties maintain a measure of independence. This is required for new variations and new adaptations to appear (and for blunders to disappear). The same should be true for the coevolution of language and the brain. Multiple users in the social domain guarantee multiple interpretive responses, which mutually adjust in social dynamics, providing new objective conventions for the brain to adapt to.

In the introduction I speculated that there exist four general categories of natural facts: physical, biological, psychological, and cultural facts. Physical facts describe the behavior of matter insofar as it is organized by information inherent in physical systems. In these systems information is preserved in forced *bonds* between subatomic particles. Biological facts describe the behavior of matter insofar as it is organized by information encoded in the DNA molecule. In biological systems information is preserved in *genes,* the units of biological inheritance. Psychological facts describe the behavior of matter insofar as it is organized by information encoded in neural assemblies. In psychological systems units of information are preserved in *engrams.* Cultural facts describe the behavior of matter insofar as it is organized by information encoded in symbols. Some changes occurring in the organization of matter are contingent on social interactions, and these interactions are themselves mediated by symbolic systems that encode information outside the body. In socio-symbolic-cultural systems, information is preserved in *memes,* Richard Dawkins's term for the unit of symbolic variation, transmission, and selection.[32] A cultural tradition is the sum total of its memes.

This chapter has followed the evolutionary emergence of a hierarchy of behavior mediation systems, including molecular, reflex, perceptual, drive, memory, emotional, cognitive and, finally, symbolic systems. Along the way it has been shown that each emergent system introduced new possibilities for adaptive behavioral responses to environmental challenges. Symbolic systems emerged from a coevolutionary process by which human brains were adapted for language competence under pressure from increasingly complex language features. This coevolutionary process may have started more than 1.5 million years ago, and continued until about 200,000 years ago when anatomically modern humans appeared on the scene. The development of symbolic competence has had profound implications for the course of human biological evolution—that is, it has very nearly stopped the process!

Evolution requires both variation and selection. Heritable traits vary as a result of mutations and genetic recombination, and the environment culls out those variants least well adapted to it. If either of these principal factors changes appreciably, then the rate of evolution will be affected. For example, if for some reason mutation rates increase, then it is probable that the rate of evolutionary change will increase as well. Or if the environment changes (as it did when the great reptiles were extinguished), there will be probable consequences for the rate of evolution. I am suggesting that when symbolic communication reached a certain point of sophistication, the coevolution of language and brain stabilized. In other words, brain and language had reached a critical level of sufficient complexity when it became both easy and adaptive to override the biases of default mentality in favor of nonintuitive mental operations. The neural infrastructure for symbolic communication was now complex enough so that any further complications would have few, if any, adaptive consequences. Once the Rubicon of symbolic reference was crossed, the rules of the adaptation game changed. The focus of selection shifted from biological variations (genes) to cultural variations (memes). Evolution would now be less about the brains that used symbols than about the symbols brains used.

To illustrate, say the average temperature was to drop in the environment of early hominids incapable of sharing nonintuitive insights. If this were to happen, then those individuals with genetic variations for extra fat or thicker fur would be preferentially selected. But in a society where ad hoc secondary rules for behavior can be shared across the group, the selective force of a chilling climate might be absorbed by the invention of new technologies for clothing or housing. The environmental challenge is the same in both scenarios, and so is the range of genetic variation for alternative physical traits. But the difference is that in the second scenario the biological consequences (preferential selection on genetic variants) were precluded by an adaptive response at the level of symbolic mediation. Cultural adaptation has preempted the expected biological adaptation.

The invention of symbolic systems meant that the evolution of behavior would henceforth be tied to the dynamics of cultural evolution. Cultures evolve as their memes change, just as species evolve with changes in the genome. The general dynamics are the same: variation followed by selection. Cultures will generate more variations of meaning than could ever be preserved in a tradition. More laws are proposed than passed, more pictures painted than hung, more songs composed than recorded, more books written than published, and so on. If a culture is to achieve even the minimal degree of coherence, it must find the means to winnow the memetic diversity in systematic ways. Failing this, a culture becomes vulnerable to the chaos of too many conflicting memes, which will ultimately countervail the adaptive value of having a symbol system for preserving nonintuitive information.

Cultures must therefore devise strategies to select for a central core of ideas about how things are (cosmology) and which things matter (morality). This central core will become the defining narrative underlying the cultural tradition, and will be used as a selective device for culling out incompatible memes. Cultures will also devise strategies to regulate the rate of cultural evolution. The virtue of having a cultural tradition is that it enables a group to adapt its behavior quickly in response to environmental challenges, but too rapid a rate of cultural evolution will generally be a destabilizing factor. The rate of cultural evolution is almost certain to be accelerated by changing circumstances in the environment. When faced with a new problem, individuals will collaborate on solutions, a process that is sure to generate meme variation. And of course, there is nothing like a new idea for stimulating new projects, or disclosing hidden discrepancies, or threatening to invalidate whole clusters of old ideas. When new ideas are introduced to a cultural tradition, they generate perplexity, conversation, interpretation, debate, insight . . . and more memes. Cultures will therefore tend to discourage novelty in order to maintain a predictable social environment. Diversity and change often create uncertainty, which always triggers anxiety and resistance. The burden of proof in a social order will almost always fall on those who want to change it.

Internal checks on diversity and change do not, however, apply *between* cultures, and while most cultures evolve slowly, they tend to diverge rapidly. If you go westward at 1 m.p.h. and I go eastward at the same speed, then the distance between us increases at a rate of 2 m.p.h. As symbolic systems emerged, and as ad hoc mental operations multiplied, there resulted a slow but steady drift away from the intuitive mentality biased by genetic inheritance. However, the additive effect on the divergence between cultures was far more dramatic.

As a cultural tradition becomes firmly established, its distinctive worldview will play a commanding role in the organization of consciousness and the mediation of behavior. The genetic biases informing intuitive mentality, however, remain securely in place, ready to do the work of mediating behavior wherever secondary operators fail to override their influence. Human nature is both uniformly shared and differentially shaped. Our nature includes genetic algorithms that bias us toward a common, species-typical intuitive mentality. But we are also endowed with potential for inventing extra-genetic algorithms, or learned biases that result in diverse, culturally specific mentalities. Every cultural tradition, therefore, represents a unique history of competition and compromise between primary and secondary operators.

HUMAN NATURE AND THE MEANING OF LIFE

This chapter has pursued a sketch of human nature by focusing on questions of how we came to be and how we carry on. In a nutshell, we are a unique combination of physical, biological, psychological, and cultural facts, endowed by nature with a plurality of interacting systems for mediating behavior. This brief summation tells us a fair amount, but it does not give us quite what we expect. Surely the point of pursuing a view of human nature is to discern the point of human existence, the meaning of life. The trouble with the meaning of life question is that it can be asked (and profitably answered) at different levels of generality. Our difficulties arise only when we assume that there must be a single right answer. The critical task for our project of sketching human nature is to ask the question at the appropriate level of generality.

Suppose we try asking the question at the most general and inclusive level. If we were to ask about the *ultimate* purpose (the grand telos) of human life, then we would have to say it is one we share with all other living beings—that is, to carry on in pursuit of reproductive fitness. If we fail in achieving this goal then the question ceases to apply. But certainly this answer is too general. It may be true enough, but asking the question at this level fails to distinguish the meaning of human life from the meaning of any and all other forms of life. The answer does not satisfy the specific requirements for a picture of *human* life.

Let us then try the question at the other extreme, at the level of concrete experienced meaning. After all, no one actually feels the deep biological imperative to seek reproductive fitness. Instead, particular men and women find meaning in more immediate teloi—like raising their children properly, or making buckets of money, or enjoying the many simple pleasures life has to offer. But answers to life's question at this lowest level of generality are no more appropriate to our task than the ultimate answer. At the level of experienced meaning there are too many answers, and it would be foolish to expect any of them to speak for all humanity. The meaning of life at the individual level is too subjective and variable for a picture of human nature.

Clearly, a portrait of human nature must seek the meaning of life at some intermediate level of generality. Religious and philosophical traditions have always understood this and have tried to develop their answers in terms of ideal teloi that apply both universally and exclusively to our species. Thus the Abrahamic traditions, for example, have insisted that the true meaning of life is to serve God's will. In Buddhist traditions the meaning of life is focused on the elimination of suffering. In the Greek philosophical tradition the telos for all humanity is to exercise reason. There are, of course, many other culturally specific variations on the true meaning of human existence. But this exercise does not get us much beyond the variability of subjective answers. And besides, cultural expressions of the ideal life are prescriptive, not descriptive. A sketch of human nature is less interested in how people *should* carry on than how they actually *do* carry on.

The meaning of human life should be expressed in terms of how our particular species pursues the ultimate telos of reproductive fitness. Like every other species, we seek the ultimate biological goal according to our peculiar nature. That is, by pursuing the many teloi that are internal to our behavior mediation systems, whether these teloi are built into the system by genetic means or incorporated into them by symbolic means. For humans there are many immediate teloi, including the biological goals inherent in our drive systems, the psychological goals implicit in our emotional and cognitive systems, and the social goals we imbibe through our symbolic systems. Human life is about whatever these goals are about.

I will claim that these immediate pursuits are about a pair of mutually depending yet mutually contending intermediate goals: *personal wholeness* and *social coherence.* Everything we do at the behest of our behavior mediation systems can be seen to contribute toward, or detract from, the achievement of one or both of these twin teloi. The claim is that the human strategy for winning the war of survival is to fight on two fronts, the battle against personal disintegration and the battle against social chaos. More positively, if we can manage to create the conditions for organizing healthy, well-integrated, robust personalities, and if we can simultaneously create the conditions for constructing coherent, cooperative, harmonious groups, we will thereby maximize the odds favoring human viability.

I am not claiming that personal wholeness and social coherence constitute genuine goals in addition to our immediate pursuits. To make good on such a claim would require one to identify specific underlying mechanisms directing us to achieve these goals, which I believe cannot be done. But this should not render the concepts of personal wholeness and social coherence useless for a picture of human nature. Surely, no one has identified specific mechanisms underlying the goal of reproductive fitness either, yet it has proven to be a useful explanatory principle. While reproductive fitness is the most general goal applying to all life forms, personal wholeness and social coherence may be said to be the most general goals applying exclusively to humans. Our understanding of the intermediate teloi of personal wholeness and social coherence is derived in part from the many immediate problems that arise when we fail to achieve them.

I have said that personal wholeness and social coherence are mutually depending and mutually contending goals. On one hand, they are interdependent values in the sense that each is a necessary condition for the other. Whole persons cannot be nurtured in a context of social chaos, nor can a coherent social order be constructed by dysfunctional individuals. It is clear that psychological problems can often be solved by social change, just as social problems may be addressed by psychological change. On the other hand, personal wholeness and social coherence are contenders in the sense that they have the potential to undermine one another. Individuals tend to lay excessive demands on social resources, while groups tend to make excessive demands for personal sacrifice.

The challenge of personal wholeness is to maximize the goods internal to our motivational systems. As we have seen, a person is a complex set of motivational systems and subsystems, each having the potential to arouse an individual to a course of action. The multiple demands of these homeostatic systems make it essential that the individual suppress some of them in favor of satisfying others. One of the ironies of human existence is that sacrificing certain desirable goods for the sake of personal integration is a necessary means to maximizing such goods. Trying to have it all is a recipe for having nothing. Whole persons are those who are fully engaged with the world, and whose motivational systems are robust yet effectively managed—persons who are able to construct agendas of sequential tasks, to anticipate outcomes, to assign priorities, and then attend to the most important matters while momentarily suppressing the demands of competing impulses. By such means the whole person is able to achieve a state of functional unity against the odds inherent in a plurality of motivational systems. "Whole" means both full and one, complex yet integrated. Our entire lives are constructed, moment by moment, through our efforts to manage the demands put upon us for knowledge, pleasure, emotional fulfillment, and self-esteem. By harmonizing these demands we maximize our achievement of goods internal to our motivational systems.

There are many ways to be a whole person. A single answer to the challenge of personal integrity does not exist. Your inherited combination of temperament, sensitivities, limitations, and abilities is not mine; and my set of environmental obstacles, opportunities, misfortunes, and advantages is not yours. Each individual is a unique combination of such variables, which means that the specific challenges to personal wholeness will be radically diverse. But if the problems are radically diverse, then we cannot expect the solutions to be any less so. The potential for human diversity is therefore sufficient to defy all norms.

And yet norms there must be, for without them we forsake the benefits of a social order. The challenge of social coherence is not unlike the challenge of multicellularity—that is, how to get a plurality of discrete parts to behave with a unity of purpose. The answer is to get the parts to conform to certain rules, to maximize conformity of behavior among members of the group. There may be many authentic ways to become a whole person, but encouraging a diversity of these ways is hardly the way to create a coherent society. A coherent society is achieved by minimizing potential for conflict, by establishing something akin to the homogeneity characteristic of an intuitive mentality.

The telos of personal wholeness is to maximize satisfaction of motives, and the telos of social coherence is to maximize conformity to shared standards of behavior. At times these twin teloi may be at odds with one another, but they need not be. It is possible to create the conditions that will be conducive to their simultaneous achievement. To create these conditions is in large measure a symbolic task, where the trick is to construct a world of memes that can inform both our therapy and our politics.

It should now be possible to recap the central features of our sketch of human nature. *Human beings are star-born, earth-formed creatures endowed by evolutionary processes to seek reproductive fitness under the guidance of biological, psychological, and cultural systems that have been selected for their utility in mediating adaptive behaviors. Humans maximize their chances for reproductive fitness by managing the complexity of these systems in ways that are conducive to the simultaneous achievement of personal wholeness and social coherence.* As we shall see, answering this difficult challenge is what religious traditions are about.

THE EDUCATION OF EMOTION

Human nature is shown by how we came to be and how we carry on. The previous chapter concentrated on the evolutionary story: We have come to be through an evolutionary process that endowed us with complex and interactive behavior mediation systems. We carry on by managing these systems in ways that enable the achievement of a wide range of short- and long-term goals. The present chapter will focus more sharply on the business of carrying on—that is, how the elements of human nature actually work together as we encounter the challenges of everyday living. The emotional systems will dominate the discussion ahead, and with good reason, for the emotions are central both to an understanding of human nature and to an understanding of religion.

The central importance of emotion studies in academic psychology has been recognized only recently. The so-called cognitive revolution of the 1960s rejected behaviorism on the principle that it ignored the key to unlocking the mysteries of human behavior—mental processes. Contrary to behaviorist doctrine, cognitive psychologists affirmed the reality of mental processes and declared them to be accessible to scientific study. The new agenda was to figure out the rules by which reasoning, problem solving, decision making, and language operated. Attention to motivational and emotional processes was omitted from the agenda, but in recent years the boundaries of cognitive science have been redrawn to recognize the significance of emotion research. Emotion and cognition are now viewed as "partners in the mind."[1] Once considered obstacles to reason, the emotions are now considered essential to rationality. As Joseph LeDoux says, minds without emotion are "souls on ice."[2] Any account of human nature that leaves emotion at the margins will be distorted. Indeed, we may agree with psychologist Richard Lazarus, who finds emotion to be our most promising window into human nature:

> From an emotional reaction we can learn much about what a person has at stake in the encounter with the environment or in life in general, how that person interprets self and world, and how harms, threats and challenges are coped with. No other concept in psychology is as richly revealing of the way an individual relates to life and to the specifics of the physical and social environment.[3]

The emotions are even more central to an understanding of the religious life. Christians typically claim that love is the essence of divine reality, and that guilt is at the heart of the human condition. Buddhists point to emotional attachments as the source of all suffering, and compassion as the chief virtue of the Buddha. Emotions are no less important for other traditions. In fact, as I shall argue, religious traditions may be viewed as schools for educating the emotions. This is true whether the teachings of a tradition are aimed at deepening or dampening human emotional experience. If our objective is to understand human nature and religion, then we must begin with the emotions.

WHAT IS AN EMOTION?

Emotions are highly complex biological-psychological-social phenomena, which probably accounts for the fact that no consensus has emerged among emotion researchers as to how they are best defined and classified. Those who favor a biological perspective will define emotions in terms of the physiological changes such as heart rate, skin conductance, hormonal secretions, brain activity, muscle tension, and so on, associated with emotional responses. There can be no significant doubt that emotional responses are distinctive biological events, and that continued research will tighten the correlation between emotions and their specific underlying anatomical structures and physiological processes.[4] Yet it is very unlikely that biological details alone will satisfy all the questions that arise concerning the unfolding of emotional events. There are clearly psychological and social factors influencing our emotional systems that cannot be fully described at the level of biological organization.

Evolutionary researchers tend to concentrate on the adaptive functions of emotional systems. In the evolutionary view emotions may be defined and classified in terms of their capacity to regulate social intercourse by way of their functions for communication and motivation. Evolutionary accounts emphasize the universal aspects of emotion and the continuity of emotional phenomena between humans and other mammals, especially the primates. The previous chapter should have left little doubt of my sympathy with the evolutionary approach. Still, it is important to see that an account of how we came to be will not fully satisfy all the questions about how we carry on. There is more to the emotions than a story of their origins in natural selection.

Closely allied to the evolutionary approach are research programs emphasizing the behavioral outcomes of emotional events. Selection can operate only on observable phenomena, like patterns of behavior. Emotional competencies should therefore be classified according to observable units of behavior. Paul Ekman and his colleagues, for example, have identified a core group of facial expressions—happiness, sadness, fear, disgust, surprise, and anger—that correlate across cultures with salient emotions.[5] A similar approach, taken by Robert Plutchik, identifies a set of eight prototypical behavior patterns that have been instrumental for survival. These patterns—incorporation, rejection, protection, destruction, reproduction, reintegration, orientation, exploration—have evolved with a corresponding set of eight basic emotions: acceptance, disgust, fear, anger, joy, sadness, surprise, and expectation.[6] Another researcher, Nico Frijda, keys human emotions not to specific expressions or behavior patterns themselves, but rather to organismic states of readiness to respond. Frijda's analysis leads him to posit ten basic emotions: desire, fear, enjoyment, interest, disgust, indifference, anger, surprise, arrogance, and humility.[7]

Cognitive psychologists are inclined to view emotions less as causes than as effects. In other words, emotions are best understood as experiential consequences of the mental processes bringing them about. Events in the world can be interpreted and evaluated in radically different ways, depending on the psychological circumstances of the individual. Imagine two children being approached by a strange dog. If one of the children has a history of bad experiences with dogs, then she will be more likely to see danger at hand, and therefore more likely to respond in fear.

Because emotions have a decisive cognitive component to them, there must be some sort of cognitive appraisal of the meaning of environmental circumstances relative to the goals and interests of the individual before an emotional event can unfold. Without such cognitive appraisals emotions do not arise. Furthermore, differences between emotions can be attributed to differences in cognitive appraisal. Emotions therefore tend to be defined in terms of the appraisal patterns that produce them. The task of cognitive theories of emotion is to specify the variables in mental processes that give rise to different emotions. It is common for cognitive theories to distinguish between primary and secondary emotions. Primary emotions (fear, anger, disgust) are thought to be simpler and more universal, while secondary emotions (guilt, love, jealousy) are more complex and variable, resulting from the more refined appraisal patterns that come with highly developed cognitive abilities.

Social constructivism is the most recent entry into emotion theory. These theorists align themselves as much with postmodernist philosophers and anthropologists as they do with mainstream psychologists. Thus it is not surprising that social constructivism tends to trivialize the role of biology and evolution in favor of a much larger causal role for social and cultural factors. Social constructivists agree that appraisals of environmental circumstances are necessary for emotional events to occur, but they go well beyond cognitive theorists in claiming that the factors governing the appraisal process are culturally determined.[8] Social constructivists emphasize the wide diversity between cultures when it comes to defining, organizing, and speaking about emotions. In this view, ethnographers are more relevant to understanding emotional realities than evolutionary psychologists are. Each culture socializes its young to distinctive appraisal patterns, which uniquely determine the meaning and the quality of emotional responses. There are no universal human emotions, and to speak of emotions in nonhumans or unsocialized humans is to be careless with one's metaphors. Even basic emotions like fear and anger involve complicated mental processes contingent on socialization within a particular cultural context. Cultures determine the objects of emotion as well as the acceptable circumstances for having an emotion. Emotions are learned cultural artifacts, thoroughly relative and infinitely variable. In the words of constructivist James Averill, "There are an infinite number of emotions. That is, societies can shape, mold, or construct as many different emotions as are functional with the social system."[9]

It is important to recognize that emotions are complex biological-psychological-social phenomena. As the foregoing survey shows, much of the theoretical work has tended to emphasize one of these domains of facts to the exclusion of the others. It seems unwise to rule out any of these theoretical approaches, for each of them is sustained by legitimate insights. Yet their differences do not promise an easy synthesis. Nevertheless, I believe it is possible to formulate a comprehensive and coherent view of the emotions that includes the best insights of these theoretical orientations. I will attempt a formulation and then follow it up with a more detailed discussion.

Essentially, *an emotion is a temporary feeling state that acquires narrative content and leads to a predisposition to act.* Emotional events are described by a process initiated by internal *biological* changes, which are appraised in terms of narrative meaning by *psychological* dynamics, which may be extensively influenced by *cultural* factors. This formulation at least gives us a start. Now for the discussion.

EMOTIONS ARE TEMPORARY FEELING STATES

Emotions are temporary feeling states, but not all feeling states are emotions. One might be in a temporary state of pain or pleasure, but these feelings do not qualify as emotions (despite metaphors about the "pain" of grief or the "pleasures" of love). Pains and pleasures can be easily identified with particular parts of the body, whereas emotions cannot. We would not know what to make of a person who reported an intense joy in his shoulder. Pains and pleasures are sensorimotor phenomena and do not necessarily involve cognitive appraisals, as emotions do. This is not to say that these different types of feeling states cannot be closely related in important ways. Pain or pleasure may play a part in the generation of an emotional event, as when an intense pain leaves one fearful that she has a threatening illness. It is also well known that emotions can modulate the intensity of pains and pleasures, as when a child's pain is mitigated by the happiness of receiving a new toy.

The relationship between pains, pleasures, and emotions is evolutionarily deep, as was noted in the previous chapter. Experiences of pain and pleasure were no doubt relevant to the selection of capacities for fear and desire. Pains and pleasures are no less important for individual development, for they play a key role in assembling the motivational prerequisites for emotional competencies.[10] Similar claims can be made for physiological drives. Hunger, for example, is a temporary feeling state that also has relevance for motivational development. Yet despite the deep and intimate relations between sensorimotor reflexes, physiological drives, and emotions, it is clear that emotions possess distinctive evolutionary and developmental stories and distinctive neurobiological underpinnings, as well as unique subjective qualities that set them apart from the temporary feelings associated with reflexes and drives.[11]

MOODS

So: Emotions are not pleasures, pains, or drives. Nor should they be confused with moods, as they often are. Moods and emotions are both temporary feeling states that can be distinguished, though not sharply, by several criteria. For one thing, moods are more stable (that is, longer lasting) than emotional states. The duration of emotional states can vary considerably, ranging from a momentary flash of anger to lingering ripples of happiness. But it is misleading to suppose that emotional responses might endure for many days, as moods often do.

Second, moods appear to have less thematic content than emotions. Unlike emotions, which arise in response to specific objects or events and culminate in an impulse for definite action, moods do not seem to be about anything in particular. We are often at a loss to say why we are in a mood, or what sort of behavior it calls for. A good mood makes everything look rosy, while a foul mood makes the whole world bleak and uninteresting. And although moods influence behavior, they do not appear to do so in systematic ways, as emotions tend to do.

Finally, in terms of subjective qualities, moods tend to be less intense and less salient in awareness than emotions. They are sometimes described as "background feelings," vaguely experienced in terms of the body's energy state.[12] Depression, irritability, and prolonged states of anxiety fall under this category.

It is important to recognize the differences between emotions and other feeling states such as sensations, drives, and moods. Equally important are the differences between emotional *states* and emotional *traits*. Richard Lazarus makes the distinction clearly: "An emotion trait means that a person possessing it has some personality characteristic that brings the emotion state about or generates it more often or more intensely than in others."[13]

TEMPERAMENTS

The concept of temperament refers to "any moderately stable, differentiating emotional or behavioral quality whose appearance in childhood is influenced by an inherited biology, including differences in brain neurochemistry."[14] Contemporary research supports the claim that we are genetically equipped with different potentials for emotional reactions, just as we are differently biased by the genes for high blood pressure or allergies. These potentials are evident in anatomical structures and physiological processes that mediate emotional responses. For example, a child is more likely to respond fearfully if he has more neural circuitry between the cortex and the amygdala, or if his brain produces high concentrations of the neurotransmitter nor-epinephrine. Thus our emotional reactivity depends in part on the wires and juices we inherit. Jerome Kagan has found that about 20 percent of infants inherit low stimulus thresholds to fear, and in about two-thirds of these subjects a pattern of inhibition persists throughout childhood.[15]

In other words, we inherit a temperamental profile which makes us more or less likely to respond emotionally: "Emotional states are biologically prepared responses released by particular events, and we believe that temperamental types differ in their susceptibility to these emotional states."[16] The fear profile is the best understood of all the emotions. One can say with confidence that some people are temperamentally more fearful than others. Research on other emotions, such as anger, happiness, and affection, is too thin to support anything but speculation, but Kagan suspects that temperamental types waiting to be described by future scientists may be as numerous as the basic categories of emotion. In other words, each of us may possess inherited biological profiles that make us more or less prone to fear, love, anger, happiness, sadness, contempt, and so on.

ATTITUDES

Biology, however, is not destiny. The inheritance of biological predispositions to emotional responses does not tell the whole story. We are also equipped to acquire emotional traits, which may be even more decisive in our emotional lives than our temperamental biases. The most important of these acquired traits are attitudes. In 1935 Gordon Allport claimed that attitude is the single most indispensable construct in social psychology. Since Allport's time some psychologists have favored eliminating the concept of attitude in favor of more basic theoretical constructs, though recent attention to the subject matter has restored plausibility to Allport's judgment.[17]

The concept of attitude should be understood as subordinate to the concept of belief. As we have seen, an important part of our nature is to be guided by reality operators in building up a large repertoire of ideas about objects, events, properties, and relations in the external world. To be human is to hold such beliefs, most of which slumber as inactive engrams, well out of awareness at any given moment. Nevertheless, they are our beliefs to the extent that we are able to activate the engrams in working memory. Some of our beliefs—most of them, probably—are value-laden. These value-laden beliefs are *attitudes*, by which I mean that their engrams are constituted in part by a positive or negative affective component. Attitudes, therefore, are *valenced beliefs* that remain relatively stable over time.

Attitudes and emotions are closely related in several ways. For one thing, beliefs are frequently valenced during the course of an emotional event. Thus, if a child is frightened by a dog early in life, she is likely to form a negative attitude about dogs. It is also the case that attitudes play a major role in the unfolding of emotional events. For example, a child holding a negative attitude about dogs is more likely to respond in fear when a dog is encountered.

But certainly not all attitudes can be traced back to episodes involving direct experience with the objects of our attitudes. Dogs are not necessary for the formation of attitudes about dogs. Many of our valenced beliefs about the world come to us through the subtle processes of socialization. Some researchers have characterized the socialization process as an "apprenticeship in thinking."[18] The idea is that from a very early age developing children actively seek to acquire attitudes about the world from their interactions with social partners. Caregivers send valenced messages to infants right from the start, and within a few weeks infants become skilled at reading the emotional meanings embedded in facial gestures and vocal intonations.

By four months of age routines of affective reciprocity have stabilized between infant and caregiver. During the fifth and sixth months infants develop an interest in attending to and manipulating objects in the environment. Before the seventh month infants appear to be single-minded, able to attend to persons or to things, but unable to coordinate their attention to both. By nine months, however, the infant is prepared to enter into episodes of "joint attention" wherein infant and caregiver focus their attention on some common object.[19] An episode of joint attention will typically include a triad composed of infant, caregiver, and object, where the object is the focus of shared attention.

At this point the child's apprenticeship in thinking begins to take off. During countless episodes of joint attention the infant regularly consults the expressions of the caregiver to pick up information about the emotional meaning of objects. If a caregiver expresses positive affect toward an object, then the infant will automatically form a positive attitude toward it. While this apprenticeship in thinking is more intense during infancy and childhood, it appears to continue throughout the life span. When you next attend a lecture, concert, or rally, take notice of how frequently adults consult the reactions of their social partners. Attitude formation is to a large extent socially mediated.

While some of our attitudes originate in the context of

emotionally charged events and others are acquired in the process of social apprenticeship, they might also arise by straightforward cognitive means. Consider how attitudes already established might prepare the way for new ones. If you have previously acquired the attitude that human life is sacred and you subsequently learn that human life is contingent on a wide range of biodiversity and a healthy ecosystem, then it is likely that you will begin to acquire pro-environmental attitudes. Why? Because affirming the value of human life together with indifference toward relevant environmental issues may now result in an intolerable level of cognitive dissonance. Consider also how we may rationalize our way into attitudes as a result of changing life circumstances. For example, participants in an arranged marriage may enter married life with indifference toward each other, but in time they become no less affectionate in their attitudes than those who married in a fit of romantic love.

Emotions are temporary feeling states that should not be confused with sensations, drives, or moods. Nor should they be confused with emotional traits such as temperaments and attitudes. Nevertheless, each of these factors may enter the picture as emotionally relevant—that is, they may play important roles in staging the onset, the unfolding, and the outcome of the emotional process. The point is that we are prepared for emotional reactions by a diversity of factors. Some of these (such as temperament) may be due primarily to genetic influence, while others (attitudes, for example) may be due primarily to cultural influences—although it should be hastily added that a person's temperaments are not closed to cultural modulation, and the proclivity to enter into social apprenticeship is ordained by the genes.

EMOTIONS ACQUIRE NARRATIVE CONTENT

Humans are narrative beings. Our lives unravel plot-like, where events have meaning in terms of how well we fare goal-wise in our relations within a physical and social environment. Emotional events are triggered whenever an event signals a goal-relevant change in these relations. In some cases the initiating event may be external (for example, a message or a transaction), and in others it may be internal (in the case of a memory or an inference). The important thing is that some element in the business at hand indicates that there is something at stake for our carrying on, something is happening that bears on our goals and interests. An emotional reaction is the process by which we determine the narrative meaning of the happening. A specific emotion is the subjective feeling that arises during this appraisal process.

A PREVIEW OF THE APPRAISAL PROCESS

I take the view, along with many cognitive psychologists, that emotions arise in the course of a subject's performing a cluster of mental operations, which can be referred to collectively as the *appraisal process*. At this point I intend merely to sketch out the components of this process, and then later in the chapter I will illustrate the process with reference to a few selected emotions. The model proposed by Richard Lazarus includes the following six components in the appraisal process.

Goal relevance. Some encounters between a subject and its environment result in emotional events, others do not. It all depends on whether the subject perceives something in the business at hand that appears to be relevant to the subject's goals. This determination of goal relevance is the initial event in the appraisal process.

Goal congruence/incongruence. It is not enough to determine that the business at hand is relevant to the subject's goal structure. The subject must also determine whether the encounter is threatening or promising. If the business at hand is construed as congruent with a subject's goals, then any one of a number of positive emotions might ensue, but if the business appears to be incongruent, then any one of a number of negative emotions might ensue.

Ego involvement. Subjects have many goals to achieve, and these will vary in terms of their relative strengths as well as in the precise manner in which the self is linked or identified with the goals. Thus a subject must appraise just what aspect of the self is at stake in the encounter. Is the subject's life or safety at stake? Social status? Beliefs? Self-esteem? Possessions? This operation in the appraisal process further limits the range of possible emotional outcomes.

Credit or blame. In some cases it may be possible to attribute responsibility for the business at hand. That is, once goal congruence has been determined, it may be possible to attribute credit to the self or some other agent. In the case of goal-incongruent appraisals, the attribution of blame may be possible. Where an appraisal of credit or blame is made, the emotional outcome will reflect it. For example, jealousy, anger, pride, guilt, and gratitude are emotional responses that depend upon attributions of credit or blame.

Coping potential. As the relevance of the encounter becomes more clear, the subject must make a rough appraisal of whether it has the resources at hand to manage the situation successfully. What are the prospects for the subject thinking or acting her way through the encounter to an acceptable outcome? In many cases a subject's emotional response will depend on this appraisal of coping potential. For example, the intensity of my fear of an attacking dog will vary in proportion to my chances of getting over the fence before it catches up with me.

Future expectations. At some point in the appraisal process it may be possible for the subject to make a judgment about what the ongoing encounter bodes for the future. How likely is it that things will change for better or worse in terms of goal congruence? The nature of this appraisal will affect the emotional outcome. I may, for example, become less upset at missing a train if I consider that another will come along soon.

It will become clear in the following discussion that subjects vary considerably in their competence to perform the appraisal operations listed above, although every subject capable of emotion will be able to perform the first three at some level of sophistication. It is in the course of a subject's appraisals that emotions acquire narrative content—that is, it is by virtue of appraisals that things and events in the world acquire meaning for the subject. I shall return to specific examples of this appraisal process later in the chapter.

Emotional analyses will never be easy, nor are they likely ever to be exact, for there are many variables to be taken into account. In addition to the background variables already discussed—sensations, drives, moods, temperaments, attitudes—much will depend on an individual's goals, which are highly variable. The concept of goals is an exceptionally messy one, ranging all the way from the goals we inherit to the ones we invent. To speak of goals is to evoke all the elements of an organism's motivational structure, a virtual Internet of linked scripts describing how the self is organized in relation to the world.

The concept of "self" is no less problematic. Just what *is* the self? What must be the case before we can speak coherently about an organized self that carries on? Does narrative content imply a narrative subject whose integrity as a goal seeker might be at stake in encounters with the world? What, if anything, does consciousness have to do with all this? And what is the relationship between consciousness and emotional experience? Any theory of emotions must hold views about these fundamental issues, and any theory of religion must be prepared to take them seriously.

EMOTIONS AND THE SELF

Neurobiologist Antonio Damasio has spent much of his career puzzling over such questions. In the book *The Feeling of What Happens,* he gives us a useful way to think about the self and emotional experience. Damasio identifies three stages in the evolutionary development of selfhood: the *proto-self,* the *core self,* and the *autobiographi-cal self.* Even the simplest organisms can be described in terms of primitive attributes of selfhood. The proto-self has the following char-acteristics: "a boundary; an internal structure; a dispositional arrange-ment for the regulation of internal states that subsumes a mandate to maintain life; a narrow range of variability of internal states so that those states are relatively stable."[20] Under this minimalist definition even the lowly amoeba qualifies for selfhood.

Proto-selves have goals. They are, in Paul Taylor's terminology, "teleological centers of life."[21] But amoebae certainly do not possess consciousness or emotions, there are no narrative meanings present in the life of an amoeba. Damasio's point, however, is merely that the proto-self is the evolutionary precursor for more elaborate forms of selfhood.

As organisms become more complex their life-maintenance goals are mediated by higher-order systems. In amoebae the teloi of life are mediated by simple biochemistry, and in insects the goals of life are elevated to mediation by reflex systems. But in animals blessed with physiological drive systems, matters become more com-plicated, involving the construction and manipulation of neural rep-resentations. Life maintenance strategies in a fish, for example, are more complicated than they are in a grasshopper. Fish are still proto-selves, but unlike grasshoppers they have drive systems that must be monitored and regulated by the brain. Fish brains receive informa-tion from internal sensors and use this information to construct neu-ral maps that register how things are faring, goal-wise, throughout the internal environment of the body.

In addition, fish brains independently construct neural signals that represent changes taking place in the world outside the body. These representations arise from consolidating information processed by the sensory systems. When a sensation occurs, events are auto-matically triggered in the internal milieu, which are immediately reg-istered by the internal mapping system. There are no object images—percepts—factored into the internal body maps of the proto-self, but merely the downstream bodily consequences of upstream sensory sig-naling. These indirect (and imageless) messages become the basis for the organism's motor responses. This means that a fish does not re-spond to a predator, or even to an *image* of a predator, but only to a set of object-less generic neural events. The fish has no experience that there is or even *might be* a predator out there—indeed, it has no sense of an external world at all. In fish the systems for registering in-ternal and external information are abstracted from one another. We might say that a fish records what is happening "out there," and it also records what is happening "in here," but it has no elaborate mechanisms to support an integration of these streams of information. It is clear that what the fish sees may be relevant to its internal milieu, but *the relevance is not registered anywhere.* The relevance is indirect, a sort of blind calibration established in the course of natural selection. If a fish darts away from a predator, it is not because it is afraid, but only because a particular sensorimotor response pattern ("see X, dart off") has proven itself adaptive in the evolutionary past. But there is no place in the neural networks where anything like a "here's-something-that-pertains-to-me" experience takes place. You can *make* fish go away, but you cannot *scare* them away.

This is not to say that fish lack the means for experience of any kind. When a fish gets chomped on by a predator, there may be a fair amount of pain going on, but it has no subjective meaning—if it has any meaning at all it can be no more significant than "tissue damage taking place at dorsal fin area." A fish may *feel* but it cannot *care.*

Many organisms, however, *do* have neural mechanisms for integrating internal and external streams of information, and in these organisms there exists a potential for subjective experience—that is, a "feeling of what happens." Damasio hypothesizes that conscious awareness (the core self) becomes possible when neural images standing for external objects are directly interfaced with neural maps standing for the internal milieu of the body. This would call for an emergent, second-order neural system that can render an image of the self as it is being causally affected by the organism's own perceptual image-making. In other words, the brain must construct an image of itself engaged in the complicated act of making an image and then passing its consequences on to the internal milieu. This global self-imaging results in core consciousness: "Core consciousness is the process of achieving a neural and mental pattern which brings together, in about the same instant, the pattern for the object, the pattern for the organism, and the pattern for the relationship between the two."[22] When this larger image incorporates a perceptual image together with its impact on the body maps, then the *relevance* of the perceived object for the body becomes immediately registered, resulting in a meaningful experience for the organism, *the experience of being something to which something is happening.* Consciousness registers the meanings of encounters in terms of feeling states that could never arise on their own from either cognitive or affective streams of information processing. The proto-self can have no such feeling of being-in-the-world. It has no unique subjective perspective, it is merely a teleological center of a life that is unconsciously calibrated to the world. By contrast, the core self is a conscious subject, an emergent reality with a completely new range of motivational and experiential possibilities.

We may say that the core self is a narrative being, but only in the narrowest sense. The conscious experience of the core self is highly circumscribed. It has an experience of what happens here and now, with no felt connections to past experiences and no expectation of enduring beyond the immediate moment. The core self is episodic, created anew with each experience of something happening. This is not narrative existence in the usual sense, where the self sustains goals over time and where today's episodes cohere, however faintly or vividly, with yesterday's and tomorrow's.

Yet there is meaning in the experience of the core self: The business at hand here and now is construed as immediately relevant to the teleological center of a life. The business at hand might bode well or ill for the carrying on of the organism. The important thing is that the core self *cares*—something is at stake, the relevance of the business for the self is *felt,* and in this feeling is a momentary burst of narrative content.

From the perspective of an external observer there may appear to be a continuous subjective entity from episode to episode, but for the core self each episode is a self-contained lifetime, a lifetime contained in the episode. The next episode calls forth the creation of the next core self in an indefinite series of discrete subjects having no continuous subjective identity. The core self is a narrative being in the sense that selections in a random collection of very short stories by very different authors are, after all, stories.

Are these bursts of narrative content what we mean by emotions? In the most elementary sense, yes. They are momentary feelings where the salient feature is a positive or negative sense of goal relevance. Are there any animals we can point to whose subjective lives correspond to this description of an episodic core self having an elementary emotional repertoire? Chickens? Pigeons? Dinosaurs? This is difficult to say. Perhaps one day the distinction between episodically conscious core selves and preconscious proto-selves will be clarified by experimental tests. For the moment, however, the distinction is hypothetical. What makes Damasio's hypothesis attractive is that it presents a plausible account of what must be the case given what we know about neurobiology, cognitive and emotional deficits, and our own subjective experience as conscious beings.

On the assumption that Damasio's story is correct, we are invited to raise the obvious question about how organisms pass from the pulsating states of core consciousness to a state of extended consciousness, where it makes sense to speak of the enduring identity of a subject. How do we get from a random collection of short stories to the epic narratives that we feel our own lives to be?

The answer to this question is incomplete at the moment but almost certainly has to do with the expansion of capacity for working memory. Working memory is the ability to juxtapose a plurality of mental objects so that various cognitive operations may be performed on them. Core consciousness appears when this capacity is at its bare minimum—that is, when a single perceptual image is held in juxtaposition to an image of body maps, making possible an integration of the two information streams.

As working memory expands to include a larger array of mental objects, so do the field of consciousness and the continuity of self-hood expand. When streams of memory images are drawn into working memory, new cognitive operations become possible and more meaningful associations are made. If a subject can recall the outcome of an episode from yesterday that resembles the happening here and now, then the assessment of what is at stake may be very different. And when memory objects can be used to construct alternative scenarios for future outcomes, then appraisals become even more complicated and the narrative meanings become more nuanced. More working memory endows a subject with the means for deeper, more powerful appraisals of what is at stake for the organism.

Working memory also enables an expanded sense of selfhood. For the chicken each episode is unique, and nothing in particular is remembered or anticipated, but for subjects with enhanced working memory there comes a capacity for distinguishing between common-place events and genuine novelty, between the routine and the surprising. Apart from the obvious adaptive value in such abilities, the sense of being-in-the-world that comes with them must be profoundly new. A sense of self-identity is born.

Damasio uses the term *autobiographical self* in referring to subjects whose conscious experience is rendered continuous by the operations of working memory. It is tempting to speculate on which animals are confined to a core self and which ones are liberated by working memory to experience an enlarged, autobiographical sense of self. The working memory of a chicken is minimal, suggesting that chickens are, at best, core selves. But ravens have by comparison a prodigious working memory, as their problem-solving abilities attest. It is likely, then, that a raven's sense of being *an enduring thing to which many things happen* is comparably enhanced. Ravens (together with mice, woodchucks, and barnyard mammals) are hardly the subjects of epic lives, but their sense of selfhood is undoubtedly more continuous and textured than a chicken's.

I imagine a continuum of selfhood among animals ranging from the pure and simple core self of, say, a chicken to progressively more continuous and richly informed states of autobiographical selfhood. This progression is marked by an increase in the duration and diversity of operative goals in the lives of subjects, as well as an expanding appreciation for meanings in the world around.

We encounter a substantial gap in the continuum when we come to consider the subjective lives of humans. Here the power of working memory is sufficient to produce genuine autobiographical selfhood, even personhood. Now we leave the world of vignettes and novellas to enter the world of sweeping sagas with convoluted subplots, refined character development, and epic triumphs and tragedies. Humans are graced not only with a *sense* of self, but also with a *concept* of self. The difference in terms of meaning is dramatic. When secondary reality operators and secondary valence operators enter the game, the mental objects drawn into working memory will include images of realities unseen and unseeable, enabling inferences about happenings and their stakes for the subject far transcending the appreciation of even the most experienced apes. Humans are not merely goal seekers, but goal *setters* as well, capable of monitoring the self, not only as it is but also as it *should be.* Humans expand the duration and diversity of their goals by constructing new types of goals, moral ones for example. Moral goals are supported by ideals for the self and various implicit standards for behavior. This function of working memory makes an enormous difference in the range of emotions open to human conscious experience.

From the foregoing it may be surmised that the onset of consciousness just is the onset of a subjective awareness of emotion. But there remains a puzzle about the evolution of this awareness. What would be the adaptive payoff in a new system enabling subjects to become consciously aware of the meaning of business at hand? Allow me to rephrase the question in terms of chickens and fish. As far as I can tell, the behavior of fish and chickens in response to approaching predators is virtually indistinguishable. Both animals react in highly stereotyped and genetically constrained sensorimotor patterns that are mediated, almost reflex-like, by the activation of limbic structures. Yet I am inclined to say that whereas the chicken is probably fearful, the fish is probably not. But here's the evolutionary puzzle: What good does the fear do? That is, what is the adaptive advantage in the extra second-order imaging loop that allows chickens to cop a momentary pulse of being in a world of meaningful objects and events? Why isn't the fish's unconscious calibration to the world good enough? After all, fish manage to escape predators just as effectively as chickens do—and more efficiently too, if we factor in the costs of extra neural imaging. The added neural wrinkle of second-order imaging seems superfluous and counterproductive from an evolutionary standpoint: doing the same with more, rather than doing more with less. So what would account for the selection of felt meanings if they cost more and don't do anything?

The answer is that meanings per se make no adaptive difference, the experience of fear itself is a no-gainer. What does make an adaptive difference is the image making, because neural images can be stored in memory. You cannot store a feeling in memory but you can store images that produce feelings. And once stored in memory they may be recalled to participate in the construction of new mental objects, whereupon they evoke something very like the original experience. And operations like this are, of course, highly adaptive. They give a subject the potential for constructing alternative scenarios of the future, and for choosing between them on the basis of their comparative emotional valence. If we did not have valenced images stored in memory, then scenarios of the future would have no meanings for us. If you were once humiliated by Professor X, then the prospect of another class from him would cause you to experience something close enough to the original feeling to assure that you would opt for Professor Y's class instead.

Such behavioral dispositions cannot be established by the calibrating devices of sensorimotor conditioning alone; they require the ability to store second-order images. It is therefore not the consciousness that is adaptive, but rather the power to create, store, and retrieve second-order images. Consciousness itself is merely an interesting side effect of this operation. At the level of chickens and fish the advantage in this new neural wrinkle is probably so small as to be almost negligible, but as capacities for memory and working memory expanded, this small advantage hit the evolutionary jackpot.

The central governing principle for the emotional life is this: *Emotional states will vary according to a subject's goals and the cognitive appraisals a subject is able to perform regarding the relevance of the business at hand for the subject's goals.* Subjects with greater capacity for working memory will develop more elaborate goal hierarchies and make more complex appraisals, and will therefore experience more differentiation in their emotional lives. At the neural level these differences involve different patterns of interaction between the limbic structures and various cortical sites in the brain. The possibilities for variation in these patterns are vast, suggesting that the differentiation of emotional states is virtually without limit.

If this is true, then it implies that our very limited categories for thinking and speaking about emotional states amount to little more than rough approximations of what is really going on in emotional events. More importantly, an infinity of emotional states would discourage the claim that emotional states are to any significant degree constrained by genetic predispositions. No consensus has been achieved on these thorny issues, but the most popular view at the moment appears to be that we inherit a relatively small number of organized emotional systems that remain open to modulating influences. But even if this were a consensus, it would be a weak one, for it leaves unresolved questions like "how small is relatively small?" and "by what means are these systems modulated?" The word "tentative" appears routinely in the literature on emotions. There is no small amount of guesswork in saying that there exist six, eight, ten, or fifteen primary and universal emotional themes upon which variations might occur. I will offer my own tentative version of an emotional typology, though I do so less to settle thorny issues than to keep the discussion of narrative content moving forward.

PRIMARY EMOTIONS. Primary emotions are feeling states that might be experienced by core selves. If we want to know what the particular primary emotions are, we should begin by asking exactly what sort of images the proto-self might bring forward to be juxtaposed with perceptual images. The answer is sensorimotor images—that is, images of the body's internal milieu actively changing in the course of established sensorimotor response patterns. Now we might ask: What feeling (if any) would be likely to arise when a core self integrates a perceptual image of an object or event with a sensorimotor image of the body freezing all motion or organizing itself to make a hasty escape? I believe the obvious answer is fear. We might next inquire: What feeling (if any) would be likely to arise when a core self integrates an image of taste or smell with a sensorimotor image of the body going into a retching response? Another obvious answer: disgust. And next: What feeling fits well with the integration of object or event images with an image of the body preparing for an aggressive attack? Anger. And finally: What feeling state would be most likely to accompany the integration of object or event images with an image of the body preparing for consumption or warmth of contact? My candidate is desire.

To these four I would add two provisional emotions: happiness and sadness. I hesitate about this because one might plausibly argue that happiness and sadness presuppose desire or fear, and therefore should be classified as secondary, or derivative emotions. Happiness occurs when a subject achieves or is about to achieve a desired goal, or when a subject vanquishes an object of anger, or escapes an object of fear or disgust (in which case we might prefer the term "relief"). Happiness therefore seems to be a response to an antecedent encounter that turns out well, where "well" is determined by the content of an earlier emotional event (desire, fear, disgust, anger). To be happy in the emotional sense, as opposed to the mood sense, a subject must make an appraisal of gains made relative to some goal other than achieving a state of happiness.[23] Likewise, sadness occurs when a subject appraises loss or helplessness relative to an earlier desired goal.

Happiness/relief and sadness might therefore be classified as "follow-on" emotions, requiring something slightly more than the minimal cognitive capacity for rendering second-order images. They appear to arise in conjunction with, or slightly consequent to, other emotions. With these provisos I will classify happiness/relief and sadness as primary emotions. Their case for primacy is somewhat compromised by their apparent contingency on other emotions, yet their feeling states are so unlike the other primary emotions that the contingency cannot be described as one of lineage or derivation.

I therefore nominate these six—fear, disgust, anger, desire, happiness, and sadness—as the basic, or primary emotions. They are basic in the sense that they presuppose no more than a minimal capacity for second-order imaging. They are also basic in the evolutionary sense of having utility for the most fundamental tasks of carrying on. These six states exhaust the possibilities for narrative content in the episodic consciousness of core subjects. At this level of primary emotions the distinction between positive affect (desire or happiness) and negative affect (fear, disgust, anger, sadness) corresponds tightly with the distinction between affiliative and dissociative behaviors.

There should be no difficulty in seeing the adaptive benefits of these primary emotions. Fear anticipates danger and motivates caution and avoidance behavior. Disgust averts animals from toxic and infectious substances, motivating them to rid their nests of harmful contaminants. Anger sends a powerful warning signal to would-be aggressors and also functions to direct and sustain the body's strength in the event of an attack. Desire focuses the body's energy resources, sustains goal-directed activity, and motivates sexual and companionate bonding behaviors. Happiness signals an openness to social bonding and has been associated with stress reduction and recuperative processes.[24] Sadness resulting from failure can induce a subject to withdraw from ill-chosen pursuits and to enter a period of inactivity, giving the body a chance to rest and repair. Sadness also sends a signal of distress, which may effectively elicit aid.

The adaptive advantages of these basic emotional strategies

would favor the selection of neural modules to mediate them. We may suppose that modules for primary emotional operations are innate to the brains of all normal animals capable of conscious experience. Our understanding of these modules is far from complete, though much has been learned in recent years, especially about the neural structures mediating the fear response.[25]

SECONDARY EMOTIONS. Secondary emotions can hardly be denied: If primary emotional systems were adaptive, then so would be any subsystems capable of modulating their responses. Such new strategies for carrying on would entail revised goal hierarchies, new appraisals of goal relevance, new complexities of memory-laden imaging, and new patterns of cortico-limbic interaction—all adding up to new possibilities for emotional experience.

It is difficult to be precise about the relationship between primary and secondary emotions. The terms suggest that secondary emotions both derive and diverge from primary emotional competencies. Primary emotions can "blend" to produce a range of secondary emotions, just as a few primary colors (red, yellow, blue) can be mixed in various proportions to produce all known colors. As information processing becomes more complex, especially in the cognitive domain, there results more refined imaging and greater differentiation in subjective experience.[26]

It is sometimes assumed that secondary emotions fall exclusively under the influence of cultural dynamics, and that only humans can have them. This is almost certainly wrong. Secondary emotions vary in complexity with the sophistication of goal hierarchies and appraisals of goal relevance. Some of the simpler secondary emotions (anxiety or affection) may occur in the repertoires of many mammals, while the most complex (jealousy, pride) occur only in humans. And there are several disputed emotions (gratitude, sympathy) which may or may not be present in a few nonhuman species.

The key to understanding these matters probably has more to do with image management in working memory than anything else. A larger working memory brings a subject closer to a genuine autobiographical self and also makes possible the use of secondary reality operators, enhancing a subject's grasp of objects, events, properties, and relations in the external environment. The construction of secondary valence operators makes possible the activity of goal setting and increases the variety of appraisals of goal relevance. As mentioned earlier, I envision a continuum of selfhood from simple core selves to more and more complex self-organization, until we come to the true autobiographical self characteristic of humans. The large gap between human and nonhuman emotional competencies is due mostly to the acquisition in early childhood of a self-concept, an event that prompts an extensive reorganization of the conditions for subjective experience and opens the door to personhood.

Late in their second year children come to master a mental schema for what persons are, as opposed to inanimate objects: Persons are agents, centers of experience who are capable of initiating behaviors. As a child's person schema becomes more refined, the schema is applied reflexively to produce the rudiments of a self-concept. This new measure of self-understanding gives a child a fresh orientation in the social world and generates intense interest in social roles. In the third year children appear to construe nearly everything in terms of narrative possibilities, a sure sign of role assimilation. As children assimilate social roles, they also internalize the goals and standards of performance implicit in those roles. Once internalized, the standards of good and bad performance are applied reflexively in a process of self-monitoring. And from the process of self-monitoring emerges an explicit and continuous self-concept.

A self-concept is a complicated thing, calling for an ample working memory capable of juxtaposing several components. An explicit self-concept presupposes both identity and agency—that is, the subject must coordinate a series of explicit memories in order to gain a sense that she is the same person today as yesterday and last week, and the subject must be able to integrate her person schema with her sense of being in the world in order to gain a sense that she is the author of her own behavior. But more: An explicit self-concept involves mental objects about "good" and "bad" as these are defined by standards of role performance, and it also presupposes the complex operation of applying these ideal standards to images representing the subject's actual performance.

Chickens cannot manage these complexities of information processing. Nor can monkeys. Chimpanzees, however, may come close enough to the mark to make their status as autobiographical selves worth debating. The acquisition of an explicit self-concept gives birth to the authentic autobiographical self, and with it come new goals (organized around the "ideal" image of selfhood) and new patterns of appraising what is at stake for them in the business at hand. In other words, along with an explicit self-concept there come new competencies for emotional experience and new possibilities for narrative meaning.

Consider guilt, for example. In order to experience guilt a subject must be able to perform the following complicated operations, which presuppose an explicit self-concept:

1. Subject must have learned certain standards of good and bad behavior.
2. Subject must have "linked" the attribution of self-worth to performing well against these standards.
3. Subject must recognize failure to meet a standard (that is, recognize, by monitoring, a discrepancy between an ideal performance and his/her actual performance).
4. Subject must form a belief that he/she *could have* behaved differently, but did not.
5. Subject must form a belief that he/she *should have* behaved differently.

Wherever an explicit self-concept exists, we may suppose a potential for these operations, and where these operations are performed, we may suppose the experience of guilt to occur.

Table 1 represents a tentative listing of secondary emotions. I hypothesize that these emotional competencies are culturally universal. There exist many universal cultural traits, such as language, religion, specialization of labor, gift exchange, marriage, and so on. It requires no special pleading to assert that there is a relatively stable core of secondary emotions shared by all humans, irrespective of cultural variations. To specify exactly what these are is a notoriously difficult and speculative task, and I reiterate that this is a tentative list. Indeed, each time I return to it I find reasons for making alterations.

The difficulty—some would say the futility—of compiling such a list is complicated by all the factors that can frustrate cross-cultural

Table I   Secondary Emotions

Presuppose minimal self-concept	Presuppose explicit self-concept
interest	hatred
anxiety	outrage
frustration/consternation	shame
affection	guilt
gratitude	envy
sympathy	jealousy
resentment	pride
contempt	grief
	resignation
	admiration
	wonder
	compassion
	alienation
	humility
	amusement

studies. In the first place there are always monumental translation problems. Does the English word for shame carry the same connotations as the German or Farsi equivalents? If not, then the words are not equivalents. Different vocabularies map onto realities in very different ways. It may well be that the English concept of shame overlaps with two or three clearly distinguished emotional terms in the lexicon of another culture.

Further, cultures differ radically in their valuation and regulation of emotions, even the less controversial primary emotions. For example, the circumstances in which anger is deemed appropriate will vary from culture to culture. When we come to the secondary emotions, matters become more slippery. Consider pride, for example. Cultures may disagree about whether or when pride is virtuous, and they may disagree about the accomplishments over which pride is warranted, and whether and how pride should be expressed. And most importantly, cultures are known to vary in the ways they define the self, and in the ways they influence the construction of goal hierarchies. All of these factors discourage speculation about a universal human core of secondary emotions. Still, one may be confident that there exists such a core without being dogmatic about a particular version of it.

The difficulties in specifying secondary emotions are real, but they should not discourage us to the point of declaring that all emotional realities are relative to particular cultural contexts. The claim that secondary emotions are universally human amounts to the claim that every human culture provides the conditions for their underlying competencies to arise in the normal course of personal development. That is, every human culture imposes standards for performance, facilitating the construction of a self-concept and the process of self-monitoring, which inevitably generates competencies for a core of secondary emotions. It is true that cultures will differ widely in their management of these competencies, but no culture has the option of not creating them. Wherever socialization occurs, we may assume competencies for a core of secondary emotions. These competencies form a large part of what I mean by the strings of human nature, upon which religious traditions play. Later in this chapter I will turn to the question of the various ways in which particular cultural contexts can modulate the emotional process.

A REVIEW OF THE APPRAISAL PROCESS

Earlier I summarized the Lazarus model of the appraisal process, a cluster of six mental operations by virtue of which emotional events acquire narrative meaning. It will now be useful to illustrate this process by analyzing appraisal patterns for a few selected emotions. But first, a qualification. The principal value of this model—as with any psychological model—is didactic and heuristic. That is, it helps us to appreciate what sorts of information processing tasks are implicit in an emotional event without suggesting that these operations always take place in the precise sequential pattern specified by the model. The actual details of the appraisal process are still obscure, but it is nevertheless possible to clarify the underlying logic of the process, which is all the model is intended to do. For this purpose the model presents formal operations in a stepwise format, from the most general to the more specific tasks. In reality most emotional responses will bear a closer resemblance to instantaneous all-or-nothing reflex patterns than to a laborious series of time-consuming computations.

Having thus qualified the model, we may illustrate it by analyzing the appraisal patterns for fear, anger, pride, and love. Fear is algorithmically the simplest of the emotions. In order for fear to arise a subject must acquire a sense of impending harm relative to an object or state of affairs. One cannot be fearful without being afraid *of* something. The appraisal pattern for fear is as follows (where X is some object or state of affairs perceived):

1. Is X relevant to the subject's goals?
   if no, then no emotion ensues
   if yes, then:
2. Is X goal-congruent or goal-incongruent?
   if goal-congruent, then positive emotions are possible
   if goal-incongruent, then:
3. Is X a threat to subject's life or safety?
   if yes, then fear ensues
   if no, then what aspect of ego-involvement is at stake?

No further appraisals are necessary for a subject's fear of X to arise. Attributions of credit or blame are irrelevant. Appraisals of coping potential and future expectations may have the effect of reducing or intensifying fear, or perhaps modulating the fear to a state of anxiety (where the threat posed by X is uncertain).

In order to achieve a state of anger a subject must have a sense that some attributable offense has caused injury to some aspect of the self. The appraisal pattern for anger proceeds as follows (where X is an event perceived):

1. Is X relevant to the subject's goals?
   if no, then no emotion ensues
   if yes, then:
2. Is X goal-congruent or goal-incongruent?

if goal-congruent, then positive emotions are possible
if goal-incongruent, then:
3. Has X resulted in damage or loss relative to any ego-involvement?
   if yes, then sadness, anxiety, or anger is possible
4. Can X be attributed to some agent or object (including the self or a scapegoat)?
   if no, then sadness or anxiety is possible
   if yes, then anger ensues, and:
5. Is retaliation for X a viable coping option?
   if yes, then anger is sustained

Both fear and anger are more likely where subjects have a temperamental predisposition to these emotional responses, or where factors such as negative mood or fatigue prevail. Anger can be facilitated by the mere presumption of blame, where no particular agent or object is identified.

Pride stands in the emotional lineage of happiness. A state of pride arises when a subject attributes credit to oneself for some valued object or outcome, or when credit is attributed to a person or group to which the subject is attached by ego-identity (as in parental pride or a swell of patriotic sentiment). The appraisal pattern for pride proceeds as follows (where X is some perceived or recollected object or event):

1. Is X relevant to the subject's goals?
   if no, then no emotion ensues
   if yes, then:
2. Is X goal-congruent or goal-incongruent?
   if goal-incongruent, then negative emotions are possible
   if goal-congruent, then:
3. Does X enhance the subject's self-esteem or reputation?
   if yes, then happiness, gratitude, and pride are possible
4. Can X be attributed to some agent?
   if yes, then gratitude or pride ensues
   if attributed to self or ego-identified agent, then pride ensues

No further appraisals are necessary for a subject's feeling of pride to arise. The feeling of pride often mingles with a sense of gratitude, especially where the credit for an accomplishment is shared with others, or where others bring the accomplishment to social attention.

The word "love" often, though not exclusively, refers to an emotional state. Perhaps even more often it refers to a social relationship or an attitude. When a person reports being "in love" with a spouse, we usually take the report to mean that he or she is in a stable relationship characterized by an enduring positive attitude toward the spouse. The condition of being in love also suggests that the person is predisposed to entering into the more acute emotional state, also called love. In the emotional sense love is a momentary feeling state that has narrative meaning relative to the person's goals and interests. It is also clear that love can arouse a physiological drive for sexual gratification, which suggests that love differentiates into romantic love and companionate love. Love stands in the emotional lineage of desire. In order to enter into a state of love, the subject must desire an enhancement of his or her ego-involvement with another person. The appraisal pattern for love proceeds as follows (where X is another person):

1. Is X relevant to the subject's goals?
   if no, then no emotion ensues
   if yes, then:
2. Is X goal-congruent or goal-incongruent?
   if goal-incongruent, then negative emotions are possible
   if goal-congruent, then:
3. Is X desired as an opportunity to deepen or extend subject's alliances?
   if yes, then affection or love ensues, and:
   if sexual arousal is added, then love is romantic rather than companionate

Attributions of credit or blame are irrelevant. An appraisal of coping potential may be relevant if there is sexual arousal, in which case love may or may not be sustained. If future expectations are positive, then love is sustained.

Once again, it should be stressed that these appraisal patterns identify formal cognitive operations that must be performed by a subject on the way toward an emotional experience. To analyze the appraisal process according to a decision tree model makes it appear far more conscious, deliberate, and abstract than it really is. In real life we seem to become emotional without cogitating about anything. For example, we often find ourselves in a state of anger without any decision on our part to become angry, and without knowing why we are angry. Examples like this are commonplace, which probably explains the commonsense view that emotions and cognition are independent: Emotions seem to be irrational and involuntary disruptions that sweep into our mental lives without warning, whereas thinking seems to be orderly, voluntary, and under rational control.

By contrast, the view I am endorsing makes cognitive appraisal an essential component of emotion, though this does not imply that these cognitive operations are always or even usually carried out at a conscious level. It is certainly possible to make judgments without being mindful of the criteria one is using. Likewise, I can get angry without sense or insight into what it is about me that makes me angry. Alternatively, I sometimes come into a state of anger by conscious and considerate means. The point is that my goal hierarchy can be brought to bear on some event in a preconscious, automatic, and intuitive mode or in a conscious, procedural, and reflective mode. But the issue of conscious versus preconscious modes of appraisal is not really critical. The important factor in appraisal is the process of cognition, not cognition of the process.[27]

It should also be noted that emotional reactions can unfold in response to internal events, such as memory or imagination, as well as external events. You may become angry when you remember how you were wronged by someone in the past, or you may get angry when you expect to be wronged in the near future. Or consider that emotions projected by actors on the stage or screen are usually the real thing, despite the fact that they lack objective stimuli. Emotions provoked in the audience are the real thing, too. Mystery novels stimulate genuine fear, passion plays inspire genuine sympathy, and obliging imaginary playmates can make a kid genuinely happy or proud. The fact that memory or imagination can initiate emotional responses reinforces the point that emotions have a cognitive component.

I have tried to indicate that an emotional event is a complicated process involving many components. The background components include several antecedent variables, such as temperament, attitude, values, past experiences, and worldview. A rich mixture of these hereditary and environmental factors is brought together in the construction of a subject's goal hierarchy. For every emotional response there is an incentive event, something happening in the environment or in the subject's mind to signal a change in the business of carrying on. The mediating components of an emotional event include appraisal pattern, action tendency, and coping strategies. The appraisal process has been described as a series of judgments by which the subject determines the narrative meaning, or rather the stakes, of an encounter. The next section will concentrate on action tendencies and coping strategies.

EMOTIONS ARE PREDISPOSITIONS TO ACT

Emotional events culminate in a temporary feeling state together with an associated urge to take some form of action. If emotions amounted to nothing more than subjective feelings, then it would be difficult to imagine how emotional systems could be selected by the dynamics of evolution. In order to be selectable, emotions had to manifest themselves in observable and patterned behaviors. This important factor has convinced some investigators that action tendencies are the defining characteristics of emotions.[28] The view I have been advancing sees action tendencies as part of a larger fabric including background components, incentive events, and mediating processes.

In order for emotional systems to evolve, their behavioral outputs would have to be stable and patterned without being rigidly stereotyped. Thus we see in emotional events that the links between subjective feelings and objective behaviors are close enough to warrant expectations, but not close enough to enable precise predictions. Generally, the action patterns characteristic of positive emotions tend to be associative, while those for negative emotions tend to be dissociative. In particular, fear is associated with a tendency to escape or avoid the object of fear, which may include behaviors as different as fleeing and freezing. Anger is normally associated with some form of aggression or attack, which may be as subtle as a threatening display or steady eye contact. Disgust is associated with a tendency to avoid, repel, or withdraw from the object of disgust. Desire tends toward approaching its object for consumption, possession, or contact. Likewise, happiness is linked to approach behavior, or expansive and interactive behavior. Sadness, meanwhile, tends toward withdrawal and inactivity.

To repeat, these are action *tendencies,* not stereotyped responses. They have all the marks of behavioral biases, governed by inherited algorithms yet open to modulation by a variety of environmental factors. The actual behaviors exhibited by a subject will reflect multiple influences, including the circumstances surrounding the incentive event, the precise course of the appraisal process, and the coping strategies employed. For example, fear of death and fear of ridicule are not likely to produce the same behavioral outcome. To begin with, they are considerably different states, even though they belong to the same emotional family. And further, individuals will cope with the two feelings in different ways. One might cope with the fear of death by slipping into a psychological strategy of denial, or one might try to preempt ridicule by creating a social distraction. Nevertheless, as different as these behaviors are, they both conform to a general rule for escape or avoidance.

Action tendencies are even more variable in the case of secondary emotions, where expectations about behavioral outcomes are much weaker. If secondary emotions are—as I suppose them to be—complex derivations from primary emotional systems, then we must assume that some of them draw upon a plurality of biases for behavior. Consider jealousy, for example, which has shades of anger, desire, and sadness. The action tendencies of these primary systems might simultaneously bias the jealous subject toward several behaviors, such as a vengeful attack of some sort against the rival, or a renewal of affections toward the loved one, or even depression and withdrawal from the scene. Any or all of these radically different behaviors might be expected to follow from the jealous state. What actually does follow in any particular case will depend on a variety of genetic and environmental factors. If the subject has inherited a strong temperament for anger, then it is more likely that an algorithm for retaliation will carry the day, whereas a strong temperament for sadness may favor depression and withdrawal. Past experiences with the rival (has he been a friend or foe?) will heavily bias the outcome as well. And so will the subject's own attitude toward jealous behavior. An important point to be seen here is that jealously is a complex state that is almost certain to generate internal conflict.

As I have described it, the emotional process begins with an incentive event, prompting the subject to integrate images of the external world with images of changes going on within the body's internal milieu. With these images juxtaposed in working memory, the appraisal process may proceed. The appraisal process amounts to a subject making narrative sense of the interface of external world and internal reactions to it. This process, I have ventured, culminates in the simultaneous emergence of two things: a subjective feeling state having narrative meaning, and an impulse or tendency to take action. In the absence of further complications the subject will act as the dominant action tendency ordains. And with this, we may say, the emotional process is complete.

But is it? Does life normally (if ever) leave matters free of further complications? If the wind blows fortuitously in my direction I am given a moment's happiness and then move on without further complications to the next encounter. But such isolated emotional events are the exception, not the rule. In real life the rule seems to be that complications abound. A psychiatrist friend once summed up his profession by saying, "People have problems because they have problems." A similar observation might be made about the emotional life: We have emotions because we have emotions. Emotional responses greatly complicate life by creating the conditions for subsequent emotional responses.

Consider jealously again. The jealous subject's emotional response was probably precipitated by an event that can be traced back to emotional encounters between his spouse and his rival. And these encounters themselves would probably not have occurred without a fairly involved history of emotions leading up to them. But now we have a new arrival, jealousy, to thicken the complications even further. Whichever action tendency gets acted out will change circumstances in ways that will undoubtedly precipitate further emotional encounters for everyone, including subject, spouse, and rival—not to mention those with whom they will interact in the course of working through the emotional pileup.

The point here is not to lament the fact that we are emotional beings. Far from it. The adaptive potential of our emotional systems is incalculable, and without them our lives would be unimaginable. I merely want to stress that emotions are powerful forces to be reckoned with and that they hold potentials for danger as well as good. It takes little reflection to realize that even slight failures to control emotional events can create havoc in personal lives and social groups.

For these reasons humans have always found it necessary to develop various strategies for managing the emotional process. I intend to say more about these strategies in the next section, but for the moment I will focus on the coping process—that is, the aspect of emotional management that is most closely associated with appraisal and action tendencies.

Implicit in a person's goal hierarchy are the following principles: (1) to ameliorate goal-incongruent encounters, and (2) to sustain or enhance goal-congruent encounters. Having these meta-goals embedded in one's self-organization is what generates the coping process. Coping with emotions includes the many things one might do or think to serve these goals in response to an emerging emotional experience. Suppose you are confronted in the woods by a hungry bear. You quickly appraise the stakes in this situation and find yourself overcome by fear and the impulse to run away. All of this belongs to the emotional event per se, and not to the coping process. The coping process includes whatever you actually do or think once the action tendency arises within you. If you decide to execute the action tendency (in other words, run like hell), then you have coped with the encounter by running. Perhaps you run because it is the option urged upon you by the action tendency and nothing else comes to mind. You cope by defaulting to the action tendency. But you might have coped with the situation differently. For example, if you had read recently that the best thing to do in such cases is to curl up on the ground and avoid making eye contact with the bear, then you might have coped in this alternative way. In either case you have coped with the fear by adopting a problem-focused strategy.[29]

Many of our coping strategies are problem-focused: for example, running from bears fearfully, combating our rivals jealously, assisting the helpless sympathetically, bragging to our friends proudly, and so on. But we also devise a multitude of emotion-focused strategies, the point of which is not to alter the world directly, but to alter some aspect of the self. One such strategy would be to alter one's goal hierarchy in some relevant way. For example, on the way home after receiving a C+ on a biology exam, a premed student might convince herself that she really doesn't want to be a doctor after all. This cognitive strategy is often used to ameliorate the sadness or shame that comes with a loss of self-esteem.

Another common emotion-focused strategy is to distract oneself from an emotional feeling by promptly pursuing some alternative goal that has potential for producing a positive emotion. If you find yourself angry at the neighbor's dog, you might head straight for the phone and call up an old friend. If you do so, the anger will abate rapidly and may not return when you hang up.

The process of reappraising the original encounter is also an effective coping strategy. If you feel a jealous impulse swelling up, you might succeed in convincing yourself that jealousy is really a childish response and that you should rather put your energies into understanding the complex motives of the offenders. Or you might pray for help in overcoming your jealousy. In taking these measures, the subject attempts to alter the narrative meaning of the incentive event. There is also the option for denial, yet another strategy for short-circuiting an emotional feeling. Denial is a common coping mechanism among patients traumatized by the news that they have a terminal illness.

There exist several so-called emotional states that might be more appropriately classified as coping strategies. Hope, for example, is widely recognized as a distinctive emotion. We might wish to say that hope is in the lineage of desire, and therefore qualifies as a secondary emotion. But hope is not consistent with the appraisal pattern for desire—that is, it is not sustained by an appraisal for positive future expectations. Indeed, it is desire in spite of negative or highly uncertain future expectations, the very appraisal feature that decomposes desire in favor of sadness. Hope is desire where there should be none. In other words, a more coherent way to see hope as a viable mental state is to view it as an artifact of the coping process.

There are other instances of what we might now call "coping states." Certain forms of love, for example, seem to qualify. The Christian concept of *agape* is understood as loving others regardless of their lovability—in fact, in spite of their *un*lovability. Christianity teaches that we should love even those who persecute us—but to do so would violate the basic algorithms that govern the appraisal pattern for love. Agape, like hope, appears to be incoherent unless we see it as a coping state resulting from a reappraisal in response to an emotional state. Many other feeling states, such as the so-called aesthetic emotions, might also be classified in this way.

My point is not to dispute that states such as these exist, or even that they have unique experiential qualities, but only to urge that we regard them as coping states and not as primary or secondary emotions. Maybe the term "tertiary emotions" would effect a constructive compromise. Tertiary emotions might be viewed as feeling states that arise in the process of coping, and may be characterized by unique blends of primary and secondary emotional properties. I will return to this concept at the end of the next section.

EMOTIONS ARE INFLUENCED BY CULTURE

Cultural traditions exert influences at every discernible stage in the emotional process. Wherever emotional variables exist we may expect to find cultural variations, which indicates that the emotional life can be manipulated by various means (including symbolic means) for achieving individual and collective goals. A comprehensive account of the many specific ways in which cultural traditions can educate the emotions is out of the question. For present purposes a general survey will suffice, but the issues will reappear piecemeal in part two, where the point will be to show how particular spiritual traditions play upon the strings of human nature. If we are correct in analyzing the emotional process in terms of background feelings, incentive events, the appraisal process, and coping strategies, then it seems appropriate to consider the influence of culture under these four headings.

BACKGROUND FEELINGS

As already observed, emotions are not to be confused with various background states and emotional traits, including sensations, moods, temperaments, and attitudes. Nevertheless, these phenomena are relevant to the emotional process and are, in varying degrees, open to cultural influences. Sensations of pleasure or pain, as well as arousal states such as hunger, thirst, or sexual excitement, can play a decisive role in the onset of emotional responses. Cultures can significantly influence the emotions contingent on these states by various means.

Pain, for example, is a universal human phenomenon, but its management is not. When I experience a headache, I typically consult the medicine cabinet for a couple of painkillers. When this socially approved strategy works, it prevents the emotional consequences associated with prolonged pain. But this pharmaceutical strategy is not followed in every culture. Similar cultural variations exist regarding the expression of sensations and drive states, and these variations may have emotional consequences as well. If a culture regards the expression of pain as a sign of weakness, then those who show pain will fail to meet a social standard, which is likely to result in shame. What particular cultures do with biologically universal sensations and drives will have much to say about the emotional life of their people.

As mentioned earlier, mood states should not be confused with emotional feelings. Moods last longer, they lack the narrative content of emotions, they are less acute than emotions, and unlike emotions they are not coupled to action tendencies. Moods are, however, directly relevant to emotional events because they affect appraisal thresholds. That is, when a subject is in a negative mood (perhaps depressed or irritable) the thresholds for negative emotional reactions (sadness, anger) are significantly lowered, while thresholds for positive emotions (joy, love) are elevated. Likewise, a positive mood will lower appraisal thresholds for positive emotions and raise those for negative emotions. When you are in a sunny mood, for example, you will be less likely to appraise an encounter as threatening or offensive.

Mood states are open to a wide range of influences, suggesting that there are many specific ways in which cultural practices might affect them, and in doing so have a profound impact on individuals' emotional lives.[30] Whenever my jogging friends miss a morning's run, they invariably report that it spoils their mood for the entire day. Whenever I am low, I find that it helps my mood to hear a good joke or to see an amusing film. Exercise and humor may, therefore, raise our thresholds to negative emotions and lower our thresholds to positive ones. How many battles, I wonder, have been prevented by court jesters who were able to elevate their monarchs' thresholds to fear or anger? But humor and exercise represent only a fraction of what cultures might do to affect moods. Art, literature, drama, music, architecture, hobbies, public celebrations, prayer, ritual, therapy, drugs—all of these bear upon our moods (and thus our emotions) in ways we cannot deny, even though their effects cannot be measured or predicted with precision.

Infants arrive in the world with emotional biases. That is, each of us inherits a temperamental profile that sets stimulus thresholds for emotional reactions. If we inherit a low threshold to fear, then we are much more likely to appraise an environmental encounter as threatening. Jerome Kagan's research indicates that about 20 percent of infants inherit low stimulus thresholds to fear. However, in only about two-thirds of these subjects does the fearful temperament persist beyond childhood. What can account for the change in behavior patterns among the remaining one-third? Kagan has found that some of these individuals retain their inherited physiology despite the changes in behavior—that is, they still have low thresholds for limbic excitability. In these subjects the change in behavior is attributed to psychological factors that compensate for a highly reactive physiological profile. But in others the physiological profile itself undergoes a change, resulting in an elevated threshold to fear. This physiological change, Kagan speculates, might be due to environmental influences.[31] The inference is that even temperament may be influenced, within certain limits, by environmental factors. Whether by psychological strategies or by subtle responses in the genomic expression, there appears to be enough slack in our temperamental profiles to suggest that cultural inputs may be significant.

Attitude formation is one of the most important ways in which cultural traditions can educate the emotions. Attitudes are valenced beliefs that we acquire and modify continuously throughout our lives. Attitudes toward persons, objects, and events originate in the course of having experiences with emotional content, or in the process of social apprenticeship. We might also generate attitudes by making inferences from previous ones. In any event, attitudes are of central importance to the emotional life because they are part of the complex fabric of an individual's goal hierarchy, the operative factor in cognitive appraisals resulting in emotional responses.

INCENTIVE EVENTS

An incentive event is anything that triggers an emotional response. An interaction with a shop clerk, a familiar melody, a broken object, a poetic image, a sudden noise—indeed, one is hard pressed to think of any observable event that might not, under the right circumstances, precipitate an emotional reaction. Defined in this way, it is obvious that what might function as an incentive event for one person might not do so for another. Each day we observe circumstances in which two people respond in radically different ways to the same stimulus. One person might be panicked by a sound that is completely ignored by the next person. The difference, of course, is entirely on the side of the individuals involved: what their different circumstances happen to be.

The differences most often show up in the course of the appraisal process or in coping strategies, but sometimes they appear earlier, in the act of perceiving an incentive event. Perception is an active, interpretive process. The circumstances in which we perceive things are often fraught with ambiguities that perceivers must resolve. Was that a blink or a wink? A grimace or a grin? Did she say "used ink" or "you stink?" Even before we get to the point of appraising the stakes in an encounter, we must achieve a coherent perception of what is happening. As William James remarked, "Whilst part of what we perceive comes through our senses from the object before us, another part (and it may be the larger part) always comes out of our own heads."[32] Our question is whether cultures can differentially prepare "our own heads" for making perceptions in ways that might systematically influence the emotional life. The answer is yes, they can.

Cultural variations have shown up in studies of visual illusion. The familiar Mueller-Lyer illusion presents subjects with two horizontal lines of equal length, one with in-facing arrows at each end and the other with out-facing arrows. The illusion effect is that subjects often mistakenly judge the line with in-facing arrows to be longer. Segall and colleagues found that American subjects were significantly more susceptible to the illusion than were African subjects.[33] The cause of this perceptual difference is not clear, but researchers agree that it has to do with strategic (that is, learned) factors rather than structural (inherited) factors. A similar strategic difference was recorded by Colin Turnbull in the following episode involving a forest-dwelling Bambuti tribesman who was unaccustomed to perceiving objects at a distance of more than a few dozen yards.

> Kenge looked over the plains and down to where a herd of about a hundred buffalo were grazing some miles away. He asked what kind of insects they were, and I told him they were buffalo, twice as big as the forest buffalo known to him. He laughed loudly and told me not to tell such stupid stories . . . We got into the car and drove down to where the animals were grazing. He watched them getting larger and larger . . . Finally, when he realized that they were real buffalo he was no longer afraid, but what puzzled him still was why they had been so small, and whether they had really been small and suddenly grown larger, or whether it had been some kind of trickery.[34]

Language differences, too, can influence perception in subtle ways. A classic experiment in perception research familiarizes subjects with a simple geometric figure and then asks them to locate the figure within the context of a more complex pattern. The duration of exposure to the original figure has been shown to have no bearing on the difficulty of identifying it when camouflaged. However, verbal labeling does have a bearing on recognition. It turns out that figures labeled with a name are much easier to locate under complicated perceptual circumstances.[35] The implication is that the semantic endowment of a cultural tradition prepares individuals for perceptual tasks. The potential herein for influencing emotional events is substantial, for cultures differ in the ways they define (and label) the boundaries of emotionally relevant factors such as social roles and relationships. It is no exaggeration to say that in certain contexts we cannot see what we do not know.

Similarly, what we do see is heavily biased by what we know best. In another study a stereoscope was used to present subjects with different pictures to their right and left eyes. One eye was shown a violent scene while the other eye was simultaneously shown a nonviolent scene. Subjects tend to resolve visual ambiguities like this by favoring one scene over the other. The results of this experiment showed that subjects with advanced training in police administration reported perceiving the violent scene twice as often as control group subjects did.[36]

Perception researchers use the term "perceptual set" to refer to the expectations and biases individuals bring to a perceptual task—as James put it, the part that comes out of our own heads. Cultural differences contribute to our perceptual sets in various and subtle ways, systematically influencing where we look, how we look, and what we look for. By educating our attention and expectations, a cultural tradition can affect our judgments about what is actually happening in the world.

THE APPRAISAL PROCESS

The appraisal process is at the epicenter of the emotional life. As we have seen, emotions are contingent on the relevance of some business at hand for an individual's goals. Positive emotions result when a subject judges the business at hand to be goal-congruent, whereas negative emotions result from appraisals of goal-incongruence. Everything depends on the individual's hierarchy of goals.

The concept of goal hierarchy has received much attention in psychological literature, perhaps most famously in the work of Abraham Maslow. Maslow developed the idea that motivation is at the core of human psychology, and that motives are organized hierarchically, the more basic needs of the individual taking priority over less basic ones.[37] Maslow's scheme is as follows:

self-actualization needs (least basic)
self-esteem needs
love and acceptance needs
safety needs
physiological needs (most basic)

According to this view, physiological needs are the most basic and will therefore normally trump our motivation to pursue safety needs. Once our physiological and safety needs are met, we are free to pursue needs for love and acceptance, and so on.

Maslow's theory correctly identifies two central facts of human psychology: We are goal-seeking beings, and our goals are organized hierarchically. But the theory overlooks another important aspect of human nature: We are goal-setting beings as well. Some of our goals are inherent in our biological makeup and others are acquired by experience. The important point, however, is that the priorities defining a person's goal hierarchy are not fixed. The ideal of a fixed hierarchy of goals may apply to animals whose behaviors are tightly organized by primary operators, but less so to goal-setting animals. A wolf, for example, might risk safety in pursuit of physiological needs, but no wolf would lay down its life for a political cause or leave the country to avoid shame. Such behaviors require an ability to inhibit default goals in favor of learned goals that are contingent on secondary valence operators.

The psychological mechanisms for transcending a fixed goal hierarchy are found in the dynamics of the self-esteem system. As we internalize standards of good and bad behavior, and as we begin to monitor our performance against images of an ideal self, we introduce a wild card into the motivational game. The dynamics of self-esteem are such that an individual may link ego needs to virtually any imaginable outcome, and may assign virtually any priority to self-esteem goals. Anything can mean everything to human beings. The result is that the goal hierarchies of human beings are broadly variable, even though the default goals will continue to exert their influence in subtle and persistent ways.

The self-esteem motive should not be confused with selfish motives. It is common for individuals to link their desire for self-esteem to performing well on standards of selfless behavior. Thus a person's goal hierarchy might subordinate selfish interests to goals that serve the interests of others. This is the paradox of the moral life: We may construct our goal hierarchies in such ways that a desired sense of self-worth may be achieved only by sacrificing self-interest. In an act of moral sacrifice I might achieve a gratifying surge of self-worth by advancing your goal of escaping some sort of harm. We both win precisely because our interests overlap.

Such overlaps of self-interest occur routinely among nonhuman animals, enforced by the logic of reciprocity, but among nonhumans the logic of reciprocity is narrowly constrained. It works perfectly well among relatives where there is a built-in overlap of genes, and it works reasonably well in small, stable populations where individuals can keep mental ledgers of favors done and favors received. Beyond these limits, however, reciprocity cannot be achieved. But in our species, where the logic of self-esteem prevails, and where secondary emotions like guilt, shame, and pride enter in, overlaps of self-interest may be achieved well beyond the default limits fixed by genetic relatedness and tit-for-tat altruism. These extended overlaps of self-interest are achieved by cultural means.

Cultural traditions encourage certain self-esteem linkages and discourage others, and by doing so they influence the goal hierarchies (and thus the emotions) of individuals. But this is not to say that an individual's goal hierarchy is fully determined by social dynamics. The assembly of a goal hierarchy is far too complicated to be determined by any single type of influence. Your goal hierarchy reflects your mood, your temperament, your attitudes, your physiological state, all your knowledge and experience—basically everything we can know about your psychological makeup has the potential for influencing your goal hierarchy. In a real sense, you *are* your goal hierarchy! And to complicate matters, a person's goal hierarchy is constantly in flux. What you considered important in childhood, or last week, or yesterday, or even an hour ago may no longer seem worthwhile.

A goal hierarchy is kaleidoscopic: At any given moment it has a definite structure, but the components are constantly and unpredictably shifting. If human behavior is inherently unpredictable, it is because our values are continuously in flux. They are not, however, chaotic. Despite constant changes in day-to-day details, the goal hierarchies of most individuals are marked by integrity over time. There remains a stable core of self-esteem linkages that continues to be fundamental in the organization of a person's life. A normal and healthy person is one whose goal hierarchy is flexible enough to enable adaptation to changing circumstances, yet stable enough to assure the persistence of a coherent personality.

Our question has been whether cultures can influence the emotional life at the level of the appraisal process. The answer is that they can profoundly affect the appraisal process by influencing the goals, values, and attitudes that generate appraisals. It is easy to see why cultural traditions have never been indifferent about goal hierarchies. William James was fond of saying that the most interesting and important thing about persons is their view of life. We might add that a view of life is nowhere more evident than in a person's goal hierarchy. A goal hierarchy tells us how a person is prepared to engage the world.

It is a person's operational view of life, where one's worldview actually meets the world. It is therefore at the level of worldview that cultural traditions are careful to exert their influence in systematic ways.

I will have much more to say in the following chapters about cultural myths and worldviews. For the moment I want to stress that they have a profound influence on the goal hierarchies of individuals, and therefore on the emotional life. It is from our cultural myths that we acquire memes for the ultimate realities and values that are reflected in our goal hierarchies. Our myths tell us how the world is made up, how it works, what its point is, what the possibilities are, and so forth. They also tell us how humans fit into the picture, what our point is, what in the world is good for us, and how we should seek to fulfill our lives. A worldview is a systematically interrelated set of beliefs whose primary function is to explain facts and certify values. When there is business at hand such beliefs will strongly influence our appraisals of what is at stake.

COPING STRATEGIES

There seems to be very little controversy over the claim that cultural factors influence the emotional life. Opinions begin to differ when we ask how much influence cultural factors have, and at what point in the emotional process their impact is most pronounced. The issues may be sharpened if we picture the emotional process unfolding in the following sequence:

incentive event → appraisal process → emotional feeling → coping strategies → behavioral expression

Causal influences entering the picture before the feeling state emerges may be called constitutive factors, while those entering after the feeling state may be called regulative factors.[38] Our question now asks whether cultural factors are constitutive or regulative, or both. They are no doubt both, but the full answer is probably much more complicated than we can presently say. I support the view that while cultural factors are potentially constitutive, it is probable that their most pronounced influence is to regulate emotional outcomes at the point of coping strategies.[39]

Cultural influences on the appraisal process do not always carry the day. It often happens—perhaps it normally happens—that initial appraisals of what is at stake reflect the primal meanings of our intuitive morality. The genetic defaults scripted into our emotional systems are formidable, and they commonly prevail despite the ministrations of culture to superimpose the ideals and goals of an override morality. Consider, for example, the famous lament of St. Paul: "For the good that I would I do not; but the evil which I would not, that I do."[40] Paul is expressing a concern about moral lapses, the failure of cultural standards to exert a constitutive influence on motives. In Paul's vocabulary such failures are attributed to the sinful condition of humanity, but in the present view we might account for moral lapses in terms of the dynamics governing a person's goal hierarchy.

We have seen that the appraisal process brings an individual's goal hierarchy to bear upon encounters with the physical and social environment. This process can never be in fact what it appears in theory—that is, some goals are inevitably excluded from the appraisal process. This is a practical necessity, for it is impossible for all the values implicit in an individual's goal hierarchy to be brought to bear simultaneously on a given encounter. The appraisal process therefore depends on which aspects of a person's goal hierarchy will actually get a hearing as the appraisal of stakes goes forward. Many factors are relevant in determining which items get a hearing, but not the least of these is the element of time. Learned values will always be at a disadvantage when there are severe time constraints because they must first be reconstructed from memory in order to inhibit and override the ever-ready intuitive values. By the time any relevant "good-that-I-would" value gets a hearing, an individual may already be off and running with a very different emotional response. When time constraints are narrow, or when the business at hand is unexpected or unfamiliar, appraisals are likely to default to the goals of our intuitive morality. But all is not lost even when appraisals lapse to intuitive goals, for the coping process gives us resources for emotional damage control.

Coping strategies include whatever a person might do or think in attempting to control the outcome of an emotional situation. It is sometimes useful to characterize coping as a process of reappraisal, where an emerging emotional feeling together with its action tendency are encountered as a new incentive event, some new business at hand to be evaluated in terms of its meaning for an individual's goals. To illustrate, suppose someone says or does something that you initially appraise as personally offensive. You immediately feel anger swelling up within, and your impulse is to retaliate. But if you have internalized a cultural bias against anger, then the onset of your anger will itself be appraised as incongruent with your goal to be vigilant against anger. Your initial response is to become angry, but then you subsequently become shameful of your anger, whereupon the whole episode takes on a different meaning. It is in coping events such as this that cultural traditions have their most decisive influence, an influence that reflects the way a tradition moralizes the emotions.

All humans have an immense stake in the process of regulating the emotional life. The reasons are obvious: Emotions mediate behaviors, and behaviors may be either conducive or threatening to the achievement of personal wholeness and social coherence. Regulating the emotional life at the level of coping strategies may be our most significant means for maximizing these twin teloi. But emotional regulation may also generate the most substantial differences between cultural traditions. Each culture has a unique history, which inevitably results in a unique perspective on the conditions for achieving personal wholeness and social coherence, not least importantly the emotional conditions. The moral education of any culture will therefore extend beyond mere behaviors to include a morality of emotion. Cultures moralize the emotional life by defining certain emotional traits as virtuous and others as reproachable.

The emotional morality of a tradition finds expression in a body of socially defined display rules, specifying the conditions under which various emotional experiences and expressions are considered appropriate or inappropriate. When is it proper for a person to feel sad or happy? Are expressions of anger ever to be tolerated? If so, under what circumstances? When should a person feel humble? Is pride a virtue or a sin? Every culture develops display rules in answer to such questions, and in doing so they issue a motivational prescription for achieving personal wholeness and social coherence.

Much can be learned about the education of emotion by examining cultural variations on the morality of particular emotions. Take sadness, for example. Carol Stearns finds that while some societies regard sadness as threatening or dangerous, others regard it as virtuous.[41] Traditional societies having a high tolerance for aggression (Kaluli, Ilongot, and Bedouin, for example) tend to see sadness as evidence of impotence and passivity, and therefore develop coping strategies to transform sadness into anger and aggression. The nonaggressive Tahitians are also wary of sadness because it drains the sense of drive and energy necessary for individual and social projects. Some societies discourage displays of sadness because they regard it as a sign of vulnerability and childishness. Societies that value autonomy have display rules against the expression of sadness because it is a sign that one is needy or lacks self-control.

In contrast to these cultural biases against sadness, there are traditions wherein sadness is regarded as an emotional virtue. In many Asian societies sadness is admired as a sign that an individual is making spiritual progress. In Iran and Sri Lanka the ability to experience sadness is a measure of a person's depth of character. In seventeenth-century England sadness was contrasted with sinfulness and associated with spiritual virtues such as wisdom, patience, and humility. Melancholic individuals were thought to be especially pensive and were much admired by intellectuals of the period. In many societies displays of sadness are encouraged as a means of achieving group solidarity in response to loss. Similar examples of cultural variation in rules regulating emotional display behavior could be offered for fear, anger, disgust, desire, and happiness.

Display rules for regulating anger reveal a similar diversity in the ways cultural traditions moralize the emotions. Among the Bedouin and the Kwakiutl anger is the approved response to the death of a loved one. The Ilongot regard anger as a positive sign of vitality and self-control. In Western societies, as far back as Aristotle, anger has been viewed as a socially constructive response to moral transgressions and a sign of personal integrity and self-respect. In these cultures anger is thought to make positive contributions to personal wholeness and social coherence. By contrast, many cultures regard displays of anger as personally and socially threatening. The Utku, for example, take such extreme measures to regulate anger that one observer was led to these remarks: "Not only do they not express anger; they do not 'feel' angry, and they do not talk about it. They do not get angry in circumstances that would surely incite us to outrage, and they do not get angry in other circumstances either."[42]

It may be an exaggeration to suggest that the Utku never experience anger, but it is clear enough that anger is far less evident among the Utku than, say, among contemporary Americans. It is hard to say why. Perhaps the Utku are blessed with genes for a high temperamental threshold to anger. Perhaps the cultural bias against anger plays a constitutive role, so that Utku appraisals seldom terminate in an angry response. Perhaps the display rules against anger enter later, at the coping stage, to redirect or suppress anger before it is gated out to behavior. Perhaps justice prevails in the Utku social order, such that there are few occasions to incite anger. Or perhaps some combination of these factors is the answer. The genetic account is the least plausible. The most likely story is that at some point in their history the Utku concluded that anger posed a threat to their individual and collective well-being, resulting in the construction of strict display rules to regulate it at the coping stage.

A few pages ago I promised to return to the concept of tertiary emotions. There I hinted that this concept might give us a useful way to think about various unique emotional states that emerge from the coping process. I will now hypothesize that tertiary emotions are highly complex feeling states that may arise as we render appraisals on emergent primary and secondary emotions. These states reflect cognitively sophisticated compromises between simpler, yet somewhat conflicting, emotional responses. We may suppose that tertiary states require much more neural looping between cortical and limbic areas of the brain than do the algorithmically simpler primary and secondary emotions. On the psychological level it would follow that tertiary states involve a lot of working memory sophistication to produce the layers of imagery implicated in emotional blending.

As the term implies, tertiary states presuppose competencies for both primary and secondary emotions. Primary emotional competencies are universal; we inherit them as do many other species of the mammalian family. Competencies for secondary emotions are also universal among humans; we acquire them in the process of socialization. Every socialized human being, therefore, is capable of the full range of primary and secondary emotions. But competencies for tertiary emotions are not universal, they are culturally specific. They presuppose primary and secondary emotional competencies, but to these they add culturally unique algorithms that derive from learning a variety of specific and complex display rules. It is possible for an individual to experience primary and secondary emotions without moralizing about the emotional life, but for tertiary states a morality of emotions is a necessary condition. And finally, it may be speculated that tertiary states are relatively difficult to enter into, by which I mean that they presuppose more than a modicum of social nurturing.

I have already proposed hope and agape as candidates for tertiary emotional states. Another familiar candidate might be nostalgia. Nostalgia is a blend of happiness, sadness, and desire. It is a bittersweet longing for the impossible return of a past forever lost. The cognitive prerequisites for a nostalgic experience are so considerable that one may reasonably doubt its existence before adulthood. A list of culturally specific tertiary emotions would be extensive, to say the least.

The Ifalukian state of *fago* may be viewed as a tertiary emotion. Fago appears to be a blend of sympathy, sadness, love, and respect. One feels fago in response to certain misfortunes befalling persons who are respectable, deserving, and well liked. Fago is thought to be a virtuous and mature emotion, typically felt by chiefs or spiritual leaders but never felt by children.[43]

Laiya might also be classified as a tertiary emotion. Laiya is especially typical among Hindu women in India and Nepal. It appears to be a complex blend of shame, gratitude, resentment, and humility.

One enters a state of laiya by successfully coping with anger at being offended. It is the feeling one gets by swallowing moral outrage in deference to a desire to uphold the social order that ultimately benefits oneself. Laiya is considered to be both a mature emotion and a supremely virtuous one.[44]

Many so-called religious emotions might be properly classified as tertiary feeling states. The oceanic feeling of oneness with God, or the sense that one is in the presence of a transcendent power, or the blessed sense of being forgiven by grace, or the serenity of release from oppressive attachments—all of these states presuppose competencies for blending primary and secondary emotional qualities. Needless to say, more attention will be devoted to religious emotions in part two.

I have said that the coping process makes it possible to inhibit emotional responses that are devalued by one's cultural tradition and to reappraise the business at hand in conformity with emotional virtues. This process is contingent on learning a wide range of display rules, which depends on a tradition of moralizing the emotions, which depends ultimately on a shared perspective regarding the conditions for achieving personal wholeness and social coherence. These elements assemble themselves in a set of culturally specific ideals of emotional and spiritual excellence, which individuals are expected to assimilate into their goal hierarchies.

I will venture the claim that the emotional virtues of a cultural tradition are its most distinctive, yet most elusive, feature. There are many distinctive elements in any cultural tradition, of course, but the emotional virtues are the most difficult ones for an alien to apprehend. The language of a cultural tradition may be strange and unique, but aliens can learn it rather quickly. The same goes for culturally distinctive technologies, cosmologies, economies, political processes, marriage conventions, and even standards for moral conduct. These phenomena have relatively little to do with the culture shock one experiences upon entering an alien cultural context. What *does* produce culture shock is a failure to pick up on the subtle but important scripts that people use to manage their subjective states. To really make sense of another cultural tradition—to achieve deep and rich insights—one must figure out not only the goal hierarchies by which people appraise whatever business is at hand, but also the emotional morality that regulates the way they cope with their own responses.

The education of emotion calls for nurturing individuals in a morality of the emotional life. The task amounts to coaxing the working memory into giving priority to the emotional virtues, ensuring that they will prevail in the coping process. Recall that the working memory is genetically biased to retrieve information that has been recently or repeatedly stored in memory systems, or has been marked by unusually poignant experience. The ultimate challenge for any cultural tradition is therefore to find the symbolic and practical means to assure that the emotional virtues will acquire these priority markers.

SUMMARY AND CONCLUSIONS

The central thesis of this book is that religious traditions are primarily about manipulating aspects of our universal human nature for the sake of achieving the twin teloi of personal wholeness and social coherence, thereby to maximize the odds favoring human reproductive fitness. The plan of the book is to present an account of human nature and then to show how various spiritual traditions have gone about the business of manipulating it.

In this part on human nature I have stressed the point that human beings are wholly the products of a cosmic evolutionary process, that we, like all other living beings, are star-born, earth-formed, fitness-maximizing creatures endowed by natural selection with a set of species-typical traits for negotiating a livelihood on this planet. Our uniqueness as a species lies in our special ability to devise symbolic systems for the mediation of behavior. This ability has loosened the constraints on our behavior, enabling a variability of adaptive strategies that far surpasses all other species. In short, we have in our nature the means for manipulating our nature.

The focus of the present chapter has been on the interactions between emotional, cognitive, and symbolic mediation systems. Primary attention has been drawn to the emotional systems because the story of how they work and how they are influenced by cultural variables provides special insight into both human nature and religious phenomena. I have supported the view that emotional competencies have an essential and substantial cognitive component. At the neurobiological level this is shown by the extensive connectivity between cortical and limbic areas of the brain. At the psychological level it is shown by the fact that emotional responses cannot occur apart from cognitive appraisals of what happens during encounters with the environment.

This model allows us to see that any developmental or evolutionary enhancements of cognitive ability will have profound consequences for appraisal patterns, and for subsequent emotional and behavioral outcomes. Thus, as the capacity for processing information in working memory was enhanced by evolutionary events, it became possible to appraise the meaning of worldly business relative to the goals of an emergent and enduring subjective reality, the conscious self. These events enhanced emotional competencies, introducing secondary emotions—that is, a range of new yet derivative motivational states, enabling a reorganization of experience and behavior. Implicit in this view is the claim that primary and secondary emotional competencies are universally shared among all normally functioning socialized members of our species. There *is* a human nature.

However, to assert that a shared ensemble of primary and secondary emotional competencies defines a universal human nature is not to suggest that these competencies are ever uniformly regulated. Variations of time and circumstance inevitably result in culturally unique perspectives on how best to organize personal and social variables. That is, humans tend to disagree about the conditions most conducive to the achievement of personal and social fulfillment. These differences, I have argued, result in widely divergent traditions for moralizing about emotions. And further, these variations show up in culturally distinctive competencies for regulating species-typical emotional competencies. I have used the concept of tertiary emotions as a way of describing the unique blends of emotional properties resulting from divergent patterns of regulating primary and secondary emotional responses. Thus we see that human emotions are highly complex bio-psycho-social phenomena, which may be parsed out in terms of primary, secondary, and tertiary competencies.

We are now at the point of shifting our attention from human nature to spiritual traditions—that is, from the adaptive meanings inherent in biological and psychological systems to the conventional meanings inherent in cultural systems. The consilience of these two foci of attention may be shown by the principle of reduction underlying the argument of this book. The principle is expressed in the following sequence:

1. The myths, symbols, and practices in a religious tradition will have a decisive influence on the mental objects featured in the working memory of individuals.
2. The mental objects (neural imaging) featured in the working memory will have a decisive influence on the cortico-limbic interactions taking place during appraisal and coping events.
3. Patterns of cortico-limbic interaction will have a decisive influence on the mediation of human behavior.
4. The mediation of behavior will have a decisive influence on the prospects for achieving personal wholeness and social coherence.

5. It is by the achievement of personal wholeness and social coherence that members of our species influence the odds favoring reproductive fitness.

It may be objected that this view reduces the conventional meanings of religious myths, symbols, and practices to "nothing but" pragmatic devices for producing certain patterns of neural stimulation. This is precisely my meaning. But I have my own (rather strong) objection to the "nothing but" language because it imputes to this view a claim that conventional meanings are completely dispensable. This is *not* my meaning. Symbols—including religious ones—are not dispensable. Their deployment is no less essential to the normal functioning of human brains than active neurons are. For our species to pursue the twin teloi of personal wholeness and social coherence apart from the conventional meanings embedded in cultural memes would be unthinkable. As Terrence Deacon reminds us, we are the symbolic species: Our brains were fashioned by natural selection to be invaded by and dominated by the conventional meanings of symbol systems. You cannot get a human brain sufficiently organized to mediate complex adaptive behaviors without the input of socio-symbolic meanings. Symbols are *real*. To say, therefore, that something is "nothing but" or "merely" a symbol is to ignore the awesome power of conventional meanings to organize neural events and inspire motivational states. The question about religious traditions is not whether we should have them, but rather how effectively and adaptively they nurture us.

THE NATURE OF RELIGION

The aim of the present chapter is to offer a generalized account of religious phenomena within a naturalistic framework. This account will begin with the thesis that all religious traditions have a common structure. Next, I will present a speculative narrative of the origins of religion. And finally, I will address questions pertaining to the functions of religion.

If there is a method behind this theoretical approach, it comes closest to the "guess-and-test" method. It goes like this. First you spend several years immersed in reading, observation, reflection, and conversation about religious phenomena. Eventually, you may get a sense that some patterns are beginning to emerge. Then you talk to several people who tend to know a lot more about many things than you do. The next step is to assemble the patterns into a series of hunches about how things work. And finally, you go back to the phenomena to see how well the hunches check out (see chapters 5 through 9).

The guess-and-test approach may sound rather unsophisticated (maybe even a bit glib) in this era of intense preoccupation with abstruse hermeneutical methodology, but it can be defended with the assertion that this appears to be the way things go in science. Good theoretical science amounts to making some informed guesses about how things work in the world, followed by a set of procedures designed to rule out the bad guesses. When scientists exhaust their own resources for ruling out their best guesses, they take a deep breath and submit their work to the scrutiny of others. The hunches in the present chapter are offered in this spirit.

THE STRUCTURE OF RELIGIOUS TRADITIONS

In the last chapter I made the assertion that humans are narrative beings. Our lives unfold as series of connected events that acquire meaning as we integrate information about objects, events, properties, and relations in the outside world with information about how things are going inside our bodies. That is, an individual's narrative meanings are constructed as the brain negotiates mergers between facts and values, between reality operators and valence operators.

Cultures, too, are narrative entities. At the core of every cultural tradition there is a story, a myth, a narrative integration of ideas about reality and value. The narrative core provides members of a culture with vital information that gives them a general orientation in nature and in history. The narrative core is the most fundamental expression of wisdom in a cultural tradition—it tells us about the kind of world we live in, what sorts of things are real and unreal, where we came from, what our true nature is, and how we fit into the larger scheme of things. These are all *cosmological* ideas, informing us about how things ultimately are in the cosmos. But the narrative core also contains ideas about *morality,* about which things ultimately matter. It tells us what is good for us and how we are to fulfill our purpose. In the narrative core of a cultural tradition ultimate facts and ultimate values are interwoven in a seamless series of connected events, in precisely the way that cognitive and affective neural images are integrated in the mental lives of individuals.

There is another important parallel between individuals and religio-cultural traditions: Both have strategies for carrying on. Individuals carry on by virtue of various mechanisms enabling survival and reproduction. Religious traditions, likewise, develop a variety of enabling strategies, or *ancillary strategies,* designed to assure that the narrative core will continue to be replicated indefinitely in the minds of individuals. Thus, it may be seen that the phenomena of any mature religious tradition will conform to a universal structure, with an integrated narrative at the core and various supportive strategies at the outside, as shown in figure 3. It is by these means—the narrative core or myth, and ancillary strategies—that a pattern of piety is nurtured in the lives of individuals.

THE NARRATIVE CORE: MYTH

Particular objects and events acquire meaning for individuals when facts and values are integrated in a valenced mental object—when im-

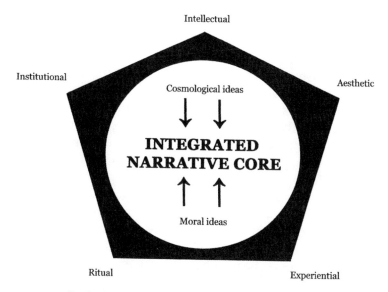

Fig. 3. A model for the structure of religious traditions.

ages from two information streams are juxtaposed in working memory. In cultural myths the entire universe acquires meaning by virtue of the conjugation of reality and value in metaphor. Reality and value are never incommensurate in mythic insight; they are merely two aspects of a unified meaning.

When the root metaphor of a mythic tradition is ingested, one apprehends that ultimate facts and values have the same source. In mythic insight, the ultimate explanation is also the ultimate validation. The root metaphor renders the real sacred and the sacred real. The force of the naturalistic fallacy—the separation of facts and values—is dissolved by the metaphors that generate myth.

Thus we see that the root metaphor of a religious tradition links cosmology to morality. In the Abrahamic traditions, for example, the root metaphor is God-as-person. God is both creator and judge, and the cosmic order and the moral order are unified under God's ultimate plan. In Greek tradition reality and value were unified by *logos,* the divine rationality inherent both in the cosmos and the human spirit. The root metaphor underlying much of Chinese myth is the Tao, the ultimate principle of balance and harmony that governs the universe. The Tao is also the Way, the moral path to human fulfillment. Opportunities for exploring such details lie ahead—the point here is to show that the deep structure of religious traditions is derived from the construction of root metaphors that unite ideas about how things ultimately are with ideas about which things ultimately matter. And this deep structure reflects the brain's integration of cognitive and affective images.

Another structural characteristic is seen in the complex relation between myth and cultural traditions. Myth integrates cosmological and moral elements of a cultural tradition, and thus it may be said that myth reflects the mentality and collective experience of a culture. But it is also true that a cultural tradition is a reflection of its myth. That is, once the myth establishes its currency within a culture, it becomes a significant cultural determinant.

Cultures become distinctive and coherent social entities by virtue of systematic and consistent decisions regulating the selection of memes. Any tradition in which innovative memes were received or rejected at random would be fickle, lacking in unity of purpose and saturated by ugliness. The members of such an unregulated tradition would be hard-pressed to find reasons to defend it, and all arguments either for or against change would be arbitrary. The most important regulating force within a culture is its myth. The meanings embedded within a mythic tradition will exert a selective influence to determine whether new meanings will be received or rejected. New ideas that do not fit the myth will be pronounced unfit, and those deemed fit will have a consistency about them for having been scrutinized by a common standard of meaning. The internal structure of a religious tradition will therefore be shaped by the myth, and especially by its root metaphor. God, one might say, is a supreme meme.

ANCILLARY STRATEGIES

Ancillary strategies provide the means by which a religious tradition may flourish. They are the strategies enabling a myth to carry on. The view presented here is that the phenomena of religion reveal five types of ancillary strategies: intellectual, experiential, ritual, aesthetic, and institutional. We may expect to find variations between traditions (and sub-traditions) in terms of their reliance on particular strategies. That is, some traditions may emphasize intellectual aspects but downplay aesthetic aspects, while others may invest themselves in ritual strategies but less so in institutional structures. Still others may reflect a balanced diversity of strategies.

INTELLECTUAL STRATEGIES. Myths are, literally, stories. They always take the narrative form and are often embellished with florid images and provocative symbolism. They also contain elements of mystery and ambiguity. These qualities have the power to excite the imagination and to draw us into the promise of deeper understandings. But myths also burden us with intellectual challenges. Despite the fact that they are entertaining, easy to remember, and broadly appealing to persons of all ages and levels of sophistication, myths are often obscure and difficult to incorporate into the concrete circumstances of everyday life. Religious myths, therefore, invite—in some sense require—interpretation. Myths excite the imagination by leaving much to the imagination. Thus we see that *mythos* (story) is typically augmented by *logos* (rational discourse) in the normal development of religious traditions.

Traditions of interpretation, normally tended by philosophy or theology, help to maintain the integrity and the accessibility of the myth by clarifying its meaning. Myths are far too important to risk any misunderstanding, so there will be a sense of urgency about setting down authorized interpretations, making it clear what the myth does and does not mean. The intellectual dimension of religious traditions naturally subdivides into metaphysics (or theology) and normative ethics, the former to clarify what the myth says about reality and the latter to make explicit its moral imperatives. As these traditions of metaphysical and moral interpretation develop, they tend to harden into doctrines of belief and codes of moral conduct. But they also tend to diversify as intellectual disputes arise, creating potentials for discord and schism into sub-traditions.

In some traditions the intellectual dimension becomes dominant, revealing a presumption that the religious life is essentially either a matter of believing a set of formal doctrines or a matter of upholding a rigid code of moral conduct. The reasons for this may be found in the particular challenges faced by religious traditions in critical historical periods. Christianity, for example, has traditionally placed a heavy emphasis on the act of belief. This is because early in its history the Christian church had to defend its proclamation against intellectual criticism from the external pagan world as well as resolve a series of divisive metaphysical disputes arising within the church. The unity of the church was at stake in these metaphysical issues, to which the church responded by developing a strong tradition of apologetic and doctrinal theology, placing a premium on the virtue of belief.

The formative challenges faced by Islamic tradition were moral in nature, not metaphysical. The Qur'an describes pre-Islamic history as a period of ignorance, by which it means an inordinate lack of moral wisdom rather than a want of philosophic acumen. Into this context Muhammad brought the ethical monotheism of the Abrahamic tradition to effect a social and political transformation of Arab culture. From the beginning, the central task of Islam was to establish a morally just society under the guidance and judgment of God. This was achieved by the development of a theocratic order, at the core of which was a concern for righteous living as prescribed by a rigorous tradition of moral law.

My point has been that religious traditions make use of intellectual resources to clarify, interpret, and defend the cosmological and moral tenets of the narrative core, thus to enhance the plausibility and relevance of the myth. In the absence of lively traditions of interpretation, religious myths are vulnerable to distortion or oblivion.

But just as myths invite interpretation, they also require reinterpretation. Despite the fact that myths are often oblique and ambiguous—or perhaps *because* they are so—they manage to endure the test of time, whereas the explicit interpretations of them tend to come and go. The intellectual custodians of religious traditions therefore station themselves at the front lines of social change, ready to guide individuals and communities through the process of adapting to fresh ideas. New experiences result in new memes, and new memes occasionally produce new fashions of thought and action, some of which mature into movements that distinguish themselves in important ways from the prevailing interpretation of the central story.

The danger is that new ideas may begin to compromise the conceptual integrity or the moral relevance of the myth. Whenever a myth faces a crisis of plausibility or relevance, it is time for the intellectuals to come up with a radical reinterpretation. In the fourth century, for example, the Neo-Platonic movement mounted a critical challenge to Christianity. The crisis was resolved when St. Augustine worked out a Neo-Platonic interpretation of the Christian faith. Later, in the thirteenth century another major intellectual crisis erupted when Aristotelian memes seeped into Christian culture. St. Thomas met this challenge by constructing an Aristotelian interpretation of the faith. Such reinterpretations restore the plausibility and relevance of the myth by demonstrating an essential compatibility between the narrative core and the new mode of thought. New modes of thought may even excite timely new insights into the timeless meanings of the myth.

The deepest possible crisis for any religious tradition would be one that challenged the realism of the tradition's root metaphor. Realism is a claim about the correspondence between a formulation of thought (a theory, model, or metaphor) and the actual, or extramental, state of reality. One qualifies as a realist about X to the extent that she believes her X-theory is an accurate and reliable description of the way things really are—that is, there really is some X existing independently in precisely the way the X-theory describes it. One is a non-realist regarding X if he denies the tight correspondence between the terms of the X-theory and the extra-mental state of reality—either because he has reason to doubt X's existence or because he has reason to doubt that the X-theory is an accurate replica of X. Anyone, therefore, who has reason to believe no God exists, or has reason to doubt that the ultimate truth about reality is reliably and accurately depicted by personal attributes, would qualify as a theological nonrealist, or a nontheist.

It is very doubtful that theistic myths, such as Judaism, Christianity, and Islam, could carry on effectively if individuals became nonrealists regarding the God-as-person metaphor. Under the conditions of theological nonrealism the God-as-person metaphor would have no currency in working memory. To illustrate: Suppose in mid-conversation I suddenly tell you to sit absolutely quiet and motionless because there is a hungry tiger standing mere inches behind you. If you are a nonrealist about this tiger (if, for example, you have good reason to believe that I am putting you on), then you will not appraise the moment in a way that will incite fear. Similarly, if an individual becomes a theological nonrealist, then traditional religious memes will not function in the appraisal process as they are intended to. Any substantial increase in nonrealism regarding the root metaphor constitutes the ultimate warning signal for a mythic tradition and its interpreters.

Theistic traditions have been faced with the ultimate crisis of creeping nonrealism for more than a century. The reasons for this are complicated, and will have to wait until part three. Of interest here is just the point that intellectual strategists have on their hands the urgent task of persuading contemporary men and women that the God-as-person metaphor is a plausible and relevant paradigm for thinking about how things ultimately are and which things ultimately matter. In some measure the fate of theistic traditions may be determined by these efforts.

EXPERIENTIAL STRATEGIES. In one way of thinking almost anything might qualify as a religious experience. If an experience is in some way contingent on the meanings of one's myth, or is in some way interpreted by the meanings of the myth, then by implication the experience is religious. But this is clearly not the way we normally speak of religious experiences. Ordinarily, there is nothing ordinary about religious experiences: They are, by conventional definition, quite extra-ordinary. But surely the conventional definition does not include all extraordinary experiences. My appendectomy several years ago was a once-in-a-lifetime experience, but hardly what we mean by a religious experience. An experience must be extraordinary in very special ways to qualify as religious. We may find it difficult to be precise in specifying the conditions for a religious experience, but we shall never achieve an understanding of religion without trying.

We might proceed by disqualifying all experiences having definite religious content but are not uncommon. For example, an individual who attends church regularly might report that she has special feelings of a religious nature each time she hears the Mass. Would these feelings qualify as religious experiences under the conventional use of the term? Probably not. Perhaps we would do well to label them "ritual experiences" to emphasize their common occurrence and to differentiate them from extraordinary experiences.

Next, we should rule out extraordinary experiences not having a direct connection with the integrated meanings of a myth. My appendectomy experience, for example, would be disqualified by this rule, as would various uncommon aesthetic experiences. We may also disqualify various momentous and life-transforming experiences that do not acquire or derive from mythic meanings. A sudden brush with death, for example, may result in profound changes in one's outlook or lifestyle, but to say it is therefore a religious experience would be misleading. Momentous experiences are especially poignant (highly valenced) and unforgettable, but this alone does not make them religious. We should also rule out many (but not all) experiences that involve radically altered states of consciousness or sense of self. One may be drunk, hallucinatory, manic, transfixed, flooded with sudden insight, ecstatic with joy, engorged with a sense of power, intensely aware, or enter into a variety of altered mental or physical states without necessarily having a religious experience.

An interesting feature of all these disqualified experiences is that they are palpably and uncontroversially *natural*. They may occur beyond the ordinary streams of experience but we are nevertheless perfectly confident that they can be explained by natural causes. But consider what happens when we combine the features of experience listed above. If someone reports having an extraordinary and momentous experience involving a radically altered state of consciousness that was utterly unforgettable and life-transforming, and was directly linked to one's integrated sense of ultimate reality and value, then we would not hesitate for a moment in labeling it a bona fide religious experience. This supports the presumption that religious experiences are natural events having thoroughly natural causes, despite any claims that they are irreducibly attributable to supernatural agency. There is every reason to believe that religious experiences are mediated by the same neural systems as those associated with ordinary experience. In fact, our understanding of the particular neural substrates of religious experiences is advancing rapidly.[1] But this is beside the point. The issue at hand is the strategic importance of religious experience in helping myths to carry on.

Extraordinary experiences are of central importance to every religious tradition—so central that many authorities have insisted that religious experience constitutes the essence of religion. There is plenty of evidence to support the double claim that extraordinary experience is both the source and the goal of the religious life. Try to imagine Judaism developing as it did without the religious experiences of Jeremiah, Isaiah, and other prophets. Would Christianity have emerged without the conversion experience of St. Paul? Is not the origin of Islam directly linked to the ecstatic visions of Muhammad? Imagine Buddhism apart from the enlightenment of the Buddha.

Religious experience appears to be equally important as the goal of the religious life. In many traditions the whole point of being religious seems to be focused on a quest for the ultimate experience. Consider the central role of the vision quest in Native American plains traditions, or the transforming power of *moksha* in Hinduism, *satori* in Zen, and the born-again experience in Pentecostal Christianity. It seems that one arrives religiously only when the ultimate experience arrives.

Much has been written in attempts to capture the varieties of religious experience in some form of typology. Thus we hear about "mystical" experiences, "numinous" experiences, and "visionary" experiences. Mystical experiences are characterized by the annihilation of conscious distinctions between subject and object, self and world. The mystic enters into an altered state of pure unified consciousness wherein all reality, the self included, is immediately and blissfully apprehended as essential oneness. In numinous experiences the subject-object distinction is preserved, even amplified, as the subject is filled with intense love and peace that come with a sense of the presence of a holy and awesome transcendent power. Visionary, or prophetic, experiences are often characterized by a trance-like state in which the seer receives a concrete message or vision communicated directly from an irresistible transcendent source.

These types of experiences are universal religious phenomena, found in all traditions at various times. They tend to share several psychological characteristics. For example, they bring one into a state of enhanced contact with reality, or immediate contact with a reality far surpassing the reality known to ordinary experience. As a result, religious experiences often have an ineffable quality about them. That is, the conventions for describing ordinary experience are completely inadequate for communicating the content of religious experiences. Further, these experiences are felt to be supremely important, typically prompting a total reorganization of individuals' lives, specifically their goal hierarchies. Finally, profound religious experiences leave a permanent mark on the memory systems. The immediate qualities of the experience are soon lost as the normal state of consciousness returns, but the reality and meaning of the experience remain indelibly etched and heavily valenced in memory.

The significance of these special experiences for the religious life can hardly be questioned, but it would be too much to claim that they constitute the essence of religion. Many individuals live out their entire lives as deeply religious persons without ever having or seeking extraordinary spiritual experiences. We shall be better served if we understand religious experience in terms of a strategic role rather than a constitutive role. That is, we should ask whether and how religious experiences might be conducive to the carrying on of particular mythic traditions. In this light we may see that religious experiences are invaluable as personal validations of a myth. Every mythic tradition encourages individuals to seek profound religious experiences because such experiences leave no room for doubt regarding the plausibility and relevance of the myth. This experiential strategy is especially important during critical periods of creeping nonrealism. If you have once had a mystical, numinous, or visionary experience of divine reality, then the probability of your becoming a religious skeptic or nonrealist is effectively reduced to zero.

Just as profound spiritual experiences reduce the probability of religious nonrealism, they also increase the probability that religious memes will get a privileged hearing in working memory during everyday appraisal and coping events. As noted earlier, working memory is biased to favor the retrieval of information that has been recently or repeatedly stored in memory systems, or has been marked by unusually poignant experience. The experiential strategy obviously exploits the latter.

RITUAL STRATEGIES. The ritual dimension is without doubt the most varied and complex of the strategies employed to support religious myths. I will use the term "ritual" in the broadest sense, to be roughly synonymous with "practice." Rituals therefore include formal officiated ceremonies as well as various informal and routine activities having explicit mythic associations. In the broad sense a ritual might be construed as any repeatable unit of behavior, the performance of which engages individuals or groups in the meanings of a religious myth or is conducive to a religious experience. A Zen monk engaged in meditation is performing a ritual, as is a rabbi reciting the Torah. Rituals include prayers, pilgrimages, fasting, feasting, hymn singing, chanting, kneeling, magic, worshipping, wedding or funeral ceremonies, sacred dances, and the like.

We may go deeper. Ritual has etymological links to Greek, Latin, and Old English words for numbering, counting, reckoning, and fitting. These associations suggest that in a ritual performance there is a concern for order, for alignment, for getting things straight, or making proper adjustments. We come close to the mark if we view ritual as some sort of performance that will effectively harmonize, synchronize, align, reconcile, or in some fashion attune humans to what is ultimately real.

The most profound insight in the history of humankind is that we should seek to live in accord with reality. Indeed, living in harmony with reality may be accepted as a formal definition of wisdom. If we live at odds with reality (foolishly), then we will be doomed, but if we live in the proper relationship with reality (wisely), then we shall be saved. Humans everywhere, and at all times, have had at least a tacit understanding of this fundamental principle. What we are less in agreement about is how we should think about reality and what we should do to bring ourselves into harmony with it. I am suggesting that ritual may be understood as practical wisdom, that is, *doing something*, performing some act that is believed to bring us into closer step with the ultimately real. A rite is for setting things aright. And what better way is there for us to achieve harmony with ultimate reality than to participate directly in the ultimate story of reality, the myth? Performing the myth, re-enacting the narrative of ultimate reality, is a way to get back on track, to hit the reset button, to get our proper bearings, to become wise.

The Mandan tribe of the North American plains achieved harmony with supernatural forces by ritual self-torture, embodying the myth that all order originates in suffering. Young men were pierced by wooden skewers and suspended aloft, while below them dancers and chanters would rehearse elements of the myth. Ancient Canaanite rituals sacrificed crops and livestock to maintain a harmonious relationship with local fertility gods. The Canaanites also practiced ritual prostitution to re-enact the earth's fertilization by the Baal of heaven, thus effectively restoring the generative powers of nature. There are literally thousands of such examples, all illustrating the close association between myth and ritual, and revealing the importance of ritual for achieving a right relationship between humans and the cosmic realities on which their fortunes depend.

Much of what we see in ritual falls under descriptions for magic or worship. These are similar in the sense that they both attempt to achieve a state of harmony with reality. Magic attempts to bring reality into harmony with the subject through an act of domination or manipulation, while worship seeks to bring the subject into harmony with reality through an act of submission or supplication.

Rituals contribute to the viability of a religious myth in several ways. For one thing, ritual is a primary strategy for transmitting the myth to the next generation—that is, rituals are educational events. They also revitalize the myth by bringing it to life in dramatic form. When individual participants engage actively in ritual performances, they are more likely to recognize how the vicissitudes of their own lives and their community are reflected in the timeless themes of the drama. This mapping of the local and particular onto the cosmic and universal enhances self-understanding and compels individuals to re-claim ownership of the myth.

In addition, ritual performances can have a validating effect for the myth. Participants in ritual events might not be moved to pro-found religious experiences, but their ritual experiences are normally meaningful in ways that reinforce realism about the myth. To illus-trate with a personal example: I once had a student confess to me that she was thoroughly agnostic except when she was performing in the college choir. For the duration of a performance she became what she described as "a true believer."

The very public nature of ritual performances often has the re-sult of enhancing a realist stance toward the myth. As we monitor the reactions of our social partners during these events, there is a predis-position to respond as they do. In other words, ritual is a form of ap-prenticeship in thinking.

And finally, ritual performances promote the viability of the myth by exploiting the working memory bias for retrieving informa-tion that has been recently or repeatedly stored in memory systems. If you attended church this morning, or if there was a time in your life when you attended regularly, then the probability is increased that mythic memes will influence your everyday appraisals and coping.

AESTHETIC STRATEGIES. The overlap between ritual and aes-thetic strategies is so extensive that it may appear pointless and arbi-trary to make a distinction between them. Ritual events almost al-ways incorporate artistic forms, and much religious art is produced expressly for ritual purposes. But despite the overlap, there remains a fundamental difference between ritual and art that should not be lost. In the previous discussion I tried to emphasize the performative na-ture of ritual. By contrast, I will here emphasize the perceptual aspect of the aesthetic dimension. The word "aesthetic," in fact, means perception. In common usage an aesthetic experience refers to one's perception of beauty in an object. The discussion ahead assumes a more specialized sense: A religio-aesthetic experience refers to one's apprehension of ultimate meaning through the natural forms of perceived objects. In such experiences ordinary sight becomes mythic insight.

To understand the aesthetic dimension of religious traditions we should first explore how art functions in the lives of those who create and encounter artistic images. It has been noted frequently that art is both expressive and impressive. That is, creating a work of art may enable the artist to express emotional feelings in uniquely satisfying ways, and encountering a work of art may engage viewers in uniquely transforming emotional responses.

If the function of art for the artist is to express emotional feelings, then we should examine artistic endeavor in light of the entire emotional process. Recall that emotional systems are homeostatic mechanisms—that is, an emotion is an arousal state that motivates an individual to take action relative to some goal. Arousal to anger will predispose one to strike out, arousal to fear will predispose one to escape, arousal to affection will predispose one to approach and affiliate, and so on. To be emotionally aroused is to be in a deficit state until some action is taken to return the individual to a state of equilibrium.

Organisms with limited capacity for working memory are destined to execute whatever action tendencies are scripted into their emotional systems. However, organisms blessed with ample working memory, and especially those capable of manipulating symbols, will be able to design a range of alternative behaviors for restoring equilibrium. In the case of anger, striking out will remain in the picture as the default behavior, but it will have to compete with many other alternatives, which are likely to take into account several goals in addition to the thwarted goal that led to an angry response.

Imagine an artist, for example, who has been aroused to anger by a critic's dismissive review. If our artist is a marginally socialized five-year-old, the default behavior may get a privileged hearing in working memory, whereupon the critic may take a kick to the shins. But suppose our artist is a mature adult whose self-esteem is linked to civilized conduct and perhaps to the prospect of future exhibitions. Kicking the critic would thwart too many cherished goals, so the kicking behavior is inhibited. Still, every time the artist recalls the hurtful review she is aroused to anger, despite various attempts to minimize or forget the offense. Then at last the artist is inspired to paint her anger, or sculpt it, or compose it in musical notation. Any of these alternative behaviors may have the effect of discharging the anger and satisfying the arousal. The point is that symbolic action is action, and while a painting or a dance may be a far cry from a kick to the shins, they may do the job equally well.

Or even better, all things considered. Remember that the artist has on board a complex hierarchy of goals, which means she must contend with a continuous stream of diverse and often conflicting emotions. If she is highly creative, her paintings may succeed in expressing several of them simultaneously, as overt actions are seldom capable of doing. It is certainly possible for a single painting or a single poem—or even a single image or metaphor—to give simultaneous expression to anger, love, and sadness. And nothing in these dynamics requires that the artist be explicitly aware of the incentive events triggering the creative process. Nor is it suggested that the artist must be aware of the influence anger had on, say, this choice of red or that choice of harsh brassy tones. All that really matters from the artist's emotional perspective is the sense that a deep urge to create has been satisfied.

Normally, of course, artists strive to be understood as well as emotionally satisfied, and thus a work of art may be intended to evoke passions as well as to express them. This is always a dicey process because the artist can never be sure how perceivers will be affected by images. From the perceiver's perspective a work of art is a piece of business that must be appraised for its meaning relative to the subject's goal hierarchy. And as we have seen, goal hierarchies are notoriously variable. This makes the artist's task infinitely difficult because if perceivers bring to the aesthetic encounter their own unique combinations of temperaments, attitudes, values, and emotional memories, then what is there to suggest that their appraisals might converge on the artist's intended meaning? Yet it frequently happens that works of art resonate deeply with masses of people despite wide differences of time and circumstance. And the reason must be that effective works of art deploy images that trigger deep biases universally scripted into human emotional systems. Master painters, sculptors, composers, poets, playwrights, and dancers, it turns out, are masters of human nature. Their own emotional experiences guide their invention of images that have a high probability of arousing emotional responses in fellow human beings. I am suggesting that elements of a universal aesthetic vocabulary lie embedded in the algorithms of human emotional systems, and that the evocative power of art is a matter of exploiting these elements by the creative use of imagery.

Snake imagery provides a useful example. Humans are genetically predisposed to be fearful of snakes, an emotional trait we share with the rest of the primate world (recall that vervet monkeys have evolved specific warning signals for snakes, eagles, and leopards). Any artist, therefore, who wishes to capture the imagination of perceivers might be well served by using snake imagery. In fact, many myths throughout the ages have featured snakes. The Judeo-Christian myth has traditionally cast the serpent as a symbol for evil, but many other traditions have chosen to venerate serpents. In India the serpent is sacred, and in Hopi tradition the serpent is a mediator between nature and humans. In some cultural myths the skin-shedding serpent is a symbol of self-transformation. The point, however, is not that the snake plays a universal role in mythology, but that it is an exceptionally common image that invariably commands awe and respect, if not outright fear. The aesthetic power of this image derives from our natural bias to be wary of snakes.

Many other genetically endowed emotional predispositions have been exploited by religious imagery. The crucifixion of Christ, for example, is a powerful image exploiting a human predisposition to feel sympathy in the presence of a fellow human enduring suffering or distress. The familiar image of Madonna and Child exploits a predispositon toward affection at the sight of an infant. The massive scale of medieval cathedrals exploits a predisposition to awe and humility when overwhelmed by size.

Here are the salient points so far:

- There is a human nature.
- Human nature includes a wide range of emotional predispositions.
- Emotional predispositions are open to manipulation by symbolic means, and thus may also be viewed as *aesthetic* predispositions.
- Every religious tradition exploits this aesthetic vocabulary in service to its narrative core—that is, to express, transmit, and revitalize the myth.

This is all very well, but certainly there is more to an aesthetic encounter than this analysis lets on. Part of the difficulty is that this picture seems to reduce the creative process to a simple matter of sublimating emotional urges, a matter of coping by symbolizing. And equally troublesome is that the perceiver appears to be reduced to a "mark," a passive victim to be puppetted about by the designs of artists. If this is to be our conception of art—and religious art in particular—then it is woefully shallow and impoverished. This is not to say that the dynamics described above are wrong, but only that they do not take us far enough. This description fails to appreciate what is involved in acts of creative imagination, both on the part of artists and on the part of perceivers.

We may take our analysis deeper by comparing a work of art—

a painting, a symphony, a poem—to a discursive theological text. As we have seen, the challenge for any intellectual interpretation of a myth is to clarify its meaning, to narrow the possibilities for understanding, to hone the story down to a systematic formulation free of ambiguities and obscurities. By stark contrast, artistic forms are multivalent, full of possibilities for variant meanings. Theology tries to identify one central meaning that all individuals can bring away from their encounter with the myth, whereas works of art try to accommodate the wide range of interests and concerns that individuals bring to the encounter. Whereas theological formulations seek to focalize the myth, artistic creations seek to diffuse it into a spectrum of meanings, some complementary, others conflicting. Yet the marvel of powerful images and metaphors is that they mingle several meanings within the constraints of singular forms. Theology expresses one thing in many layers, but art expresses many layers in one thing.

The challenge for the artist is to give expression to a deeply felt urge to create. If the emotional texture of the artist is univalent, then the expression is likely to be inelegant—like poster art, which is only slightly more contrived than a kick to the shins. But when the emotional texture is complex, the artist faces an immense imaginative challenge: How to make the many one, how to show everything as one thing, how to express this urgent yet recalcitrant swarm of possibilities in a satisfying, beautiful form. There is an agreeable etymological tie between "beauty" and "bounty": Beauty, one might say, is emotional bounty contained by sensuous form.

The aesthetic encounter is no less demanding for the perceiver. The imaginative challenge for the perceiver is to achieve a unifying insight by blending a bounty of emotions aroused in the encounter. Consider the emotional possibilities inherent when a Christian encounters the image of crucifixion. As already observed, the crucifix exploits a predisposition to sympathy upon witnessing the suffering of innocent fellow humans. But appraisals of the crucifix will be multivalent for those already acquainted with the myth. In addition to sympathy there may be several other emotions aroused. The crucifix will remind the perceiver that Christ endured his suffering as a sacrifice for the benefit of all humanity. This recognition will bring gratitude into the picture.

The image is also a reminder that the sacrifice was a necessary consequence of human sinfulness—that is, the perceiver himself is the one deserving punishment on the cross. This attribution of self-blame will arouse a sense of guilt. A further step in the coping process might arouse the sense of eternal hope made possible by the sacrifice. Or one might be aroused to the feeling of being loved, or to the feeling of anger at the injustice of executing an innocent man. Many additional emotional responses are possible, depending both on the aesthetic qualities of the particular crucifix and on the concerns brought to the encounter by perceivers.

Obviously, none of these emotions arise as a Christian hurries past a hospital lobby crucifix to visit a sick friend. But in the context of a worship service, or in prayer, or even in a reflective pause while leaving the hospital, such responses would not be surprising. The point is that when emotional complexity like this does arise, the perceiver is challenged to cope with it by achieving a coherent response. How can so much be at stake in this one image? How can all of these feelings be gathered into harmony? What does this object *mean?*

Many Christians are never bothered in these ways by the crucifix, and among those who are, the incidence of unifying insight may be very low. And the insights of those having them are sure to vary considerably. But we may be confident that achieving a blended emotional response to the image will forever transform its meaning for the perceiver, and this transformation will amount to a deepened insight into the narrative integration of ultimate reality and value. And this assures, in turn, that the meanings of the myth will become far more likely to enter into working memory during the appraisal and coping events of everyday life. Art revitalizes the power of myth.

INSTITUTIONAL STRATEGIES. Nothing is more vital to a cultural tradition than its myth, its integrated vision of reality and value. It follows that decisions affecting the myth—its interpretation, its transmission, its embodiment in morality, ritual, and art—will be of fundamental importance to religious traditions. Every tradition must establish strategies for making these decisions in an orderly manner. These strategies constitute the institutional dimension of religion. In many cultures the familiar Western distinction between religious and non-religious phenomena makes no sense, so virtually all institutions may be regarded as religious. But even in these cases it is possible to identify special leadership roles in which authority regarding the myth has been institutionalized.

Strategies for regulating the care and feeding of mythic traditions vary considerably, depending on the history and complexity of the particular social order. In some traditions religious institutions are highly structured and formalized. The institutional structure of the Roman Catholic tradition, for example, is monarchical and hierarchical, patterned as it was on the political order of imperial Rome. By contrast, Calvinist tradition, while formalized, is democratic. One step further, Quaker institutions are both informal and egalitarian.

Some traditions, such as Hinduism, select religious leaders from a designated social class, while others determine leadership issues by a process of apprenticeship. Still others leave authority to chance, deferring to charismatic spiritual leaders who arise spontaneously from the ranks. Most traditions reveal a history of strict gender discrimination regarding positions of leadership.

Every religious tradition maintains institutional strategies for transmitting the myth to future generations. In some cases this function is carried out in the context of ritual and ceremonial events, and in other cases there will be formalized instruction in schools, churches, or monasteries, or by attachment to a guru. Missionary traditions, such as Christianity, Islam, and Buddhism, have developed special institutions for bringing outsiders into the story tradition. On occasion, unfortunately, missionary zeal has expressed itself in the form of military conquest.

Institutional authority also extends into the domains of interpretation, ritual practice, and aesthetic expression. As noted earlier, metaphysical and moral interpretations of mythic traditions have a tendency to diversify. Sometimes disagreements over modes of thought or norms of behavior grow serious enough to jeopardize group solidarity and cooperation. Aberrant forms of the myth must be dealt with, else the truth about ultimate reality and value will be compromised by dissipation. Traditions must therefore have institutional policies and means for resolving internal intellectual conflicts.

Religious traditions normally institutionalize the regulation of ritual events, for someone has to decide when, where, and how often to perform which rituals. And someone, not just anyone, has to officiate. There is often far too much at stake in the performance of rituals to risk their being handled improperly by ill-qualified persons. To institutionalize a ritual is to endow it with solemnity and efficacy. When members of a religious community lack confidence in their rituals, they become vulnerable to uncertainty and anxiety. Even today, for example, there are many Christians who will not allow their children to be baptized by a woman out of fear that the sacrament will not be effective.

The aesthetic dimension has always had special significance for the official keepers of mythic traditions. It matters greatly what people regularly see and hear. Thus religious traditions exercise institutional strategies for regulating symbolism and art. Artworks, like rituals, are incentive events, and a good share of religious leadership has to do with initiating incentive events. Cultures with robust aesthetic traditions will saturate public spaces with monuments and images having mythic content. Renaissance Europe, like contemporary India, was awash with sensuous reminders of myth. This was no accident, for the institutional church was actively involved in promoting and regulating the arts. In religions where the aesthetic dimension is emphasized, consciousness is continuously assailed by mythic images.

These regulatory activities are not without inherent danger—that is, religious institutions have a tendency to become authoritarian regimes that brook no dissent from any quarter. This tendency becomes obvious when one considers that no human beliefs, values, or activities escape the purview of mythic authority. If religious leaders, therefore, fancy themselves authorities on the myth, they might take the next step to presume authority on everything imaginable. An absurd presumption, of course, but not one without instance. Many traditions have discovered—some the hard way—that it is possible to institutionalize prophylactics against such presumption. The protestant commitment to democratic process is one strategy (Luther's "priesthood of all believers," for example). Another strategy is to decentralize authority, as Buddhism has done. Yet another is to institutionalize religious pluralism, as was accomplished by the U.S. Constitution. It is possible, of course, to guard too closely against the authoritarian tendency, to the point where diversity itself undermines prospects for sharing memes about ultimate reality and value. In the end, every cultural tradition must feel its own way between the Scylla of absolutism and the Charybdis of relativism.

It is always tempting to capture the essence of complex phenomena in a concise definition, or in this case to define what religion essentially is. I tend to agree with the anti-essentialist view that religious phenomena are too complex and variable to submit to one-line definitions. Nevertheless, as I have tried to show, it is possible to reduce the complexity to a pattern of elemental components. I have proposed a structural model to describe all particular religions as narrative traditions or myths, which are formulated and revitalized by a set of ancillary strategies. These strategies (intellectual, experiential, ritual, aesthetic, institutional) may be seen as overlapping dimensions that collectively shape the religious life.

Each of these strategies of a religious tradition is sufficiently vital that it is possible to construe any one of them as the essence of the whole. One might argue, therefore, that the essence of religion is myth, a narrative integrating ideas about how things ultimately are and which things ultimately matter. But see how this formulation might be plausibly reduced to any of the five dimensions.

Religion is intellectual. If religion can be identified as myth, and if myth is defined as ideas about reality and value, then the real essence of any religion is belief in the most explicit and systematically coherent interpretation of those ideas.

Religion is experiential. If religion can be identified with ideas expressing the essential unity of reality and value, then the real essence of religion must be the source of these ideas—that is, the original and immediate experience of apprehending the Ultimate.

Religion is ritual. Ideas about the integration of reality and value are just that: ideas. But ideas are essentially instruments of practice. Myths themselves originated as narrative accounts of ritual practices, which constitute the real essence of religion.

Religion is aesthetic. Myth is narrative, and ritual is dramatic performance, and both are essentially forms of aesthetic expression. Ideas about reality and value have their ultimate source in the creative imagination. The real essence of any religion is objectified in its art.

Religion is institutional. Each of the above may have a point. But at the end of the day the real essence of any religious tradition is what its authorities say it is.

The superficial plausibility of these reductions suggests not that religion is essentially one of these elements, but that it is essentially a collaboration of them. Anyone who sets out to develop a general theory of religion makes the assumption that behind all the complex and variable phenomena there lurks one kind of thing, waiting to be described. The view suggested here is that the one kind of thing to be described is a socio-symbolic system of interdependent elements. We may suppose that some components of religious systems were already on the scene in some form before they were stabilized in the systematic patterns we now recognize as religion. For example, certain forms of ritual and aesthetic expression were no doubt practiced before anything resembling myths appeared. Their functions, however, would have been slightly different than those described above.

I believe that socio-symbolic systems (such as religious traditions) are natural entities, like atomic systems, molecular systems, organ systems, neural systems, and the rest, which have emerged by natural causes in the creative process of cosmic evolution. In this section I have tried to summarize the general anatomical structure of religious systems. The next two sections will focus in turn on their evolutionary origins and physiological functions.

THE ORIGINS OF RELIGION

The previous discussion proposed a structural definition of religion: At the core of every religious tradition there is found a narrative vision, a myth unifying ultimate reality and value, a story that is expressed, transmitted, and revitalized by a variety of ancillary strategies. The challenge now is to show how such socio-symbolic systems might have emerged in the natural course of evolution.

Any attempt to answer the question of religious origins will be highly speculative and full of debatable assumptions. I do not pretend to have a final answer. The extent of my claim is this: If religion is properly identified as mythic tradition, then the following account may be taken as a plausible way to think about its origins in human evolution.

THE NATURE OF THE QUESTION

If our question is not of the sort that lends itself to final and decisive answers, then what sort of question is it? Does the question lend itself to naturalistic answers? Many accounts of the origins of religion insist that it does not. It is a common feature in many traditions to offer accounts of their own origins in supernatural events, the implication being that the tradition cannot be made intelligible apart from a recognition of supernatural causes. In these cases the question about religious origins is treated as a theological question. Islam, for example, makes a claim for its origins in the divinely inspired visions of Muhammad; Judaism is unintelligible apart from God's initiation of a covenant with Israel; and Christianity cannot be understood apart from the incarnation of Christ and the work of the Holy Spirit.

Some theological accounts allow for naturalistic explanations of false religions, but insist that the true way has divine origins. Both Isaiah and Jeremiah, for example, attribute their neighbors' beliefs to ignorance and delusion, while their own beliefs are based on God's revelation.[2] John Calvin held a similar view, suggesting that humans were initially created with the capacity to apprehend divine truths, but because of original sin they fell into ignorance and dupery, which can be cleared away only by divine intervention.[3] A modern version of this view is found in the work of John Hick. Hick believes that primitive religions might be given a naturalistic account, but the sudden emergence of the great religions of the world in the so-called Axial Period (circa sixth century B.C.E.) can be understood only on the assumption of divine self-disclosure.[4]

There are no compelling non-circular arguments against treating the origins of religion as a theological question. In the introduction I suggested that the only viable way to justify a naturalistic theory of religion is to produce one. That is, if the naturalist can provide a plausible and satisfying story of religious origins, then competing supernaturalist accounts will be left to the mercy of Ockham's razor. Even if we agree that the origin of religion is not a theological question, it remains unclear what sort of question it is. In some sense, for certain, it is a historical question, or more precisely an archaeological one. But if we treat it exclusively as an archaeological question we cannot expect much in the way of an answer, because the archaeological evidence pertinent to religious origins is notably scarce and vague. In the end archaeology can do little more than place constraints on speculations generated by other fields of inquiry. For example, there is no firm archaeological evidence to suggest that religious traditions might have existed before the great leap forward in cultural evolution that took place about forty thousand years ago, and there is plenty of evidence for their existence after that point.[5] Any hypothesis about the origins of religion must be consistent with this time frame.

Many theories of religious origins have assumed that the question is psychological in nature. The general idea is that religion emerged as a strategy for coping with particular psychological challenges, either cognitive or emotional. Some theorists have regarded religion as a cognitive enterprise, a way of thinking that satisfies a psychological need to make coherent sense of human experience. Like a sort of primitive science, early religious ideas functioned to explain what was happening in the world. Just as theoretical physicists today posit elementary particles to account for physical phenomena, so primitive peoples posited spirits and gods to explain their observations. Stewart Guthrie labels these theories of religion intellectualist or rationalist.[6]

Edward Tylor and James Frazer were influential writers who developed cognitive theories of the origins of religion. Tylor envisioned the primitive mind as bewildered by the phenomena of death and dreams. How could such things be explained? To solve these mysteries the primitive mind developed a theory of soul or spirit, the principle of animation. Persons live, grow, act, and speak by virtue of possessing a soul. When they die, their souls depart. Dreams and illness were explained by the principle that alien spirits were capable of entering the human body. Nature, too, is animated. The sun, the moon, the forest—all the phenomena of nature were infused with anthropomorphic spirits.

James Frazer argued that the impetus for religion had less to do with understanding natural forces than it did with controlling them. He thought the ultimate origins of religion were found in magic—primitive techniques for coercing nature into compliance with human goals. Magic evolved into religion when primitive peoples became disillusioned with its results. The point of religion is to perform rituals designed to please the gods, not to manipulate them.

Since Tylor and Frazer, most psychological theories of the origins of religion have focused less on cognitive challenges and more on emotional ones. Sigmund Freud's speculations on the psychological origins of religion have been the most influential and controversial.[7] In Freud's view, religious ideas have little to do with providing a rational interpretation of the way the world works. Quite the opposite: Religious ideas create a world of fantasy that makes the hopelessness of human life more tolerable. Religion provides succor for the psyche, an irrational strategy for hanging on to the security and freedom of childhood in the face of natural terrors and the moral constraints of the social order. Religion, says Freud, is an illusion invented to compensate for the anxiety-producing realization that our deepest wishes cannot possibly be fulfilled. Religion began when the primitive mind projected a fatherly God into the heavens as the ultimate guarantor of wish fulfillment.

Psychological theories of religious origins have attracted criticism from various perspectives over the years, yet they continue to find supporters. Even if all the particular theories fail, however, it would be difficult to say that the matter of religious origins is not in some fundamental sense a psychological question. Clearly, before any religious tradition could emerge in human evolution, there would have to be in place certain prerequisite cognitive and emotional competencies. Moreover, it is obvious that loss of certain psychological functions will have consequences for an individual's religious life, and that the acquisition of a religious orientation will have a bearing on psychological function. There can be no doubt that whatever happened about forty thousand years ago to give rise to religious traditions had something profoundly to do with human psychology. The issue is not whether the origin of religion is a psychological question, but whether it is exclusively so.

In addition to psychological theories, various accounts of religious origins have concentrated on social factors. Here the general idea is that religion first emerged as a strategy for legitimating and maintaining the social order. The most notable proponent of the sociogenic theory of religion has been Emile Durkheim.[8] Durkheim believed it was social experience that first gave rise to notions of divine beings. Religion is not about interpreting natural phenomena or coping with insecurities: it is essentially about group survival.

Durkheim was impressed by what he perceived to be a fundamental fact underlying every religious tradition: a distinction between the sacred and the profane. In every society some things have sacred status while others are considered non-sacred, or profane. When these things are carefully analyzed, it becomes clear that sacred objects (whatever they happen to be) always represent non-optional, absolute social values, whether good or evil. In Durkheim, when you see "sacred," read: "vital interest of the group." By contrast, profane objects signify socially optional values, relative to the interests of individuals. Religion is always focused on the sacred—that is, always concerned with what matters for the coherence of the group. Religion originates in this distinction between the sacred and the profane, between what is vital for the group and what is not. The gods are to be understood as mere symbols personifying the transcendent reality of the group. The central activity of religion is found in its ritual life, which creates social solidarity and preserves the social order by reinforcing group consciousness.

Whatever one thinks about the details of Durkheim's claims (or any other sociogenic theory of religion), it seems impossible to deny the fact that religious beliefs and practices are of decisive importance for social solidarity and cooperation. Furthermore, it is normally difficult, if not impossible, to understand religious doctrines apart from the social circumstances influencing them. It seems therefore unreasonable to withhold a consideration of social dynamics from any account of what happened forty thousand years ago to stimulate the formation of religious traditions.

Despite the fact that theoretical attempts to generalize and naturalize religious phenomena are notorious for their failures, I remain convinced that such a theory is possible. This chapter is premised on the claim that the quest for a general and naturalistic theory may be satisfied by describing the structure, the origins, and the functions of religious traditions. The question of origins is a complex one, as we have just seen, involving both psychological and social dynamics. I believe this question may be answered to the extent that it is possible to reconstruct the psychological and social circumstances prevailing among our ancestors forty thousand years ago when religion first appeared. Any such reconstruction will be highly speculative, to be sure, and must be content to share the stage with virtually every hypothesis that does not run counter to our best accounts of the evolution of human nature.

I have already argued for the view that every religious tradition features a narrative core, a myth that integrates cosmology with morality by the unifying power of a root metaphor. If this view is correct, then the story of religious origins may be told by describing the developmental process that prompted the articulation of a mythic vision. The following account describes this process as unfolding in two stages. The first stage involves a gradual transformation in the way early humans thought about the external world. The second stage involves a relatively abrupt change in the way they thought about the moral order. The second stage, I will argue, was triggered by a fundamental change in social organization. These two revolutions in human thought—one in cosmology, the other in morality—established the prerequisites for mythopoesis—mythmaking.

STAGE ONE: BEYOND INTUITIVE SCIENCE

In chapter 2, I argued for a distinction between intuitive mentality and ad hoc mentality. Intuitive mentality is the default mentality of our species, governed by inherited rules for constructing and manipulating mental objects. If all humans were limited to these intuitive rules for processing information, then we would all think and act pretty much alike, as chimps do. But of course we do not all think and act alike because we are capable of inventing extragenetic rules, enabling us to transcend our common intuitive mentality in favor of diverse ad hoc mentalities.

I have also argued that our worldviews break down into ideas about how things are in the external world and ideas about which things matter for human fulfillment. These ideas about reality and value may be either intuitive or ad hoc. Thus an intuitive mentality would include an intuitive science and an intuitive morality. Likewise, we may speak of ad hoc science and ad hoc morality, emerging from the constructions of secondary reality operators and secondary valence operators. Of course we never observe intuitive science or morality in their pristine manifestations because contemporary mentality everywhere shows the influence of ad hoc rules. Nevertheless, if we could travel backward in time we would doubtless find the thinking of our prehistoric ancestors to be less and less influenced by ad hoc categories and concepts, eventually arriving at a prelinguistic period when intuitive mentality dominated. I now want to suggest that a reconstruction of the path forward from intuitive mentality to the rise of religious traditions will tell a story of two great transitions in human thought. The first was a gradual transition in our thinking about the world, and the second was a sudden transition in thinking about ourselves.

No one can say for sure what language was like in its formative stages, but we may suppose that the use of true symbols was preceded by the indexical mode of reference—that is, sounds or gestures standing for objects and events in the world. I speculated earlier that even *Homo erectus* might have invented a sizeable indexical vocabulary, perhaps to refer to various plants and animals, impending events, individuals, and maybe even a range of behavioral routines. Higher order symbolic reference came to build on this semantic foundation, but would have been very slow in developing. From about 1.5 million years ago until 200,000 years ago, it may be supposed that our ancestors were increasingly talkative, but probably did not engage in high abstractions. Sophistication in semantic and syntactical properties would have accelerated after 200,000 years ago. For the next 160,000 years or so—until 40,000 years ago—we might envision the development of ad hoc scientific ideas. That is, various sophisticated categories and concepts pertaining to the objects, events, properties, and relations making up the physical environment.

We may even be fairly confident about the nature of this emergent ad hoc science: It would have been thoroughly anthropomorphic. Bear in mind that communication is a social enterprise, and that brains selected for their symbol-using powers would be highly specialized for attending to the behavior of conspecifics in order to ferret out their motives and intentions. Brains specialized for social intercourse would be especially adept at reading the minds of others. Indeed, brains unable to judge the motives and intentions of others would hardly be worth talking to. It seems reasonable to suppose that mind-reading capabilities and symbolic competence developed in tandem.

The mind-reading expertise of prehistoric *Homo sapiens* would have been gainfully employed in extrasocial cognitive challenges as well, such as predicting the motives and behavior of predators and prey, based on the analogy to human experience. Thus it seems likely that the specialized mind-reading brain of humans, when turned to the task of comprehending objects, events, properties, and relations in nature, would attempt to read the mind of nature—that is, to look for various motives and intentions underlying natural phenomena.

Contemporary humans tend to be wary of the fallacies inherent in anthropomorphizing nature, but in the infancy of our species this practice probably worked out rather well as a first approximation of how things actually work. We cannot know for certain what it is like to be a rabbit, but it is sure to be much more like being a human than a stone or a river. It may even be practical to conceive of storms and wildfires by analogy to human agency.[9] To make humankind the measure of nature, as I suppose prehistoric ad hoc science did, was no doubt a highly adaptive intellectual strategy.

In this view, the emergent ad hoc science of the Middle Paleolithic period (125,000 to 40,000 years ago) came to regard the natural world as brimming with verve and motive. It is therefore reasonable to agree with Edward Tylor's attribution of an animistic worldview to our prehistoric ancestors. This is not to suggest, however, that humans prior to forty thousand years ago were in possession of a religious orientation. They presumed the reality of spirits or gods embodied in natural phenomena, but the function of such concepts was too limited to qualify as religious. Gods and spirits functioned as quasi-scientific models, theoretical constructs that rendered intelligible the otherwise bewildering and sometimes frightening phenomena of the natural order. There were probably stories told about various gods and their powers, but these would have been anecdotal and explanatory, lacking the ultimate dimensions of myth. Granted, beliefs about superhuman agency in nature have familiar analogues in known religious traditions, and they might even be regarded as preliminary to mythic insight, but they are not in themselves a sufficient condition for a religious worldview.

It is entirely reasonable that an emergent ad hoc science would have included notions of immortality as well as beliefs about nature gods. Consider how mystifying death is as a purely intellectual problem. In one moment you are interacting with a person who talks, smiles, and answers questions, and in the next moment you are reduced to interacting with an inanimate mound of flesh. Such a change is intellectually no less mysterious than the disappearance of a material object into thin air. When material objects appear to vanish into thin air, we are left with some serious explaining to do. Such things just don't happen. Nevertheless, in the case of death something real has clearly vanished. But what? When material objects appear to vanish we rescue our realism about them by finding an alternative explanation. Death presents a similar intellectual problem.

One may suppose, then, that as ad hoc science began to appear among our prehistoric ancestors, life and death would have been high on the agenda of conceptual problems to be solved. Humans are innately realist about persons, no less than about material objects. From a purely intellectual perspective, therefore, it would have been more reasonable for early ad hoc scientists to posit some sort of personal transformation (say, into the spirit realm) than to suppose that persons just ceased to exist. This picture of an early origin for ideas about immortality is consistent with the archaeological record, which indicates that modern humans as early as 100,000 years ago were systematically burying their dead with provisions for a presumed journey ahead.[10]

I have been preparing the way for a thesis about the origin of religious traditions. Religion presupposes a cultural myth—that is, a shared narrative integrating an explicit cosmology with an explicit morality. My thesis holds that cosmological ideas and moral ideas became explicit independently, as two stages in cultural evolution. The first stage—roughly from 200,000 to 40,000 years ago—saw the gradual articulation of an anthropomorphic cosmology, abstract categories and concepts for describing natural phenomena. Included in this ad hoc vocabulary were ideas about gods and immortality. These were proto-scientific ideas, not religious ones. Nor would they become religious ideas until the dawn of mythic insight at the time of the great leap forward in cultural evolution.

STAGE TWO: BEYOND INTUITIVE MORALITY

The mystery surrounding the great leap forward is this: If human beings were capable of language and abstract thought from at least 200,000 years ago, then why does the archaeological record show an absence of cultural diversity and sophistication until as recently as 40,000 years ago? In my judgment the most convincing answer is that prior to about 40,000 years ago social groups were too small, too fluid, and too uncomplicated for individuals to engage in the sustained patterns of interactions necessary for creating progressive traditions. As new patterns of social organization developed and stabilized, the social conditions for the great leap forward were met.

For most of human history our ancestors subsisted as hunter-gatherers, under the conditions of minimal social organization. Hunter-gatherers typically lived in small groups ranging in size from ten to forty individuals. Normally individuals would spend much of their time dispersed over a fairly large territory, periodically congregating in a base camp. The occupancy of a base camp might vary from a few days to a month, or perhaps longer if resources were abundant. As food supplies became scarce, the group would move on to a new base camp. Observations of contemporary hunter-gatherers indicate that groups might change campsites for other reasons as well. If a large animal has been killed it might be more efficient to move the camp than to transport the meat. Or a group may change sites simply because someone has fallen ill or experienced a bad dream.[11]

Hunting-gathering groups lived at very low population densities, seldom exceeding three individuals per square mile. The rate of population growth was near zero, for several reasons. For one thing, the death rate was high due to accidents and disease. One analysis of archaeological remains estimates that individuals had no better than a fifty-fifty chance of surviving beyond the age of twenty.[12] In addition, ovulation was reduced whenever women were nursing or when their body fat declined. It is also likely that early hunter-gatherers practiced both abortion and infanticide, as some contemporary nomadic groups are known to do.

The key to understanding these groups is that the social glue

holding them together was genetic. That is, interactions between individuals were governed by the algorithms of kin selection and reciprocal altruism. Kinship ties were no doubt the most important, but
we may suppose that some groups were initially composed of exiles
and strays who came together by accident, and were then held together by the tit-for-tat dynamics of reciprocal altruism. The important thing is that each group, whether a family or a clique, was a free-
hanging social mobile, twisting and swirling according to the rules
of intuitive morality. Primitive hunter-gatherers lived, literally, from
hand to mouth. It was a life based on instant gratification: "They
gather what they need and consume it immediately; they exchange
services straightaway and employ a simple technology that involves
a small investment."[13] Everything was scaled to the campsite: They
practiced a campsite economy, supported by a campsite technology,
and regulated by a campsite morality.

In such groups there would be no sense of history and no gran-
diose ambitions for a future life. Hunter-gatherers may have been
consummate campsite problem solvers, but they were not reflective
about matters of human nature, origins, and destiny. Nor would they
have felt a sense of group identity. The group itself was entirely incidental and of no account. They achieved genuine solidarity and cohe-
siveness, for sure, but these were based on personalized attachments
between individuals rather than a sense of belonging to a transcendent whole. Moral obligations were determined by reciprocity and
emotional exigency—you shared everything with everyone either because you felt affection, sympathy, or gratitude toward them, or
because you feared reprisals. Gut feelings were adequate to regulate
social relations, so there was no need for articulate norms of moral
conduct. There may have been gods hovering around the campsite,
but they had nothing to do with the moral order. That was the province of the genes.

The hunting-gathering way of life was sufficiently adaptive for
many millennia, yet it imposed serious hardships and constraints that
might have been alleviated by the formation of larger groups. Larger
groups would have the advantage of greater productivity, enabling
the buildup of food surpluses to hedge against temporary shortages.
Larger groups would be more efficient hunters and would be able to
tackle larger prey with predictable success. Importantly, larger groups
would have an extended knowledge base. They would also be less
vulnerable to raids by other groups, and more capable of executing
their own raiding ambitions. In short, larger groups would have enjoyed greater prosperity and security. The real puzzle, therefore, is
not why our ancestors started forming larger groups at about the time

of the great leap forward, but why they did not do so much earlier. What was the obstacle to larger social systems?

Social systems are just like other systems in this respect: The more components in the system, the more complex the rules must be for coordinating them in an orderly manner. Consider a simple game of catch between two players, Peter and Paul. This is a two-particle system having very simple rules. Peter follows the rule "catch ball from Paul, then throw ball to Paul." Paul follows his own version of the same rule. Suppose, now, that Mary enters the game. It will still be a game of catch, but the addition of a third player necessarily complicates the rules. When Peter catches the ball he must now have a rule to determine whether to throw it to Paul or Mary. To keep the game orderly he must introduce a more complex rule. One thing is clear, however: If Peter and Paul continue playing by the old rule, the three-particle system will collapse. Mary will get upset at being excluded and will probably pack off—or worse, she may vent her anger at Peter and Paul.

Early attempts at large group living would have been faced with a similar challenge. For thousands of generations the game of sociality had followed the rules of kin selection and reciprocal altruism, the gut reaction morality of the species. These rules mandated the campsite scale of social organization. If hunting-gathering groups managed to increase in size during boom periods, the social mobile was sure to become disorderly and eventually fission into independent subgroups. Any experiment in large group living would have to introduce new rules for behavior that were consonant with the new scale of sociality. But what were the rules for maintaining harmony in a large group? No one would have known, of course.

But more seriously, the very idea of having rules to regulate social behavior would have been unintelligible to hunter-gatherers. It is not that the social order of the campsite was unregulated, but that the regulation was implicit, automatically enforced at a preconscious level by inherited algorithms. To acquire new rules for enlarged group life would require one of two things: either a fortuitous mutation in the algorithms underlying intuitive morality, or the ad hoc invention of articulate rules to override the sirens of intuitive morality. What happened, of course, was the latter. But, again, what were the appropriate ad hoc rules that needed to be imposed? And furthermore, what could possibly induce individuals bent on instant gratification to comply with them?

We may suppose that our ancestors engaged in many failed attempts to pass through the bottleneck from intuitive morality to ad hoc morality. To describe how they made this breakthrough is a bit like describing how biological systems emerged from chemical systems: There are many plausible ways in which it might have happened. I propose the following.

In the course of their migrations hunting-gathering bands would inevitably encounter other small groups, perhaps at watering holes where large fauna swarmed. Encounters like this, where there was more than enough game for everyone, would tend not to be hostile. On the contrary, they might even be exhilarating and memorable. In addition, individuals might spontaneously cooperate in the kill, perhaps learning a thing or two, and share a meal together. Such meetings might even result in the exchange of a few females. One can only guess at the duration of such events, but it is likely that tensions would eventually swell to a flash point and outbreaks of hostility would end the party as abruptly as it began. Lacking the social means for resolving conflicts on this larger scale, groups would be relieved to part company with good riddance.

How often would events like this occur during the lifetime of an individual? Impossible to say, of course, but one might safely assume that the probability of such chance occurrences would vary with ecological fluctuations. One can easily imagine circumstances under which a few small bands might begin to regularize such meetings by reuniting on an annual basis. Doing so would constitute a major advance in social organization.

Periodic reunions must have been common during the millennia before the great leap forward. It was probably not unusual for coalitions of this sort to be ended once and for all by the demise of one of the groups. And presumably an exceptionally ugly incident during the course of a meeting might result in two groups remaining permanent enemies, each of them careful to avoid returning to the scene. But what would follow from a pattern in which two groups reunite each year for several years running? During the months of separation individuals in each group would reflect on the previous reunion and anticipate the next one. They might even prepare for it by making special gifts. When reunions took place, the tempo of social life would rev up to a bazaar-like pitch as the groups feasted together, exchanged gifts, and selected marriage partners.

In time the reunion events would become stereotyped as participants would know what activities to expect and when. A series of proto-rituals would begin to develop. The annual exchange of members (especially women) would eventually result in a shared language and a shared cosmology, not to mention genetic ties. Lots of personal information would be exchanged, including accounts of what had become of non-returning participants. A good share of time spent in reunion events would be devoted to stories and fond recollections of reunions past. In the end, however, the atmosphere of these events would thicken with stress and restlessness, and everyone would be happy to return to the more comfortable scale of meandering camp life.

Allowing for these kinds of chance encounters enables the following points. First, the formation of episodic yet stable coalitions between hunting-gathering groups may have been the most radical social transformation ever in the history of our species, preparing the foundations for subsequent transformations to tribal alliances, chiefdoms, and nation states. Each of these transformations may be described in the language of emergent systems. The change from the intuitive sociality of hunting-gathering bands to episodic coalitions is analogous to the emergence of multicellularity from unicellular life forms.

Furthermore, it is striking to realize how quickly such a transformation might have occurred, and how quickly it might be lost to the default sociality. It seems natural to suppose that wherever ecological conditions were favorable, such experiments in episodic coalitions would have occurred. The duration of these coalitions might have varied considerably, depending on environmental stress factors. In especially stressful periods the reunions might have lasted longer, or component groups might have split up but remained in close proximity to enable intergroup monitoring.

Second, episodic coalitions of hunting-gathering bands offered the best of both worlds to our Paleolithic ancestors. That is, the external support provided a measure of prosperity and security that isolated hunter-gatherers could not achieve. But at the same time, these coalitions were minimally demanding in terms of individual investment and sacrifice. When the costs of social investment escalated there was always the exit option, yet when you needed help there was somewhere to go.

Third, the formation of episodic coalitions would call for the invention and enforcement of ad hoc rules for behavior. A more complex social order requires more complex rules for maintaining it. We all know that certain behaviors tolerated at home with the family are simply not allowed in public. When you're at church or at Grandma's house, you're expected to be on your best behavior. The same discrepancy would have held between normal camp life and the more structured social context of reunion events—only worse, for it would have felt like an absolute difference between rules and no rules. The intuitive morality of camp life came naturally because there were no explicit rules to learn and to obey. Campsite morality was governed by gut reactions, whereas reunion life was unnatural, counterintuitive. In the reunion context there were things you could not do, no matter how intensely you felt like doing them. And there were things you had to do, despite your inclination to refuse. In the reunion context everyone had to exercise a lot more deliberate control over their behavior, and this would have been very difficult and stressful for individuals accustomed to a life of instant gratification. Nevertheless, they somehow managed . . . more or less.

There is no way to determine the nature of the first articulate ad hoc morality, but we may be confident that the specific rules were generated by the following algorithm: Thou shalt not do whatever it was that caused trouble at the last reunion. This would be an excellent general strategy for kick-starting ad hoc morality, but the simplicity of this algorithm is deceptive. For one thing, it would not always be clear exactly what caused trouble at the last reunion. Maybe it was this behavior, but then it could have been that one. To be on the safe side there should perhaps be rules against both behaviors. Furthermore, we can be sure that different behaviors caused trouble in different years. Thus we can easily see how this general algorithm, simple as it appears, could begin to generate a rather elaborate set of explicit rules. Each year, therefore, the psychological costs of reunion life would escalate. But the upside would be that if the right rules were introduced, then each year the reunion event would be more satisfying and harmonious.

It might also be noted that the introduction of ad hoc moral rules would have a decisive bearing on the dynamics of self-concept and self-monitoring. To learn explicit rules for behavior is to acquire performance standards for an ideal self against which the individual compares his actual behaviors in the process of self-monitoring. During reunion events individuals would be engaged in a radically new way of thinking about themselves.

And finally, the formation of episodic coalitions would have generated a sense of group identity. Isolated bands were identifiable groups, too, but individuals would hardly be aware of this fact. The camp group was a negligible by-product of the real substance of sociality, which was all about personal attachments. But the discrepancy between the atomic life of the band and the molecular life of the coalition would have made "grouphood" a salient social fact, no longer to be taken for granted. One now entered into a group in a deliberate manner, and on the inside everything would feel strange. The group was suddenly a thing, an entity that one might think about, a reality about which one might bear an attitude, a presence to which one seems to be obligated. Yet finding a satisfying way to think about the molecular group would be extraordinarily difficult. One could not actually see this mysterious entity, although its reality could not be denied. What kind of reality is it?

MYTHOPOESIS AND RELIGION

For a very long time our ancestors coupled an ad hoc cosmology to an intuitive morality. From the time of *Homo erectus* we were coming to think about the world at a level of sophistication that far transcended the cognitive abilities of our simian cousins. Yet we continued to organize ourselves socially in roughly the same way they did. Chimpanzee politics and *Homo erectus* politics were pretty similar. But at some point along the way (or perhaps at many points along the way), *Homo sapiens* made an important breakthrough to ad hoc morality and post-intuitive sociality. They got into it in a small way at first, on a part-time experimental basis, and we cannot assume that the results were uniformly positive. It is conceivable that some episodic coalitions failed due to environmental factors, some due to disease, and in other cases the experiment might have failed because of poor insights about which rules to enforce, or because the means of enforcement were too harsh. But eventually the enterprise stabilized and progressive traditions emerged.

We will fail to understand the depth of this social transformation if we do not appreciate how the sheer novelty of it generated a crisis of self-understanding. I have suggested that the introduction of ad hoc values triggered changes in the way individuals thought about themselves and the group. Ultimately this would have been the case, but initially it is probably more accurate to say that it triggered confusion and perplexity. At the very least, the new social circumstances would have generated both ambivalence and dissonance in the minds of individuals—ambivalence because the larger group was both freeing and limiting, both exhilarating and frustrating; and dissonance because the movement between social contexts would burden individuals with potential for internally conflicting goal hierarchies. Ambivalence and dissonance are threats to personal wholeness, and individuals experiencing these states would hunger for the means to resolve them.

The deep point is that the new experiment in post-intuitive sociality would have left individuals confused and perplexed about fundamental personal and social realities. We may summarize the crisis of self-understanding in the following set of questions:

What kind of reality is a human being?
What is good for a human being?

What kind of reality is a social group?
What is good for the social order?

Obviously, the questions would not have appeared in anything like these precise forms, but these were nevertheless the pressing concerns.

Something had to be done. As long as there was no satisfying way to think about the nature and the nurture of personal and social realities, any experiment in post-intuitive sociality would remain inherently unstable. These were the psychological and social circumstances that prompted mythopoesis, or mythmaking. Our ancestors found themselves in a state of desperate need for a story that could tell them who they were, where they came from, what the group was, how it came to be, and why they should follow the new rules.

But who could tell such a story? Where would the vocabulary for such a narrative come from? My suggestion is that the early thinkers who took on these difficult issues mimicked the ancient biological trick of putting old adaptations to new uses. That is, they borrowed the language of ad hoc cosmology, already well developed to explain the phenomena of nature, and expanded it to include the nature and nurture of self and society. The cosmic order and the moral order were thus unified by the anthropomorphic vocabulary of gods and spirits. The powers that send the rain and command the wind and replenish the earth were now understood to be the very same powers that created people, brought them together, and commanded them to obey a particular set of rules. As these first myths emerged from the imaginations of storytellers, we may suppose that reunion events acquired a new focal point: the re-enactment of the primal narrative. The first religious traditions arose as these stories found their distinctive strategies for carrying on.

THE FUNCTIONS OF RELIGION

Having explored the anatomical structure and the evolutionary origins of religion, we now turn to its physiological functions. We may begin by taking an insight from the science of physiology itself: The best physiologists are pathologists. That is, if you really want to know how organisms work, then figure out why they quit working. We know that the life process has ultimately failed when the rate of an organism's chemical decomposition overtakes the rate of chemical composition, and the job of the pathologist is to trace this reversal to injury or malfunction in some organ system. Analogously, we might seek to understand how religions work by beginning with symptoms of their failure, and then tracing these symptoms back to malfunction in some subsystem.

What, then, are the telltale symptoms that there is injury or malfunction somewhere in the anatomy of a religious system? I propose this: We may suspect a breakdown in religious function whenever we observe systematic losses in the achievement of personal wholeness and social coherence. For example, whenever solidarity, cooperation, security, and harmony appear to be decreasing, or whenever social animosity, discrimination, injustice, and conflict appear to be increasing, we may begin to suspect a failure of religious function. And likewise, whenever happiness, tolerance, generosity, and forgiveness appear to be giving way to depression, aggression, obsession, and repression, we may wonder about the religious life.

Is this claiming too much for religion? Surely it cannot be said that religion is responsible for all the gains and losses in personal wholeness and social coherence. An earthquake, a flood, or even a brain tumor might result in considerable losses in personal wholeness and social coherence, but what has religion got to do with these? Nothing. A religious tradition cannot be blamed for losses such as these, of course. But what if a religious tradition failed to make any appreciable difference in coping with or restoring these losses? Might we not suspect religious malfunction then? I am not claiming that all losses in personal wholeness and social coherence are attributable to a failure of religion. What I am claiming is that religious traditions are sufficiently relevant to achieving, maintaining, and restoring these goals that we may at least suspect religious malfunction whenever they are noticeably in decline.

Does the claim that the principle functions of religion are to enhance personal wholeness and social coherence amount merely to an arbitrary assertion? Why should this claim displace alternative ones, like the assertion that religion functions to bring us closer to God, or that it frees us from sin, or informs us about what God is like and what God wants, or that it helps to overcome the illusion of selfhood, or (for that matter) that it takes our minds beyond mundane concerns like personal wholeness and social coherence? I have no compelling answer to this question. It is relevant, however, that every religious tradition includes clear and forceful notions about the human predicament and what should be done about it, and these notions leave us in no doubt about the personal and social dimensions of the human condition. Every tradition provides us with resources for therapy and politics, which is ultimately what I mean by the vocabulary of personal wholeness and social coherence. The real question is whether the religious life could offer anything *more* than personal wholeness and social coherence—that is, whether the language of salvation (communion with God, eternal glory, blessedness) effectively reduces to the language of personal wholeness and social coherence.

There is much to be said for the claim that talk about salvation actually derives its meaning from the language of personal and social well-being. Imagine, for a moment, that all our longings for personal wholeness and social coherence have been fulfilled. What would the goal of salvation, or coming into glory, add to that? One might say that coming into glory is something more to be longed for, and that one cannot be fulfilled spiritually without this promise. No doubt there are many men and women who believe this way. But consider that spiritual fulfillment of the sort one expects from the promise of glory will always be a part of what is meant by personal wholeness. It is therefore difficult to miss the point that the promise of glory is a means to personal wholeness, and not something more to be longed for. It is the promise of fulfillment that counts, not the fulfillment of the promise. Indeed, it is difficult to escape the more general point that the entire language of theology is absorbed without trace into the vocabulary of therapy and politics.

It may help to be more precise about the nature of personal wholeness and social coherence. The simultaneous achievement of these goals is very simple to describe in general terms. Personal wholeness exists when an individual is able to maximize the satisfaction of self-interests with a minimum of mental conflict. Social coherence exists when there is a maximum amount of solidarity and cooperation among individuals in a group. The secret to achieving these goals simultaneously is to construct the conditions for maximizing the confluence, or overlap, of self-interest among individuals.

This process may be easy to describe, but it is exceptionally difficult to achieve. We have seen that on the scale of hunting-gathering bands, overlaps of self-interest are effectively established by an intuitive morality scripted by the genes. But when the social order ratchets upward to episodic coalitions, tribal alliances, chiefdoms, or nation states, then overlaps of self-interest get out of whack and intuitive morality must be overridden appropriately by an ad hoc morality commensurate with the new scale of social organization. Ultimately, then, the secret to a simultaneous achievement of the twin teloi in a post-intuitive social context is to construct the ad hoc foundations for group-wide overlaps of self-interest. Failure in this enterprise invites a return to the default form of social organization. In fact, breakdowns in personal wholeness and social coherence may be perceived as outbreaks of intuitive morality.

But how are the ad hoc foundations for overlaps of self-interest in large groups to be constructed and maintained? They are constructed and maintained by educating the emotions—that is, by nurturing the mentality of individuals under a comprehensive vision of how things ultimately are and which things ultimately matter. In particular, it calls for nurturing of background feelings, the appraisal process, and the coping process. A mythic tradition will encourage certain moods and discourage others, thereby changing thresholds to various emotional responses. For example, if an individual surrounds herself with religious art and music, and if she performs rituals regularly (such as prayer, meditation, and worship), then her moods will tend to become patterned under the influence of the myth.

Religious traditions also nurture the formation of attitudes. By the dynamics of social apprenticeships, and by virtue of experiences deliberately induced by orchestrated incentive events, individuals will acquire a distinctive set of valenced beliefs that will bias their emotional responses. For example, if you take the attitude that all members of the group are your brothers and sisters, then you will be more likely to cooperate with them.

The appraisal process may be patterned by mythic influences as well. This influence begins as individuals learn to perceive objects, events, properties, and relations according to the categories and concepts embedded in the cosmology of the myth. Appraisal patterns develop as individuals reconstruct their self-esteem linkages to reflect the goal hierarchy endorsed by their tradition. For example, if individuals link their self-esteem to performing well against ad hoc moral standards (like the golden rule), then overlaps of self-interest will expand.[14]

Religiously informed goal hierarchies will also pattern the coping process, as individuals will be equipped to inhibit certain emotional displays and to reappraise the meaning of their encounters to reflect an ad hoc morality of emotions. When individuals are nurtured in a hierarchy of spiritual virtues, then emotional events that begin in anger might well result in displays of compassion.

When the ancillary strategies of a mythic tradition are effective, then individuals will acquire and share a distinctive pattern of piety. That is, when the mythic vision is interpreted plausibly, when it is performed ritually, when it is objectified aesthetically, when it is regulated and administered socially, and when it is validated by subjective experience—when these things happen, a pattern of piety will emerge as a dominant factor in modulating background moods and attitudes, and in guiding perceptual, appraisal, and coping events from moment to moment in the lives of individuals.

This description makes the education of emotion appear as a

once-and-for-all event, which it is certainly not. The religious life is a life of constant revitalization. The myth is under continuous critique and reinterpretation, the rituals must be performed regularly, the art must be ubiquitous, the social administration must be vigilant, and the subjective validation must be deepened. If the point is to bias the working memory toward mythically informed mental objects, then the education of emotion must be unceasing. The spiritual life must be tended daily, even hourly, for the minions of evil—by which one can only mean the whisperings of intuitive mentality—are relentless.

The ultimate function of a religious tradition is to enhance personal wholeness and social coherence by nurturing the conscious and unconscious lives of individuals. I have suggested that this ultimate function has both a therapeutic focus and a political focus. The therapeutic function of religion is to transform the individual from an orientation of self-centeredness to one of reality-centeredness. The integrated meanings of a mythic vision offer me a way to apprehend the deep meaning of the cosmic order. The entire realm of being, ultimate reality itself, is presented to me as an incentive event to which I respond in newly positive ways nurtured by the story. The universal story now becomes my own particular story. My self-understanding is transformed, for now I apprehend my true nature, my ultimate origins, what is genuinely good for me, and how I might be fulfilled. I now have a new perspective by which to order my aspirations and obligations within the larger scheme of things, including the social order.

The selfish goals I once pursued have been displaced by new commitments and my self-esteem is now linked to projects that advance the common good. There may continue to be moments of mental conflict in my life, but as these arise I will return to the story to recover my bearings in the world. The story is always available to me, to absorb my uncertainties and to conquer my temptation to default back to narrow self-interests. As the challenges of everyday life erode my personal integrity, I may restore my soul by revitalizing the story.

But the saving grace of myth goes beyond therapy to exercise a political function as well. When a mythic vision is ingested widely among members of a group, a synergy of parts emerges in service to the whole. Differences between individuals are trivialized by deeper, common interests. When you and I share a myth, we affirm a common origin, a common nature, and a common destiny. We have the same perspective on things in general, we are committed to the same goals, we share the same attitudes and expectations, and our orientation within the cosmic and moral order is the same. Ultimately, what is good for me is also good for you. There may continue to be moments of conflicting self-interest between us, but the community provides occasions for transcending these differences to achieve a unity of purpose. When the ongoing competition of social life wears away at overlaps of self-interest, then the collective affirmation of a shared story may restore them at the level of ultimate truth and value. By the power of myth I become whole and we become one.

All of this assumes, of course, that our social context offers us a single mythic vision. A few pages ago I suggested that breakdowns in personal wholeness and social coherence might be perceived as outbreaks of intuitive morality. But this is not the whole picture. Failure to achieve and maintain the twin teloi may also result from a plurality of myths. If we differ on ultimates, then we are ultimately different. If you have one vision of ultimate reality and value, and I have another, then we have nowhere to go when we encounter deep conflicts of interest. If we lack a shared orientation in nature and history, and if we lack a common understanding of who we are and what we should strive for together, then we lack the means to transcend our differences when the social chips are down. We will be left to think that the other just has it wrong.

And further, we may be inclined to believe that the other's way of thinking constitutes a serious threat to the true way of thinking. At the end of the day, if we cannot reconcile our differences by ingesting a common story, one that transcends both our myths, then any overlaps of self-interest will be accidental and unstable. Then as conflicts arise, as they inevitably do, our only alternatives will be to resort to brute force or to part company.

But it should be clear that these are impossible options in a world simultaneously bloated with instruments of warfare and depleted by overpopulation and insatiable consumption. As our global environmental crisis continues to worsen, we shall be challenged as never before to construct the ad hoc foundations for human solidarity and cooperation. We are called—here, now, urgently—to the task of mythopoesis. These troubling issues will meet us again in part three.

ON
SPIRITUAL
TRADITIONS

Our primate ancestors had a good thing going: They were instructed by a common genetic playbook to think, feel, and act in ways that kept their social mobile relatively stable and tolerably satisfying. The gut reactions of their social emotions were sufficient means to achieve overlaps of self-interest, assuring that individual and collective goals could be met simultaneously. The circumstances of modern humans haven't changed so much that we cannot recognize the same basic dynamics. We, too, are social creatures, and our social mobiles, albeit much larger and more complex, still depend on achieving overlaps of self-interest. The big difference is in the playbooks. Theirs was fixed, constraining them to go on playing on the same scale and by the same innate instructions no matter what. Ours, however, is a compendium of genetic defaults overlaid by ad hoc algorithms, enabling human sociality to transcend the limitations of familial bands.

I have maintained that overlaps of self-interest commensurate with an enlarged social order can be achieved when individuals come to share a myth, an integrated vision of ultimate reality and value. When the members of a group are nurtured by the meanings of a shared myth, then their behavior mediation systems will be coordinated by a common set of algorithms for thinking, feeling, and acting—in generally the same way that our primate ancestors were governed by their genetic playbook.

Part two of this volume has been assigned the task of showing how particular religious traditions play on the strings of human nature to make all of this happen—that is, how religious traditions attempt to nurture individuals in the conventional meanings that might enhance overlaps of self-interest. Five religious traditions will be considered: Judaism, Christianity, Islam, Hinduism, and Buddhism. Each will be given its own chapter, organized as follows by a common five-part outline.

- *Historical context.* Each chapter will show how the tradition developed in historical circumstances where needs for personal wholeness and social coherence were not being met. In each case the response was to construct a narrative tradition having the conventional meanings that were necessary for re-educating behavior mediation systems.
- *Myth.* Each chapter will summarize the central myth of the tradition and will clarify the manner in which the root metaphors integrate cosmology and morality. The cosmological component tells us what the myth expects followers to be realists about, while the moral component tells us about the goal hierarchy they are expected to carry into their encounters with the world.
- *Emotional appeals.* Each chapter will show how salient elements in the myth appeal to human emotional systems, thereby influencing the attitudes and goal hierarchies of individuals.
- *Ancillary strategies.* Each chapter will show how the tradition has developed intellectual, institutional, experiential, aesthetic, and ritual strategies to support the inculcation and maintenance of attitudes and goal hierarchies.
- *Personal wholeness and social coherence.* Each chapter will show how the ministrations of the tradition result in self-esteem links that may be expected to enhance prospects for achieving overlaps of self-interest.

5

JUDAISM

HISTORICAL CONTEXT

The Late Bronze Age was a period of political turmoil in the eastern Mediterranean region. Egyptian authority was in decline, the Hittite empire to the north had fragmented into small warring city-states, and Dorian invasions were displacing groups from the Aegean coasts. Widespread hostilities and instabilities brought migrations of diverse peoples into the territory of Canaanite civilization. Archaeological evidence suggests that waves of immigrants settled in the eastern Mediterranean region during the period from 1300 to 1000 B.C.E.[1]

The most notable fact about the newcomers was their diversity: Some were farmers, others were semi-nomads, mercenaries, and refugee slaves. They converged from all directions on a Canaanite city-state system that was ill prepared to prevent the influx of foreigners. When Egyptian hegemony ended in the region, the Canaanite cities became independent feudal entities, each with serious problems of internal stability. The foreign immigrants were therefore relatively unopposed as they constructed villages and settled down to farming and herding in the territories between cities.

The social and political dynamics of the region were complicated. In general, the geographical and cultural boundaries between tribes, as well as those between tribes and their neighboring city-states, were in considerable flux. At certain times differences hardened into open warfare, but at other times they softened into solidarity against a common military threat. Some rural tribes gradually assimilated into the indigenous feudal system, while others remained aloof or were refused.

The religious complexion of the region was no less confusing. Religious differences, too, would have hardened and softened—and even blended—in the course of local interactions. The indigenous religion of the agrarian Canaanites was Baalism, which centered on nature gods and fertility rites. Ritual practice safeguarded the cycle of fertility and the harmony of the natural order. Periodic reenactments of the myth of Baal's death and resurrection would assure the fertility of the land and its people. Most of the immigrants readily adopted Canaanite fertility traditions, presumably because it was considered unwise agricultural practice not to do so.

But the acquisition of new gods did not necessarily displace the influence of old ones. Groups unfamiliar with the agrarian way of life would have brought with them non-agrarian ideas about gods. Semi-nomads and mercenaries would come bearing traditions of patron gods, gods of warfare who intervened in the events of history to protect the group during episodes of crisis. A uniquely attractive variation on the patron-god idea was Yahwism, a tradition associated with refugee slaves from Egypt. This tradition held that Yahweh had delivered his chosen people from a life of bondage and had led them to Canaan through the hardships and dangers of an unforgiving wilderness. Like a magnetic core, this narrative was embellished by attracting fragments of legends from a variety of immigrant traditions.

Eventually the oral tradition of Yahweh's patronage grew until it dominated the historical consciousness of a majority of immigrant tribes. Nevertheless, the forces of nature were not to be ignored, which meant that fertility cults were not dispensable. Gradually, the religious diversity of the area winnowed to a pair of coexisting myths: the militaristic Yahwist myth of history and the agricultural Baalist myth of nature. The discrepancies between these religious paradigms could be overlooked. After all, battle and harvest are completely different spheres of human activity, so there should be no problem in construing them as the provinces of different gods.

By the eleventh century B.C.E., Palestine had become a pastiche of ethnic, cultural, and religious diversity, slowly tending toward homogeneity. Meanwhile, the tribes had begun to develop military alliances in response to a persistent threat posed by the Philistines to the west. By late century the threat was growing in scope and intensity as the Philistines unleashed their ambition to dominate the entire eastern Mediterranean region. It soon became apparent that the tribal alliance could not resist the Philistine threat unless they unified their military resources under the central leadership of a monarch. This was a highly controversial proposal, for some of the tribes insisted that allegiance to a king would alienate the very God, Yahweh, who had protected them from hostilities in the past. To swear allegiance to an earthly king would demonstrate lack of trust in Yahweh, an act no less foolhardy than going into battle unarmed. In the end, however, a king was elected, the Philistine threat was thwarted, and Israel found itself in the commanding position of being a formidable, if accidental, nation.

Just as there is nothing like the fear of defeat to cement the collective resolve of a nation, so there is nothing like the pride of victory to galvanize its collective identity. Momentarily united by a veneer of military success, the young nation was left to invent the cultural means for a deep and enduring sense of solidarity. One thing was clear: There was no going back to a loose confederation of tribes. Any remaining elements of tribalism would somehow have to be consolidated to reflect the new reality of a unified monarchial state. King David (1000–961 B.C.E.) took many practical measures to unify the realm, but it was not until the reign of Solomon (961–922 B.C.E.) that the new social order acquired ideological legitimation.

Solomon commissioned a court historian to write an official account of the birth of the nation. The result was a comprehensive narrative drawing together many strands of oral tradition to demonstrate how the experience of the past was continuous with the new social order. The author—dubbed "the Yahwist" by biblical scholars—made it possible to see the emergence of Israel as a matter of manifest destiny, implicit in God's purpose from the moment of creation. For the Yahwist any separation of nature and history into distinct realms was inconceivable. There is only one God, Yahweh, who is creator and lord of both the primordial order of nature and the moral order of history. The organizing theme for the Yahwist's epic was God's unswerving patronage toward his chosen people, Israel. The basic terms of the covenant were these: If the people willfully trusted God, then he would bless them unconditionally with peace and prosperity. The deliverance from bondage in Egypt, the safekeeping through the wilderness, and the conquest of Canaan were all presented as acts performed by God to fulfill his promise to Israel—and these were all prologue to the final blessing of nationhood.

Yahwism, however, was not Judaism. The decisive events giving rise to Judaism came at the *end* of the nation's history, not at the beginning. Nationhood ended abruptly in 586 B.C.E., when Babylonian forces utterly devastated Israel. The invaders torched Jerusalem, killed most of its residents, and left a mere remnant (fewer than five thousand) to be exiled to Babylon. A catastrophe of this magnitude had been anticipated by some of Israel's prophets, but their forewarnings had met with contempt and hostility. And what good were such warnings now? Here was a crisis of cosmic proportion, and one that could not be made intelligible by the Yahwist's triumphalist ideology. Yahwism, it was clear, had been brutally refuted by the facts of history. The circumstances of the exiled remnant were not exactly conducive to the achievement of personal wholeness and social coherence, as the Psalmist reveals:

> By the rivers of Babylon—there we sat down and there we wept when we
> remembered Zion.
> On the willows there we hung up our harps.
> For there our captors asked us for songs, and our tormentors asked for
> mirth, saying, "Sing us one of the songs of Zion!"
> How could we sing the Lord's song in a foreign land?
> (Ps. 137:1–4)

By the rivers of Babylon the Israelites experienced what Camus called "the incorrigible sorrow of all prisoners and exiles, which is to live in company with a memory that serves no purpose."[2] Personal wholeness and social coherence might have been better served by burying the past and assimilating into Babylonian culture, which in fact many of the exiles did. But there were some among them who found the inspiration to reexamine the past and to assemble a new vision of hope for the future. These were the architects of Judaism, the priests and prophets who initiated what was perhaps the most creative and productive period in Hebrew history.

It was during the exile that strands of oral and written tradition were recast under a new vision. It was during the exile that legal traditions were rediscovered and given a central role in the new venture. It was during the exile that the insights of pre-exilic prophets were fully appreciated. And it was during the exile that the synagogue was founded, where a new pattern of piety began to emerge. Judaism emerged in response to the despair of the exile, nurtured by the prophetic insistence that God's covenant with Israel was not unconditional, but fiercely conditional.

THE MYTH OF COVENANT BLESSING

The mythic tradition associated with Judaism was given its distinctive expression during and after the exile, though it bore important elements of continuity with pre-exilic traditions. The essential elements of the myth of covenant blessing may be summarized as follows.

At the beginning of time God created an orderly world from a watery chaos, placing the first humans in a garden of peace and prosperity for their enjoyment. Adam and Eve were created in God's image, as whole persons free to choose and to create within their own sphere. The garden paradise was a world without evil, adversity, or death. All that was, was good, for nothing in the world was contrary to God's will. God imposed a single moral constraint on the freedom of Adam and Eve: They were forbidden to eat fruit from the tree of knowledge of good and evil. If they disobeyed God's command, they would be subject to death.

Failing to resist temptation, Adam and Eve ate the forbidden fruit and were driven by God from the garden and left to face the challenges and adversities of a fallen order of nature. The progeny of Adam and Eve continued to displease God, until God flooded the earth, eliminating all forms of life except for a remnant spared on Noah's ark. God then initiated a new covenant with Noah, but this, too, was marked by a failure of obedience. Then came Abraham. If Abraham obeyed God's will, then God would bless Abraham by making him father to a great nation. The covenant with Abraham was renewed at various points throughout the biblical narrative, the most decisive renewal event following a miraculous escape from Egypt.

The descendants of Abraham had been enslaved by the Egyptians and held in bondage for 400 years. God took pity on the suffering slaves, delivering them from captivity and leading them through a wilderness to Mt. Sinai, where the terms of the covenant were revealed to Moses. The terms were clear: If the Hebrews obey God's will, as expressed in the laws given to Moses, then God will bring them to the promised land of Canaan where they will flourish as a peaceful and prosperous nation. But if they disobey the laws, then God will allow them once again to suffer at the hands of hostile nations. Implicit in the covenant was the idea that all forms of adversity and suffering befalling the people of Israel were properly to be construed as evidence of their failure to love and please God. In the face of adversity, therefore, the remedy was clear: Repent for the sins of disobedience and return to the straight and narrow path of doing God's will. This remedy became a persistent theme among the prophets of the headstrong nation.

God fulfilled the promise of nationhood by leading the Israelites to victory against the Canaanites and the Philistines. A glorious kingdom flourished, but then began to falter as Israel's kings persisted in disobeying God's will. God therefore punished Israel by allowing the nation to suffer defeat and exile by Babylonian forces. After fifty years in exile the Jews were allowed to return to the homeland, there to struggle with the Law as they continued to hope for enduring peace and prosperity.

The genius of the myth of covenant blessing—indeed, any myth—lies in its effectiveness to condition the appraisal process. And this effectiveness is contingent on realism about the central metaphor of the myth: *God as person*. Suppose someone informs you that there is poison in the soup you've just been eating. If you are a nonrealist about the poison, you will remain emotionally unchanged by the information, but if you are a realist about it you will be instantly gripped by fear. The genius of Judaism exploits the same principle, and it cannot work unless one is a realist about God as person. In fact, operationally speaking, realism just is a cognitive predisposition that decisively affects the appraisal process. Pious Jews are therefore theological realists. This does not preclude sophisticated nonrealist modes of appropriating and valuing certain elements of Judaic tradition. It only means that such modes are marginal to the religious life.

The root metaphor, God as person, achieves an integration of fact and value, cosmology and morality. On the cosmological side it declares how things ultimately are in the universe. Ultimately, God is the creator, sustainer, and absolute ruler of everything that is. And on the moral side the covenant myth declares what ultimately matters: In the end the only thing that matters is what God wants. God does not want certain outcomes because they happen to be good. Rather, outcomes are good if and only if God wants them. The will of God is therefore both the ultimate explanation for any fact and the ultimate justification for any value.

The Judaic myth also presumes realism about the covenant blessing. One must believe that God has ordered creation such that good outcomes for the people of Israel are contingent on their obedience to God's commands. The covenant, therefore, is both a metaphysical principle and a moral one. Inherent in this view is a presumption about historical realism as well. That is, the covenant is the ultimate hermeneutic for historical understanding. Events in history can be made intelligible only on the assumption that God has a plan, and that the execution of this plan relies on the willful obedience to God's law. The facts of history confirm the logic of the covenant.

Finally, Judaism presumes realism about the law. It insists that the law is a true and reliable formulation of God's will—not a guess about what God wants, not a list of inferences from historical experience, not a mere reflection of the interests of the ruling class, but the real thing, a revelation from God, an authoritative transmission of the terms of the covenant. If the Jews are expected to obey God's will, then they must be free of doubt over what God's will actually is.

If one is a realist about these central elements of the Judaic myth, then something follows about human goal hierarchies. The terms of the covenant amount to an expression of the conditions for carrying on, a set of algorithms for survival and fulfillment. The entire force of Judaic tradition has been concentrated on conditioning and transforming the cognitive and emotional lives of individuals so that they become more likely to obey God's will. The pious Jew is one who has been nurtured to appraise all of life's encounters in a manner that reflects what God wants, one whose working memory is dominated by the covenant myth. This is what the tradition has meant by "walking with God." The impious are those whose goal hierarchies reflect either a default to intuitive morality or seduction by falsehood. Obedience to the sovereign will of God is central to Judaic piety, and on this essential point there has been no significant debate or development in the course of the tradition. Everything else about Judaism belongs to the realm of detail.

EMOTIONAL APPEALS

The myth of covenant blessing is brought to life by concrete images and events that appeal directly to human emotional systems. In particular, the covenant myth plays on our emotional competencies for awe, humility, fear, gratitude, guilt, love, and hope. These emotions arise from encounters with the myth, culminating in predispositions to act. When we are moved as the story intends, then nothing makes more sense than to respond as the story bids us. The myth compels individuals to reconstruct their attitudes, goal hierarchies, and self-esteem linkages in the light of these responses.

AWE AND HUMILITY

The biblical narrative is emphatic about the power and sovereignty of God. God is almighty, the creator and lord of heaven and earth, whose strength is beyond human comprehension. A single breath from God was sufficient to annihilate the Egyptian forces:

> In the greatness of your majesty you overthrew your adversaries;
> you sent out your fury, it consumed them like stubble.
> At the blast of your nostrils the waters piled up, the floods
> stood up in a heap. . . .
> You blew with your wind, the sea covered them;
> they sank like lead in the mighty waters.
> (Exod. 15:7–10)

The biblical narrative is bloated with imagery of God's almighty acts, and the authors of the texts count on these images to provoke a sense of awe, reverence, and humility in the audience. The action tendency associated with these emotional responses is to cower and submit to the will of an overwhelming divine presence. This behavioral outcome is precisely what the biblical authors aim for—that is, a disposition to relent and obey.

The Psalms are excellent resources for telling us which emotions are appropriate marks of piety in response to God's sovereignty. Here is one of many endorsements of awe and humility:

> For the Lord is a great God, and a great King above all gods.
> In his hand are the depths of the earth; the heights of the mountains
> are his also.
> The sea is his, for he made it, and the dry land, which his hands have formed.
> O come, let us worship and bow down, let us kneel before the Lord,
> our Maker!
> (Ps. 95:3–6)

FEAR

The mythmakers of Judaic tradition were astute observers of human nature. They knew nothing of the language of genetically predisposed emotional systems and their inherent action tendencies, but they nevertheless effectively exploited the algorithms for threats to induce fear, and for fear to trigger a response to escape or submit. In other words, they realized that fear, like awe and humility, could function as a constructive emotional virtue. The idea was to frighten listeners into submission and obedience.

Indeed, plucking on the strings of fear became something of a specialty among biblical authors. God is frequently portrayed as jealous and wrathful, ready to vent his anger on those who disobey: "But if you defy me, you are doomed to die. I, the Lord, have spoken." (Isa. 1:20) The Bible is awash with God's open threats against the people of Israel, many of which are carefully crafted with vivid imagery to chill the spine and scorch the memory:

> The Lord will afflict you with the boils of Egypt, with ulcers, scurvy, and itch, of which you cannot be healed. The Lord will afflict you with madness, blindness, and confusion of mind; you shall grope about at noon as blind people grope in darkness, but you shall be unable to find your way; and you shall be continually abused and robbed, without anyone to help. You shall become engaged to a woman, but another man shall lie with her. You shall build a house, but not live in it. You shall plant a vineyard, but not enjoy its fruit. Your ox shall be butchered before your eyes, but you shall not eat of it. . . . The Lord will strike you on the knees and on the legs with grievous boils of which you cannot be healed, from the sole of your foot to the crown of your head. . . .

All these curses shall come upon you, pursuing and over-
taking you until you are destroyed, because you did not obey the
Lord your God, by observing the commandments and the decrees
that he commanded you.

(Deut. 28:27–35; 45)

GRATITUDE

All normal humans are wired to experience gratitude under certain
conditions. When we receive a favor or a gift, and when this positive
outcome can be attributed to someone else, then we automatically
feel grateful and become predisposed to the action tendency of offering thanks or reciprocating the favor. The biblical authors exploited
the mechanisms of gratitude to achieve their central purpose—that is,
to increase the probabilities for obedience to God's will. The Bible
takes pains to emphasize the many close calls and fortunate events
that occurred in Israel's history, but it insists that these events were
acts of God, and in no way could they be attributed to the people of
Israel. Whatever good happened was exclusively attributed to the generosity of God, and never to the merits of the people or to the power
of fertility gods.

Do not say to yourself, "My power and the might of my own hand
have gotten me this wealth." But remember the Lord your God,
for it is he who gives you power to get wealth, so that he may
confirm his covenant that he swore to your ancestors, as he is doing today.

(Deut. 8:17–18)

Judaic piety has always been clear about the importance of re-
sponding in humble gratitude to God's gracious acts by serving his
will. Pride is always without virtue because it displaces gratitude. If
there is any credit to be given, it must go to God: "Not to us, O Lord,
not to us, but to your name give glory, for the sake of your steadfast
love and your faithfulness." (Ps. 115:1)

GUILT

The appraisal pattern for guilt involves self-blame for negative outcomes. If something bad happens as a result of my behavior, and if I
believe that I could have and should have behaved differently, then
blame for the bad outcome falls to me, whereupon I will be aroused
to guilt. The action tendency for alleviating guilt is to act in ways that
might undo, repair, or compensate for the negative outcome.

It is easy to see that the biblical view of the world encourages the experience of guilt. Cosmology and morality are so tightly integrated in the biblical concept of divine sovereignty that anything happening in nature or history may be given a moral construction. Furthermore, the moral construction typical of biblical authors is that all good outcomes are attributable to God's generosity, while all bad outcomes are attributable to Israel's disobedience. All suffering and adversity, whether resulting from human agency or natural forces, are ultimately construed as evidence of human wrongdoing. If *anything* goes wrong, it is our fault.

The biblical authors serve up many specific indictments of human wrongdoing. Adam and Eve were charged with disobeying God's prohibition against eating from the forbidden tree (Gen. 3); the descendents of Adam were indicted for having consistently evil thoughts (Gen. 5:5); accusations of grave sins were brought against the people of Sodom and Gomorrah (Gen. 18:20); the refugees from Egypt were charged with worshipping false gods (Exod. 32; Deut. 9:16); and time after time the people of Israel and their rulers were accused of violating specific terms of the covenant. The prophetic literature is harsh and pointed in laying blame on God's chosen people for the suffering and adversity that resulted from these transgressions. Hosea's image of Israel as an unfaithful whoring wife is especially caustic (Hos. 16). The biblical narrative leaves no doubt that from its beginning, the history of the covenant has been characterized by God's faithfulness and Israel's failures (Ps. 78; Jer. 2:1–13).

In addition to the many specific indictments, the Hebrew Bible addresses the universal condition of human sinfulness. There is no one on earth who does not sin (1 Kings 8:46; Eccles. 7:20); no one is entitled to a clear conscience (Prov. 20:9); all humans are equally guilty (Ps. 53:3); even the best of their actions are evil (Isa. 64:6); and it has been this way from the very beginning (Isa. 64:5). The general condition of humanity is clear: When left to their own devices, the people of God, both individually and collectively, are capable of nothing but evil deeds resulting in disorder and suffering.

The Garden of Eden narrative shows that the action tendency of running away to hide will not work. The only remaining option for persons aroused to guilt is to seek repair—that is, to repent, to beg for God's mercy, and to reorganize their goal hierarchies: "Let the wicked forsake their way, and the unrighteous their thoughts; let them return to the Lord, that he may have mercy on them." (Isa. 55:7)

LOVE

The biblical authors vigorously exploit an inherited human predisposition to experience love and affection under certain circumstances. We are by nature predisposed to feel affection for close kin, especially parents, and to love those who demonstrate their concern for our well-being. The action tendency triggered by affection is to reach out for intimate contact with the beloved, and to respond to their expressions of love by doing what pleases them.

The Bible variously portrays God as shepherd, king, warrior, husband, judge, craftsman, and teacher. But by far the most commonly used metaphor is the image of father. The biblical writers want us to think of ourselves as children of an attentive and benevolent father whose love will never fail us, even though we have a special talent for abusing his love. The Bible is content to leave God's love in the realm of mystery. Nothing, not even pity (Exod. 33:19), can adequately explain the depths of God's love. At times divine love turns to wrath and punishment, but this is completely intelligible because of Israel's rebellious and headstrong nature. What is not intelligible, however, is God's patience and long-suffering with a people so undeserving of his love. Time after time God pardons the unpardonable and embraces them with his steadfast love: "For you, O Lord, are good and forgiving, abounding in steadfast love to all who call on you." (Ps. 86:5)

For an audience already genetically predisposed to love their kin, the fatherly image by itself might be enough to arouse love for God. But such a father! Divine love and devotion to Israel far exceeds all reasonable expectations even for parental love. In the presence of a love that transcends human understanding, the biblical authors leave us no emotional alternative but to respond in love. We must love God because he has loved us beyond measure: "You shall love the Lord your God with all your heart, and with all your soul, and with all your might." (Deut. 6:5) And to love God means to serve his will: "So now, O Israel, what does the Lord your God require of you? Only to fear the Lord your God, to walk in all his ways, to love him, to serve the Lord your God with all your heart and with all your soul." (Deut. 10:12)

HOPE

I have already described hope as a complex human emotion: It amounts to desire for a positive outcome in the face of discouraging prospects. The biblical authors have constructed the narrative of

Israel's history in such a way that it inspires hope for the future. No matter how bleak things look at the moment, the past was even bleaker. Yet, even then, God in his mercy and steadfast love came to Israel's defense. Remember the Exodus! The odds against escaping Egyptian forces were overwhelming, but look at what God did for us then: "He is your praise; he is your God, who has done for you these great and awesome things that your own eyes have seen. Your ancestors went down to Egypt seventy persons; and now the Lord your God has made them as numerous as the stars in heaven." (Deut. 12:21–22)

The appraisal pattern for hope requires a belief in the possibility of a positive outcome, even though the probability is low. No matter how poor the odds are for winning the lottery, I may hope to win as long as I believe it is possible. The story of Israel's past is presented by biblical authors as a license to hope, serving as a pep talk for those who are inclined to despair. The appraisal pattern inherent in the biblical narrative inspires hope and displaces despair.

But there is a catch! Just as one cannot hope to win the lottery without purchasing a ticket—for then the odds would be zero—so Israel cannot hope to prosper or escape doom without keeping the terms of the covenant. With God on their side Israel might hope against any odds, but without their obedience even hope is lost. By arousing their audience to hope, the biblical authors create a disposition to obey.

I have said earlier that the secret to any post-intuitive social order is to create overlaps of self-interest, or—what amounts to the same thing—a coincidence of goal hierarchies between individuals. I have been trying to show how the mythmakers of Judaic tradition were masters in the art of creating the symbolic conditions for such overlaps. The rhetorical challenge for these mythmakers was to transform the attitudes and goal hierarchies of their audience to conform to God's will. They accomplished this by making a unified appeal to cognitive and emotional systems.

On the cognitive front they presented a coherent worldview in which living in harmony with reality meant living in harmony with the lord of nature and history. And on the emotional front they aroused awe and humility before God, fear of God, gratitude to God, guilt for betraying God, love for God, and hope in God. The action tendency triggered by each of these emotions is precisely the same—that is, to do what pleases God. Any one of these emotional responses might be effective on its own, but when they are aroused simultaneously, or serially, or when there is a blending effect, then they create a pious mental state in which nothing could appear more desirable, more reasonable, or more advantageous to self-esteem than serving God's will.

The pious Jew *wants* to serve God: "With my mouth open I pant, because I long for your commandments." (Ps. 119:131)

ANCILLARY STRATEGIES

It is all very well to show how encounters with a myth engage the kinds of emotional responses that shape attitudes and transform goal hierarchies, but such effects are notoriously ephemeral. I once heard a journalist confess that his attitudes were mostly a reflection of his most recent interview. And goal hierarchies are fluid realities, too, changing from moment to moment as we encounter new items of business. It must also be remembered that goal hierarchies have a strong tendency to default to the persistent demands of intuitive morality. While I am engaged by the covenant myth—while I am into the story—it may be likely that my will is in harmony with God's will. But how do things look when I am *not* into the story? Is there any way to assure that what matters to God will matter to me when I meet up with an old foe, get cut off in traffic, miss my flight, lose my job, get mugged, or even win the lottery? If a religious tradition fails to influence appraisals in situations such as these, then it seems to me it fails absolutely.

Ancillary strategies are designed for the care and feeding of the myth, to enable and support its transmission, to maintain and strengthen its influence. That is, they aspire to command our behavior by dominating the systems that produce it. In this respect, perhaps, they resemble clever marketing strategies and political campaign strategies, but with this important difference: Spiritual traditions ask for more than sales and votes—they ask for our entire lives. The ancillary strategies of Judaism, like those of other traditions, apply something like a full-court press to the soul. They do whatever they can to keep us into the story, to saturate us with the meanings of the myth. As Jacob Neusner claims, "To be a classical Jew is to be intoxicated by faith in God, to live every moment in his presence, to shape every hour by the paradigm of the Torah."[3]

INTELLECTUAL STRATEGIES

Religious traditions vary in their use of intellectual resources for the purpose of clarifying, interpreting, and defending the cosmological and moral components of their mythic core. Intellectual energy in Judaic tradition has always concentrated more heavily on ethics than on metaphysics. The reason for this is well-expressed by Moses Maimonides, the influential philosopher of the twelfth century: "All attributes of God that are found in God's books . . . are attributes of God's actions, and not at all descriptions of God's essence."[4]

God so far transcends human understanding that metaphysical speculations and disputes about his ultimate nature are fruitless and beside the point. What God is like cannot be known, but what God *wants* can be known from his dealings with Israel. And these dealings must be known. The premise of intellectual endeavor in Judaism is that knowing the Law, and remembering it, is an absolute prerequisite to obeying the Law. You cannot do what is right unless you first know what is right. Judaic tradition has therefore, from its very beginnings, stressed diligence in learning the Law: "Never forget these commands that I am giving you today. Teach them to your children. Repeat them when you are at home and when you are away, when you are resting and when you are working." (Deut. 6:6–7)

Knowing what is right, however, is never as easy as it sounds. It is not so simple a matter as committing to memory the 613 laws specified in the Torah. This is never discouraged, of course, but it is never sufficient. To understand fully the meaning of these laws and what is implied by them, and then to figure out how they apply to the many particular situations of human life—all of this requires careful study, reflection, analysis and interpretation. Preoccupation with the Law and its ramifications has generated an exceptionally rich tradition of scholarship and discourse on ethics within Judaism.

I will not attempt to summarize the historical and literary characteristics of this bulky tradition, but simply indicate that there exist layers upon layers of rabbinical commentary and practical instruction on both ritual and social obligations imposed by the Law. The material of these traditions ranges from the erudite and technical to the folksy and humorous, but all of it is aimed at comprehending what God's will actually is and how best to obey it.

Pious Jews are expected to immerse themselves in the study of their intellectual heritage. It has functioned as an important element in educating the emotions, for the more time and energy spent considering aspects of the Law, the more likely it becomes that its moral content will influence working memory during everyday appraisal and coping events, and therefore the less likely it becomes that individuals will default to the self-centered inclinations of intuitive morality. Virtue becomes second nature in those who know the Law:

> Every one of these good qualities becomes habitual with him
> who studies the Torah; for that study indeed leads to the forma-

tion of a noble character. Therefore, happy is he who toils in the Law! For this gracious toil fix daily times, of long or short duration, for it is the best of all works that a man can do.[5]

EXPERIENTIAL STRATEGIES

It is common for religious traditions to encourage individuals to seek extraordinary experiences as occasions for personal validations of their worldviews. Intense subjective religious experiences serve a double function: They tend to preclude doubts about the intellectual plausibility of the myth, and they exploit the bias of working memory to retrieve information marked with strong affective valence. Given these important properties—and given a veneration for the prophets, for whom spiritual ecstasy was the norm—it may seem odd that mainstream Judaism has been relatively unenthusiastic when it comes to ecstatic forms of piety.

But there are non-trivial reasons why the tradition might be wary of extraordinary experiences of all three types, visionary, numinal, and mystical. Once the canon of the Hebrew Bible became fixed, in the first century C.E., no further revelations or prophetic visions were expected from any quarter. God's revelation was now complete, and it was all right there in the scriptures. There was good reason, therefore, to regard any further prophetic visions to be either superfluous or subversive.

Profound numinal experiences of divine presence had negative potential as well. The subjective immediacy of these experiences might compromise the view that God presents himself most fully in the scriptures. Moreover, numinal experiences tend to inflate the importance of individual spirituality and to diminish the communal sense of religious identity.

It is also difficult to see how the extraordinary experiences of mystics would have significant appeal to mainstream Judaic piety. Such experiences are said to obliterate the distinctions between creator and creation, a distinction that has always been central to Judaism. And furthermore, mystics typically claim that the temporal quality of ordinary experience is simply annihilated in a blissful sea of eternity—in other words, time as we know it is ultimately an illusion. Such talk would not go down well for a tradition in which time is considered both real and sacred.

Despite these considerations, it would be misleading to suggest that Judaic tradition has been insouciant about all manner of extraordinary experience. Napoleon, it is said, once asked the scientist Laplace whether he could cite any evidence for the existence of God.

After a moment's thought Laplace replied, "The Jews!" This story captures the Judaic sense that no experience could be more extraordinary or compelling than Israel's historical experience. The truest religious experiences are not found in subjective ecstasy, where the individual is removed from the temporal order (ecstasy = "stand out"), but rather in the apprehension of God's real presence within the sacred history of his chosen people.

So far mainstream Judaism has been described in terms of its dual emphasis upon intellectual pursuits (to the exclusion of intense subjective spirituality), and moral issues (to the exclusion of metaphysical speculation). But it would be a distortion of Judaism to overlook minority traditions that developed in part by reaction against these emphases. Chief among these minority traditions are the Kabbalah and Hasidism, both of which have developed strategies to nurture intense subjective experience.

The Kabbalah is a body of metaphysical doctrine that inspired an esoteric and mystical interpretation of Judaic tradition. Flourishing in medieval Spain, the sources of this speculative movement are obscure.[6] Kabbalistic metaphysics attempts to explain how the infinite God, a supernatural transcendent unity, could create a finite sensual world of material diversity. Creation is explained in terms of a series of divine emanations—God's spiritual essence radiates ten spheres of divine reality, forming a descending ontological chain from a unity of spirit to a diversity of matter. These ten spheres also describe the spiritual path of holiness by which the human soul may perfect itself, ultimately to achieve a blissful union with God. Human souls are conceived as fragments of the divine spirit, which may ascend to reunion with God by adherence to strict disciplines of moral righteousness, contemplation of the Law, and deep yoga-like meditation. The operative dynamic behind Kabbalistic teaching and practice is that the promise of ecstatic experiences of divine consciousness (or the heavily valenced memory of them) would be sufficient to keep one's goal hierarchy well in line with the attitudes and values of Judaic tradition.

Modern Hasidism emerged in the eighteenth century as an eclectic reaction against the intellectual elitism and arid spirituality of rabbinic scholarship. To the extent that Hasidism had a metaphysical component, it was drawn from Kabbalistic doctrine. In the process of creation, sparks of the divine essence descended and were encased within material "husks." The point of religious practice was to ignite and release these divine sparks, thereby enabling their reunion with God. But unlike the Kabbalah, Hasidic Judaism bordered on antinomialism—that is, it rebelled against the idea that the only way to reach God was through ascetic discipline and contemplation of the Law. Instead, Hasidic piety educated the emotional life by charismatic means. Hasidic cults formed around charismatic leaders (the *Tzaddiqim*), who were venerated as miracle workers and emulated as spiritual paragons. The followers of these charismatic figures were typically poor and uneducated peasants, the very ones who were spiritually disenfranchised by an overly intellectual and elitist religious establishment. The Tzaddiqim taught that God is omnipresent and immanent, revealed not merely in scripture but in all forms of nature, and most particularly in their own flamboyant presence. Their authority was God's authority, giving them the power to command the goal hierarchies of their followers. When charismatic leaders embody God's will, there is little need for scripture as an instrument for educating the emotions.

The Tzaddiqim were themselves spiritually demonstrative, given to open displays of enthusiasm and ecstasy. Hasidic worship was expansive and celebratory, marked by vigorous dancing, joyous singing, and even violent physical exertion. The role of the Tzaddiqim in worship events was nearly that of an emotional puppeteer, able to manipulate followers and excite them to extraordinary heights of mystical experience. The point was to stimulate a state of ecstatic self-forgetfulness:

> Forget everybody and everything during your worship; forget yourself and your needs; forget the people of whom you have need. . . . Before you can find God you must lose yourself. In him who is full of himself there is no room for God.[7]

The overall effect of Kabbalistic and Hasidic mysticism was to introduce a kind of spiritual pragmatism into Judaic tradition. If the intellectual and moral emphases of mainstream rabbinic tradition failed to educate the emotions, then room could be made for esoteric and charismatic strategies that nurtured intense subjective experiences.

AESTHETIC STRATEGIES

Every religious tradition invents ways to influence emotional systems by aesthetic means. These efforts range from strategies aimed at modulating background states to those aimed at exploiting the biases of working memory. Works of art, music, dance, theater, and so on are notoriously effective in formulating attitudes and in triggering moods that raise or lower thresholds to particular emotional responses.[8] Aesthetic strategies also affect memory systems, assuring that elements of the myth will be readily accessible to working memory. Not least of all, works of creative imagination have potential for generating poignant insights that strengthen the valences of memories associated with a myth. These modulations of neural systems have significant consequences for the process of appraising and coping with everyday encounters with the environment.

The exploitation of aesthetic means for educating the emotions has been minimal in Judaism when compared with most religious traditions. The source of this underdevelopment was the biblical prohibition against idolatry:

> When the Lord spoke to you from the fire on Mount Sinai, you did not see any form. For your own good, then, make certain that you do not sin by making yourselves an idol of any form at all—whether man or woman, animal or bird, reptile or fish.
>
> (Deut. 4:15–18)

Rabbinical scholarship consistently interpreted this injunction in the broadest possible sense, thereby fostering an attitude of suspicion toward the fine arts in general. Thus, from the biblical period right up to the Enlightenment, artistic renderings of the covenant myth were minimal in both scope and scale. The most remarkable exception has been music:

> The prohibition of fine arts . . . channeled Jewish talent into the minor arts of silver jewelry, embroidery, manuscript illumination and illustration, and the like. Quite similarly, the prohibition of instrumental music resulted in a concentration on vocal music, on singing, which came to play an exceedingly important role in Jewish life.[9]

Music, of course, has played an important role in all cultures at all times, precisely because it has proven to be instrumental for the achievement of personal wholeness and social coherence. As the bards tell us, music "soothes the savage breast" and "hales souls out of men's bodies." We settle our babies into sleep with lullabies, we rouse our troops to patriotic courage with marches, we drown our sorrows with funeral dirges, and we collect ourselves into meditation with mantras. A simple melody can unlock a storehouse of vivid memories and transport us to particular times and places in the past.

It has been well-known from antiquity that music has the power to heal by modulating background states and enhancing emotional responses.[10] It has been equally well-recognized that music creates solidarity between individuals, perhaps by providing a shared focus of attention as well as a common structure of temporal experience.[11] According to Deryck Cooke music has such command over the human spirit because it is the closest thing we have to a genuine language of the emotions.[12] Anyone interested in manipulating the strings of human nature would do well to begin with music.

Like most elements in Judaic tradition, the emphasis on vocal music has links to a preoccupation with scripture. The Talmud scorns unsung readings of the Torah, and Kabbalistic tradition teaches that the gates of heaven can be opened only by song. Every word of the Hebrew Bible has been assigned musical notation, a convention nearly as old as the canon itself.[13] Musical embellishment of any text carries a potential to guide and intensify emotional responses. Imagine viewing *Psycho* or *Shindler's List* without the emotional tutoring provided by the music. The melodic intonation of biblical texts exploits our emotional competencies in the same way. Jewish cantorial art has been careful about selecting modal forms to accompany biblical texts. A variety of leitmotif structures evolved for chanting scriptures, each mode intended to evoke appropriate moods and responses.[14]

In Hasidic tradition music was essential, even to the point of rivaling scripture in importance. Chant modes, cantillation motifs, and hymn themes were freely elaborated and embellished. As befits mysticism, Hasidic music emphasized spontaneity and creative improvisation. The soul, it was thought, could not take flight without music. Some Hasidic groups dispensed with texts altogether, relying exclusively on musical effects. The soul was elevated to higher levels of ecstasy by progressive stages of melody and movement.[15]

There is obviously more to say about the use of aesthetic strategies in promoting Jewish piety. A thorough treatment would consider the role of architecture, painting, manuscript illustration, and various forms of symbolism and craftwork. But it was the poetic quality of vocal music that dominated the aesthetic dimension and reached into the hearts of pious Jews. It was this tradition that provided vital spiritual resources through the centuries, resources that could be tapped into when circumstances demanded.

A forceful testimony to the depth and adaptive value of this aesthetic tradition is found in Gila Flam's book *Singing for Survival*, an account of how vocal music enabled Jewish life and hope in the Polish ghettos during World War II. Song provided resources for both personal and collective endurance:

> So I was without a job, without school, and without friends and people to keep me company. A human being always strives for something. So I found the songs. . . . At that time I did not understand the significance of these songs. Now I do; when one listens to a song he escapes from everyday life, he escapes from his despair.[16]

During weekdays we sat together, and in the evenings we sang songs by the light of a small burner. Every meeting we learned a new song. . . . [We] sang many songs in Yiddish, such as "Above my window, two doves are standing." This song is full of hope and romance, and this is what nourished us. Because of our singing we could spiritually escape the ghetto.[17]

RITUAL STRATEGIES

The classical Jew, in Neusner's phrase, is "intoxicated by faith in God." To be constantly mindful of God in this way means that the individual's goal hierarchy will be continuously monitored and reoriented to God's will, so that each encounter with the environment will be appraised and coped with in a manner that reflects Judaic piety. Ritual practices play an important role in this process of nurturing patterns of appraisal and coping.

The sanctification of time in Judaic tradition is nowhere more evident than in its ritual life. Ritual practices provide a framework for measuring the passage of time. One might even say that ritual is the ultimate timekeeper for the pious Jew. The passage of time in a day is reckoned by prayers; the week progresses from Sabbath to Sabbath; the year is clocked by seasonal festivals; and the lifetime of an individual is punctuated by rites of passage.

Prior to the destruction of the second temple in 70 C.E., Jewish worship life was centered on ritual sacrifices. But once the temple was gone, the focus of worship shifted to offerings of prayer in synagogues as well as in the home, both of which are considered sacred places. Daily prayers are an exceptionally important aspect of Judaic piety. Each day begins and ends in prayer, and prayers are sung before and after every meal. The morning prayer is the longest and, in a sense, amounts to a rehearsal of emotional virtues. The morning prayer renews the covenant relationship with God and includes expressions of praise, humility, gratitude, love, guilt, and hope for God's continued intercession. This prayer amounts to a daily reorientation of the individual's goal hierarchy and a preparation for the day's encounters. The evening prayer sums up the day and asks forgiveness for failures of obedience.

The week is defined by the Sabbath, a day of rest and celebration for God's act of creation. The Sabbath is "the day on which we are called upon to share in what is eternal in time, to turn from the world of creation to the creation of the world."[18] The Sabbath is observed in the home as well as in the synagogue. In the home, the Friday Sabbath meal is the best and most elegant meal of the week, and is usually adorned by prayers, candles, spices, wine, song, and much conversation about the religious life. The Sabbath continues on Saturday with a worship service in the synagogue, where families congregate for lengthy prayers and, especially, for readings from scripture.

The Sabbath is a special event, one to be both remembered and anticipated as the week wears on. More than anything else, the Sabbath is a sign of the covenant and a gift from God, to be received with praise and thanksgiving. When the Sabbath is faithfully honored, it leaves a residual sense of gratitude that is likely—because of its recent and frequent storage—to bubble into appraisal events throughout the week.

The major festivals of the Jewish calendar are the three pilgrimage festivals (Pesach, Shavuot, Sukkot) and the High Holy Days (Rosh Hashanah and Yom Kippur). The pilgrimage festivals are so named because these were the three occasions during biblical times when all the men of Israel were expected to be in attendance at the Jerusalem temple. These were originally agricultural festivals marking the spring (barley), summer (wheat), and autumn (fruit) harvests, but as fertility practices were displaced, these festivals were changed to commemorate decisive events in Israel's sacred history. Taken together, the pilgrimage festivals commemorate the Exodus, the most decisive and formative event of all. Pesach (Passover) recalls the escape from slavery in Egypt. Shavuot (Feast of Weeks) hallows the day of revelation when God gave the commandments to Moses on Mount Sinai. Sukkot (Festival of Booths) remembers the hardships of the wilderness when the vulnerable refugees lived from hand to mouth and slept in makeshift huts.

These annual rituals are more than mere commemorative events. They are occasions for reliving the past, designed to educate the emotions by recreating the actual experiences of the Exodus. If these rituals are even slightly effective, then worshippers will simulate the experience of absolute dependence on God that was felt by the runaway slaves, and will consequently feel a deep sense of gratitude for God's acts of kindness. Ritually induced experiences like these influence appraisal events by reorienting goal hierarchies and enhancing the emotional valences of memory objects.

The High Holy Days (also called "awe-inspiring days") are given in Judaic tradition as a ten-day ritual event focused on self-examination and spiritual renewal. The cycle begins on Rosh Hashanah, the day of the Jewish New Year. This day commemorates the creation of the world, but it is also a solemn day of moral reflection and repentance. According to Talmudic tradition, three books of judgment are opened on Rosh Hashanah: one for the perfectly righteous, a second for the completely wicked, and a third for the vast majority of human-kind who fall between these extremes. The perfectly righteous are immediately acquitted and their reward is sealed. The completely wicked are immediately condemned and their punishment is likewise sealed. But the vast majority are held in suspense—neither convicted nor acquitted—until Yom Kippur, the Day of Atonement.

On Rosh Hashanah the ram's horn is sounded as a wake-up call for Jews to engage in serious moral reflection: "You who are asleep, wake up! Search your deeds and repent. Look into your souls, you who indulge all year in trifles. Amend your ways; let each one of you give up his evil course and purpose." [19] For the next ten days individuals are left to think about the kinds of lives they have been living during the past year, and to commit themselves to better lives in the year ahead. This is also a time for asking and granting forgiveness among fellow human beings for wrongs done in the past.

And then, on Yom Kippur, the reckoning comes. This is the most solemn of all Jewish holidays, a day of fasting, continuous prayer, and self-disclosure. On this day penitents make a full confession of their sins: "We are guilt laden, we have been faithless. . . . We have committed iniquity, caused unrighteousness, have been presumptuous. . . . We have counseled evil, scoffed, revolted, blasphemed." [20] And finally, with repentance complete, comes a plea for redemption: "May it be your will that I may no more sin, and forgive the sins I have already committed in your abundant compassion." [21]

When sincerely undertaken, this ten-day preoccupation with moral self-examination makes a profound contribution to educating the emotions. This is no glib exercise in making New Year's resolutions; it is rather the utterly serious business of moral and spiritual transformation. The whole point of these High Holy Days is to dismantle and reconstruct one's goal hierarchy before God and community. The process involves much humility, anxiety, and guilt, but it also inspires gratitude, love, and hope. The whole pattern of Judaic piety is revitalized by these intense ritual means.

Religious traditions commonly ritualize the salient events of an individual's lifetime, and in this way they invest personal biographies with mythic significance. The events typically marked by rites of passage are birth, puberty, marriage, and death. In Judaic tradition newborn children are received into the social group by a name-giving ceremony. For girls, the name-giving takes place in the synagogue on the first Sabbath following the birth. Boys are given names on the eighth day after birth as part of the rite of circumcision, an unforgettable tangible sign of God's covenant with Israel.

Passage into adulthood is recognized in the Bat Mitzvah ceremony for girls (age twelve), and the Bar Mitzvah ceremony for boys (age thirteen). From this rite forward, youths are considered "sons and daughters of the commandment," falling under the full obligations of the Torah. Much more than a simple recognition of puberty, these events are solemn occasions for reflecting on the personal freedoms and moral responsibilities that come with the adult life, and for committing oneself to a goal hierarchy. The special recognition conferred by the ritual ceremony makes it memorable, thus biasing appraisal events in favor of the traditional goal hierarchy.

Marriage, too, is intended to be a memorable occasion for moral reflection and commitment to traditional values. As a couple joins in marriage, they undertake new goals with a publicly declared commitment to serve them faithfully. For the young couple a wedding ceremony is an emotionally poignant event that will be recalled annually, revitalizing the personal commitments. For the social group a wedding restores hope for the continuation of God's chosen people.

Judaic tradition treats death in a simple and straightforward manner. The body is washed and buried within a day or two—no embalming, no cosmetic adornment, no open casket, no wake. Mourners remain at home for seven days and continue to offer special prayers for eleven months. If death is anticipated, then the individual will make a full confession of sins, using parts of the liturgy for Yom Kippur. The practice of issuing an "ethical will" is especially relevant to the process of educating the emotions. For many centuries inheritance of worldly goods was regulated by civil law, so last wills and testaments were uncommon among Jews. More typical was the preparation—sometimes well in advance—of moral instructions to be left to descendants. It was not uncommon for these ethical wills to undergo several revisions, finally emerging as polished moral treatises. I find it difficult to imagine that any genre of moral discourse could carry more weight or be more emotionally significant than the deathbed wishes of parents for the moral character of their children.[22] Here is an excerpt from a typical ethical will:

> My children. . . . Beware of deceit and envy. Work diligently and acquire wisdom. Lead a life of sincerity. Have pity on the poor and the weak. Be compassionate toward all persons and animals. Anger is blind and does not permit one to see the face of another as it really is. Do not become angry when someone speaks against you; do not become vain when you are praised. Speak the truth to your neighbor, and love each other with a true heart. Hatred is evil; it makes small things appear great. Put envy out of your souls, and love one another with singleness of heart. Be patient with one another's faults and overlook them. A good man shows

mercy to everyone, even to sinners. He is neither envious nor jealous of others, but rejoices always in their good fortune.[23]

INSTITUTIONAL STRATEGIES

Religious traditions have developed a range of symbolic strategies for nurturing and modulating behavior mediation systems. Rational discourse, extraordinary experiences, artistic forms, dramatic reenactments—each plays a role in educating the emotions, thereby to enhance the prospects for achieving personal wholeness and social coherence. To these we may add institutional authority—that is, the legitimate use of power to influence the willful behavior of individuals.

The institutional dimension of classical Judaism was shaped under the conditions of political domination. For twenty-five centuries Judaism persisted merely as an ethnic and religious reality, not a political one. Wherever Jews lived—Palestine, Africa, Asia, Europe—they were caught between their accountability to the dominant social order and their distinctive identity as God's chosen people. Under these conditions Jewish leadership came to have a dual role. In one role they were the official point of contact between the Jewish community and their host government, and in another role they were expected to regulate the internal affairs of the local Jewish community. This dual function had already taken shape during the period of the second temple (536 B.C.E.–70 C.E.), when the high priests of the temple were in command.

By the first century C.E., however, the authority of the old order priesthood was giving way to the vigorous moral leadership of a rising class of religious scholars. This transition of authority from priest to scholar, from temple to synagogue, was sealed with the destruction of the temple in 70 C.E. From the first century forward, Judaism became synonymous with the supreme and almost unquestioned authority of the rabbis.

The institutional pillars of Rabbinical Judaism were the synagogue and the academies. The synagogue—simultaneously a house of prayer, assembly hall, meeting place for business people, courthouse, charity center, and elementary school—emerged as the nerve center of Jewish social life. The day-to-day operations of the synagogue normally fell to wealthy elders of the community. The rabbis prepared, appointed, and advised the judges and local administrators who supervised activities and legal proceedings in the synagogue. The academies were centers of education and legislation, which is to say they were places where the Torah was carefully studied and interpreted for purposes of social policy and moral instruction.

It is no exaggeration to say that the institutions of classical Judaism saw to it that local communities were literally saturated with the authority of the Torah. All decisions regarding personal and social life—including education, worship, legal and moral duties, business, neighborly relations, marriage, child rearing, healthcare, work and leisure—were carefully scrutinized for their compliance with God's word. To say that these institutions had a profound bearing on the goal hierarchies and working memories of individuals would be an understatement.

PERSONAL WHOLENESS AND SOCIAL COHERENCE

We humans maximize our potential for survival by creating the conditions for simultaneously achieving personal wholeness and social coherence—that is, by achieving a high degree of social solidarity and cooperation without sacrificing opportunities for individuals to satisfy their own desires. This can be achieved by creating overlaps of self-interest, so that when individuals advance their own interests they coincidentally advance the interests of others. And these conditions, I have argued, arise from the process of linking self-esteem to prosocial outcomes.

In the deepest sense this is what Judaism is all about. If I achieve a sense of enhanced self-worth whenever my behavior conforms to the Law, and if Law-abiding behaviors actually do make contributions to the well-being of others, thereby stabilizing the social mobile, then the conditions for human viability will be met and Judaism will have done what it was designed to do.

Much has been said in this chapter about the importance of the Law, but only now do we come to the point of asking whether and how the Law makes a contribution to personal wholeness and social coherence. Everything depends on the details of self-esteem links. Nothing is gained if I link my self-esteem to obeying the Law, but then do not know what the Law entails. And nothing is gained if obeying the Law does not result in overlaps of self-interest. This is why a proper understanding of the Law is so heavily stressed in Judaic tradition. But expert knowledge of the whole Law is an overwhelming undertaking, well beyond the practical reach of most pious Jews. At a minimum, all Jews are expected to link their self-worth to the core and the essence of the Law. The core of the Law is embodied in the Ten Commandments, which every Jew is expected to know by heart and to take to heart. The self-esteem of pious Jews will be linked to performing well against these commands, which can be summarized as follows:

1. Worship of gods other than Yahweh is prohibited.
2. Worship of idols is prohibited.
3. Making wrongful use of God's name is prohibited.
4. The Sabbath must be honored and observed.
5. Parents must be honored.
6. Murder is prohibited.
7. Adultery is prohibited.
8. Theft is prohibited.
9. False witness against one's neighbor is prohibited.
10. Coveting anything belonging to one's neighbor is prohibited.
    (Exod. 20:1–14; Deut. 5:6–18)

Abiding by the core of the Law is a necessary condition for Judaic piety, but it has never been considered sufficient. One must go deeper, to embrace the essence of the Law. The essence of the Law is found in the prophetic admonitions to do justice. "The Lord is a God of justice" (Isa. 30:18), who delights in "steadfast love, justice and righteousness in the earth" (Jer. 9:24), who wants humans to "hold fast to love and justice" (Hos. 12:6), and to "let justice roll down like waters" (Amos 5:24). Micah best sums up the essence of the Law in this familiar passage: "He has told you, O mortal, what is good; and what does the Lord require of you but to do justice, and to love kindness, and to walk humbly with your God?" (Mic. 6:8).

Daniel Maguire has observed that the biblical ideal of justice (*sedaqah*) bears little resemblance to the concept that has dominated Western culture since antiquity.[24] To the Western mind justice is dispassionate, unbiased, blind to circumstance, and driven to protect the universal rights of individuals. But to the biblical mind justice is impassioned, heavily biased, informed by circumstance, and driven to achieve social solidarity and peace. In sum, Western justice amounts to rights-based impartiality, while biblical justice reflects needs-based partiality. The biblical concept of justice is biased against the "haves" in favor of the "have nots." Whereas Western justice responds to evil deeds, biblical justice responds to human needs. The central importance of human needs, relative to individual rights, is made abundantly clear in the coverage of case law in the Torah (Deut. 12–25). To do justice, therefore, means to care for the needs of the poor, the sick, the orphans, the aliens, the widows, the homeless, the hungry and the oppressed. Where these needs are unmet, justice is not served, and where justice is not served, peace will not prevail, for "the effect of justice will be peace" (Isa. 32:17).

On the surface it may appear that these actions on behalf of social justice call for personal sacrifices, but in fact they do not. To the contrary, the call to do justice is an opportunity to advance personal wholeness and social coherence at the same time. Again, everything depends on self-esteem links. If my agenda is God's agenda, then my self-esteem, my personal fulfillment, will be achieved only by doing what is just. In liberating the oppressed, in feeding the hungry, and in sheltering the homeless I find occasions to make myself whole. When our self-esteem is linked to justice, then any sign of injustice to you will be appraised as a personal threat to me, and any injustice to me will be a threat to you. In this way, overlaps of self-interest are constructed by making prosocial self-esteem links.

This principle works with emotional virtues as well as behavioral virtues. If your self-esteem is linked to steadfast love, walking humbly with your god, and not coveting the fortunes of others, then whenever contrary feelings begin to swell up, you will sense a discrepancy between what is and what should be, and this will prompt you to reappraise the business at hand, perhaps safely before your behavior gives you away.

The ultimate objective of the myth of covenant blessing, together with its ancillary strategies, is to bring individuals to the point of making these prosocial self-esteem links. And in the end, Judaic wisdom is committed to the wager that herein lies the salvation of our species.

CHRISTIANITY

HISTORICAL CONTEXT

Jesus of Nazareth was born at a time when many Jews were giving up on history. Too often, and for too long, they had been rudely disappointed by historical events. Disappointments became serious in 586 B.C.E., when Jerusalem was destroyed and the Jews were exiled to Babylon. Fifty years later they were liberated by the Persians and allowed to return to their homeland, only to spend the next five centuries in and out of vassalage to a series of foreign rulers.

The worst humiliation came in the mid-second century, when the Seleucid monarchy outlawed the practice of Judaism and established Hellenistic cult worship in the restored Jerusalem temple. This indignity provoked a successful Jewish revolt in 166 B.C.E., bringing the Jewish community a hundred years of relative autonomy under the leadership of the Hasmonean dynasty. But the ostensibly Jewish Hasmoneans were hardly more Jewish or any less brutal than Seleucid rulers.

By 63 B.C.E. bitter resentment had given way to civil unrest, forcing a beleaguered Hasmonean leadership to seek assistance from the Romans, who promptly took advantage of the situation to declare Palestine an imperial province. Roman policy only worsened religious, political, and economic strife in the Jewish community. For post-exilic Jews, history had become "a descent into hell." [1] It is no wonder, then, that many Jews were tempted to look beyond history for the promise of covenant blessing.

Resources for transhistorical hope came in the form of radical new ideas about nature and history. Following the Seleucid desecration of the temple, many Jews were especially drawn to the apocalyptic worldview, a peculiar amalgam of pagan, Persian, and Hellenistic thought. The key elements in Jewish apocalypticism included a dualistic cosmology, claims to revealed truth, a preoccupation with eschatology and messianic speculation.

Cosmological dualism. Apocalyptic thought makes sharp distinctions between this world and the other world, nature and supernature, history and eternity, matter and spirit, earth and heaven—that is, it creates a dualistic cosmology. The heavenly world is ultimately real, while the earthly world is derivative and transitory. The earthly world of nature and history is not an illusory realm, but it is finite and in every way contingent on the heavenly world. Events in the earthly world are mere effects of primary events taking place in the heavenly world, in the way a puppet show is determined by an off-stage puppeteer. Heaven is populated by spiritual beings who are locked in a cosmic battle of good versus evil. The earthly realm is temporarily under the control of the Evil One and his angels, but God is soon to put an end to evil—including nature and history—and will then establish an eternal realm of goodness and light.

Revelation. The primary events of heaven are completely inaccessible to earthly beings by any natural means. The only way earthlings can know about heavenly events is through the gift of apocalypse (Greek: "revelation"). Apocalyptic literature therefore claims to record revelations from the inaccessible world, revelations that are exceedingly important because their highly symbolic imagery portends events about to unfold on earth. As if drawn up into heaven for a comprehensive preview of the future, the apocalyptist writes a symbolic guide to help earthlings prepare themselves for the momentous events to come. For those who are uninitiated to the revelation, the realm of history will remain clouded and bewildering, frightful and discouraging. But to those fortunate enough to receive the heavenly guide, the events of history will be suddenly clear and intelligible, full of hopeful signs of glories to come.

Eschatology. From its beginnings Judaic tradition affirmed that history has a point, which helps to explain why the relentless disappointments of the post-exilic period were so hard for pious Jews to comprehend. How could God's promise of covenant blessing be reconciled with the discouraging facts of history? Apocalyptic thought provided the ultimate explanation, pressing the meaning of history to its extreme: The point of history is its climactic end, the *eschaton.* Far from being absurd and random, history is driven by a tightly scripted plan. The outcome is certain: At the appointed time the forces of goodness and light will prevail over darkness and evil, and curtains will then be drawn across the stage of history. There is no cause for anxiety or alarm, for God's promise of blessing is soon to be fulfilled beyond all imagination. All that remains is for the pious to stand fast and endure the decisive last days in strict obedience to God's will. This faithful remnant will be richly rewarded in the new heavenly order. Pre-exilic Judaic belief stressed that salvation was terrestrial, communal, and historical, but for Jewish apocalypticism the unit of salvation was the individual soul, and the destiny of the saved was the transhistorical realm of heaven.

Messianism. The messianic theme within Jewish apocalyptic was perfectly consistent with ancient Judaic tradition. Throughout Israel's history God was seen to make frequent use of special agents who were anointed to execute some particular function. In Jewish apocalyptic literature "messiah" ("anointed one") serves as a title for the agent(s) who will preside over the climax of history and inaugurate the transhistorical realm. There were, however, several speculative variations on the messianic theme. Some groups expected a prophet, others expected a cosmic figure with metaphysical credentials, others expected a priest, others a king in the Davidic lineage, and still others expected a series of messiahs to play out specialized roles.

Apocalypticism represents the kind of edgy thinking that engenders extremist movements and creates sharp divisions within communities. This was certainly the experience of the Palestinian Jewish community during the heyday of apocalyptic, from the mid-second century B.C.E. until its rapid decline in the second century C.E. Long before Jesus' birth, Judaism had splintered into a spectrum of entrenched parties, each with a distinctive diagnosis of current events and a prescription for the future. The Sadducees were old-order priestly Jews, the unpopular aristocrats of Jerusalem society whose principal concern was to preserve the status quo. They rejected the new-fangled ideas of apocalyptic, including belief in personal immortality, and held fast to a literal interpretation of the Torah.

The Pharisees were a populist party, standing in the tradition of the ancient scribes. They were mostly active in synagogue life, where they were widely respected as scholars and teachers. The Pharisees accepted much of the apocalyptic worldview, including a variation of cosmological dualism, personal immortality, and hopes for a Davidic messiah. They stressed meticulous observance of the law as a condition for salvation.

The Zealots had much in common with the Pharisees (from whose ranks they originated), but they were convinced that there was nothing to gain by waiting in purity for God's intervention in history.

They were militant activists and apparently believed that their role in history was to initiate the great cosmic battle by taking up arms in a holy war against the Romans. God was expected to join in the battle at the decisive moment to bring about the final denouement.

The Essenes appear to have originated as disenfranchised priests who broke away from the corrupted temple to lead the monastic life of study and ritual purity. They expected two messiahs: first a priestly messiah who would sanctify them for the end times, and then a warrior messiah who would lead them in a holy war against the Gentiles, marking the beginning of the end.

The Christians were yet another Jewish apocalyptic group, having much in common with the Essenes, although they were predominately lay people who rejected the concept of holy war. The Christians believed that their crucified teacher, Jesus, would return from heaven transformed as the messianic Son of Man to execute the final judgment. As they waited for his return, the Christians exhorted one another to care for the needy and spread the word of repentance and salvation to the ignorant.

Like other apocalyptic sects within Judaism, Christianity originated as a spiritual endurance program in response to conditions of intensifying religious, political, and economic oppression. This was a period of increasing uncertainty, suspicion, and anxiety, when the conditions for personal wholeness and social coherence were breaking down. Christians waited in earnest for the imminent return of the messiah, who was expected to bring about the cataclysmic end of history. These expectations were obviously misconceived, leaving the Christian community stranded on earth to face the social and spiritual challenges of historical existence. As they came to terms with these challenges, a distinctive myth of God incarnate took shape.

THE MYTH OF GOD INCARNATE

The challenges of historical existence became palpable to the Christian community toward the end of the first century. A decisive turning point centered on the events of 66–70 C.E., when a Jewish uprising brought a devastating response from Roman forces. Jerusalem was besieged, the temple was destroyed, and several Jewish groups—the Sadducees, the Zealots, the Essenes—were routed. Leadership of the Palestinian Jewish community fell to a group of Pharisees who established an academy at Jamnia, where they labored for the elements of a self-understanding that would ultimately reunify Judaism.

The Christians, meanwhile, were left with identity problems of their own. What was the nature of their claim to be the People of God now that they were being rejected by the new leaders of Judaism? Why had Jesus failed to return in victory as expected? What was the real meaning of Jesus' life, work, and death? How should they understand their belief that God was active in the person of Jesus? What is expected of them now? The myth of God incarnate grew out of attempts to answer such questions.

The Christian myth was given a definitive summary in the Nicene Creed, as follows:

> We believe in one God the Father All-Sovereign, maker of heaven and earth, and of all things visible and invisible;
> And in one Lord Jesus Christ, the only-begotten Son of God, Begotten of the Father before all the ages, Light of Light, true God of true God, begotten not made, of one substance with the Father, through whom all things were made; who for us men and for our salvation came down from the heavens, and was made flesh of the Holy Spirit and the Virgin Mary, and became man, and was crucified for us under Pontius Pilate, and suffered and was buried, and rose again on the third day according to the Scriptures, and ascended into the heavens, and sitteth on the right hand of the Father, and cometh again with glory to judge living and dead, of whose kingdom there shall be no end;
> And in the Holy Spirit, the Lord and the Life-giver, that proceedeth from the Father, who with Father and Son is worshipped together and glorified together, who spake through the prophets;
> In one holy Catholic and Apostolic Church;
> We acknowledge one baptism unto remission of sins. We look for a resurrection of the dead, and the life of the age to come.[2]

The centerpiece of the Christian myth is the assertion that Jesus is Lord. Initially this claim probably meant only that Jesus was an agent of God, the Messiah (that is, the Christ) who was chosen to fulfill a special apocalyptic role. But the identification of Jesus with divine reality soon took on metaphysical dimensions as well as historical ones. Jesus was the embodiment of divine wisdom, the metaphysical Son of God, a heavenly being who entered into nature and history by supernatural means. He was the ultimate self-disclosure of divine reality in human form. Christian doctrine made the extraordinary claim that Jesus was fully human, yet he was "of one substance with the Father."

This bold identification with the Lord of nature and history is the key to understanding the remarkable events of Jesus' life, as well as the hopes invested in him by the Christian community. The point of the incarnation was to bring God's purpose in history to fulfillment—that is, to enact the final judgment of human righteousness and to bless eternally those who remained faithful to God's will. The incarnation was God's ultimate act of grace. Repeatedly throughout Israel's checkered history God had intervened to act graciously on behalf of his chosen people, despite their failures to fulfill the conditions of the covenant.

But now, in the Christ event, God was settling Israel's account once and for all—the entire balance of obedience owed to God was being forgiven, the slate of all sins was being wiped clean by God's boundless act of grace. Herein lies the explanation for Jesus' death: As the wages of sin are death, the Messiah had to pay them with his own death. Jesus went to the cross burdened with the sins of all humanity, there to satisfy all the terms of the covenant, there to atone for every last sin of disobedience. Now, finally, the temple and all its fastidious obsession with laws and rites of purification was rendered obsolete.

The resurrection of Jesus was the decisive validation that the wages of sin had been paid—that the check, so to speak, had cleared the accounts. Jesus had returned to heaven as the first fruit of the harvest of salvation. The path was now open for all humans to enter freely and humbly into the Kingdom of Heaven. The only condition was a spiritual one: to recognize and to accept these momentous events for what they were, and to express faith in the saving grace of God in Christ. Expressions of Christian faith normally included baptism and participation in communal meals of celebration, but they also included willful acts of Christ-like charity and an eagerness to share the good news of salvation with others.

The Christian myth forecasts a victorious return of Jesus to resurrect the dead and bring the final cataclysmic end to nature and history. In post-apocalyptic understandings, however, this final denouement was deferred to an indefinite time in the future. The second coming is still a sure thing, but its timing is not. Meanwhile, Christians are left in the era of the historic church to follow the guidance of God's Spirit in bringing the gospel of salvation to the world. The Christian mission brought a new sense of urgency to the universalist vision of the prophet Isaiah. God's forgiveness and blessing applies to all humanity, no less to the Gentiles than to the Jews.

Previously, I offered an operational definition of realism, suggesting that one may be considered a realist about some X (an entity, event, property, or relation) if cognitions involving X have the power to produce decisive effects on one's appraisal processes. The Christian myth cannot function properly in the absence of realism about its most central claims. Like Judaism, Christianity presumes realism about its root metaphor, God as person.

The critical difference is that the Christian understanding of God is conditioned by the language of incarnation. Pious Christians are theological realists, and they are also realists about the two natures of Christ—that is, he was simultaneously human and divine, a historical figure and a heavenly being. Christianity also presumes realism about the atonement, the connection between the person of Christ and the salvation of the world. Realism about the atonement implies realism about the crucifixion and resurrection of Christ, the sine qua non events of the Christian proclamation: "If Christ has not been raised, then our proclamation has been in vain." (Rom. 15:14)

The efficacy of the Christian myth also depends on realistic attitudes concerning cosmological dualism and the eschaton. The earthly realm of nature and history is a temporary instrument for realizing the eternal kingdom of heaven (and the eternal thralldom of hell). The world of nature and history is therefore subordinate to a higher, perfect order of reality. At the appointed time Christ will return to the earthly realm for the final judgment and transformation of creation. Individuals with a compromised sense of realism about these features of the ancient apocalyptic worldview will be less susceptible to the emotional effects aimed for by the Christian myth.

EMOTIONAL APPEALS

Christianity affirmed its spiritual continuity with Judaic tradition right from the beginning. Indeed, Christians declared that they, and not recalcitrant Jews, were the true heirs to the covenant blessing. Judeo-Christian continuity is most tangible in the fact that Jews and Christians share a common spiritual resource in the Hebrew Bible. The authority of these ancient scriptures remained intact for the authors of the New Testament, who would have seen no point in rewriting the sacred history of God's dealings with the world. Moreover, the New Testament writers continued to address the same rhetorical task undertaken by the ancient scriptures—that is, to bring individuals to transform their goal hierarchies to conform with God's will. The central message remained the same: repent and obey!

Still, there was much more to say, since God's dealings with the world had taken a dramatic and decisive turn in the Christ event. The parting of the ways between Judaism and Christianity was a complicated affair, marked by several factors that help to explain both subtle and dramatic differences between their patterns of piety. Whereas Judaism recovered from its flirtation with apocalyptic thought, Christianity did not. Apocalyptic emphases on cosmological dualism, eschatology, revelation, and messianism remained central to the Christian myth, even as these underwent various metaphysical reinterpretations. Rabbinical tradition, meanwhile, withdrew from metaphysical speculation to concentrate on the moral life. That is, Judaic piety concerned itself with orthopraxis (right conduct), while Christian piety focused on orthodoxy (right belief).

Another important difference is that Christianity developed a universal mission, investing much of its energy in the Hellenistic world, while Judaism continued to foster nationalistic sentiments. A correlative difference is that Christianity became more explicitly individualist, whereas Judaism retained shades of collectivism. In other words, for Christians salvation was thought to be inclusive and individually realized, but in Judaism the divine promise was intended for an exclusive group.

Far more important than these, however, was a fundamental difference in the centering of the two traditions: Christianity became radically christocentric while Judaism remained steadfastly nomocentric. For Judaism God was incarnate in the Law, but in the Christian myth God was incarnate in the person of Christ. The contrast was not merely incidental, for it entailed profound differences in the imaging of divine reality. That is, whereas Jews and Christians continued to proclaim the same God, they differently imagined what God was like and what God willed. For Christians God was like Christ—indeed, God *was* Christ—and God's will is that humans become Christ-like.

The emotional appeal of the christocentric myth represents a shift from history to biography. The narrative of the Hebrew Bible engages the emotions in response to great sweeping events in history—the Abrahamic saga, the bondage in Egypt, the Exodus and wilderness experience, the fulfillment of God's promise of nationhood, the exile and restoration, and so on. The Christian myth does nothing to displace any of this, but its distinctive appeal is concentrated in images of Christ's birth, ministry, suffering, death, and resurrection. These biographical images seek to shape a prosocial piety among Christians by appealing to capacities for sympathy, guilt, gratitude, hope, fear, and love.

SYMPATHY

The Christian myth exploits a prosocial predisposition to feel sympathy for the sorrow and suffering of other persons under certain

circumstances. When the suffering person is appraised as blameworthy for a negative outcome, then a sympathetic response is normally
blocked, but where there is no attribution of blame, sympathy is
likely to result. In the gospel narratives Jesus is portrayed as a kind
and gentle man, a blameless and willing servant of God. Yet this innocent man was unjustly accused and brutally tortured, left helpless
and forsaken to die on the cross. The normal response to this scene
is sympathy, for which the implicit action tendency is to step forward
and take measures to alleviate the suffering. It is in the nature of sympathy to want what the victim wants, namely relief for the victim.

   Where no action directly relevant to the suffering is possible, the
sympathizer is rendered helpless but still aroused to do something for
the sake of the victim. In the case of anyone moved to sympathy by
the crucifixion, there is no question of stopping the suffering, but
there remains a next-best thing, which is to validate the suffering by
doing what Jesus would want done. The imagery of Christ's suffering
and death, therefore, has the effect of predisposing the sympathizer
to serve the will of God.

GUILT

Guilt arises when individuals transgress what they have internalized
as moral imperatives. When I do—or even wish to do or accidentally
do—what I believe to be wrong, then I am likely to feel guilty. Moreover, the intensity of guilt increases to the extent that other persons
are brought to harm by the transgression. I am likely to feel mildly
guilty if I carelessly run a red light, but I am certain to feel intensely
guilty if my transgression causes an accident. The action pattern
for guilt is to take sacrificial measures that might compensate for the
transgression. These, of course, will vary with the circumstances but
might include an apology, performing a favor, seeking punishment, or
even taking measures to reform one's life.

   Feelings of guilt are unavoidable for anyone who has internalized
social standards because we often find ourselves faced with moral
dilemmas forcing us to transgress one value for the sake of another.
Christian tradition has used the language of original sin to explain
universal guilt (by Adam's sin we are born guilty), but it seems more
appropriate to say simply that guilt is an inevitable consequence of
socialization.

   The Christian myth exploits guilt even more aggressively than
does Judaic piety. Not only are humans blameworthy for generations
upon generations of disobeying God's law, but they are additionally
to blame because their transgressions were the occasion for the inno-

cent suffering and death of Jesus: "Because of our sins he was given over to die." (Rom. 4:25) The measures one is prepared to take in atoning for such a burden are likely to be extreme, and indeed, the measures aimed for by the Christian myth fall nothing short of a radical transformation of one's life.

GRATITUDE

The role of gratitude is no less important to Christian piety than it is to Judaic piety. Recall that individuals are genetically predisposed to feel grateful when they receive a gift or favor from another, and this experience arouses them to expressive actions such as offering thanks or reciprocating the favor. And recall that the intensity of gratitude (and therefore the magnitude of response) will vary with the degree to which the receiver expects or deserves the favor. We have already observed how the Judaic myth boosts the intensity of gratitude to God by insisting that the saving events of the Exodus and the settlement in Canaan were wholly undeserved.

The Christian myth brought a new interpretation to the sinful and undeserving nature of humanity. St. Paul's view is that Adam's disobedience infected human nature with sin, a sort of metaphysical virus that renders us incapable of doing God's will, and therefore intrinsically undeserving of God's blessing.[3] This account of human nature makes Israel's dismal historical experience more intelligible, and it also heightens the sense of gratitude that comes with receiving the free gift of salvation through the Christ event. The pious Christian, aroused to gratitude, responds in acts of worship and service to God's will.

The element of sacrifice inherent in the Christian myth achieves an additional emotional boosting effect. Consider that the intensity of one's gratitude will vary with the value of the gift or favor, and that several variables may factor into the evaluation. One of these factors is the sacrifice made by the donor: the greater the sacrifice, the deeper the gratitude. I have a colleague who displays on his wall a jigsaw puzzle that was assembled and framed for him by a former student. Not a very impressive gift, until one considers that the student was blind. A similar gift from a sighted student would have inspired far less gratitude. Now consider the astonishing claim made for the gift of salvation by the Christian myth: The benefactor sacrificed the life of his only son in order to bring this gift to a selfish and undeserving beneficiary. This is what one might call a story of extreme grace, which is intended to provoke an act of extreme reciprocity: "So then, my brothers, because of God's great mercy to us I appeal to you:

Offer yourselves as a living sacrifice to God, dedicated to his service and pleasing to him. This is the true worship that you should offer." (Rom. 12:1)

HOPE

One might suspect that the Christian myth was crafted for no other reason than to inspire hope, for the dynamics of hope are epitomized in the Christ narrative. If hope is a desire for a positive outcome despite negative prospects, then what better images could one find than the crucifixion and resurrection of Christ? Here we have a story about a great liberator, the Messiah, the promised one who embodies promise. And what becomes of him? He is ridiculed, harassed, whipped, strung up, and murdered. How bleak could things get? The prospects for a dead man to lead God's people to victory are grim indeed. By comparison, the odds favoring a band of runaway Egyptian slaves to become a great nation look pretty good. But death is not the end of the story, for the Messiah is restored to life and ascends to heaven in glory, there to prepare an eternal home for the faithful. The difference between outlook and outcome could not be more astonishing. Christians encountering this story would be fortified with hope.

The New Testament writers understood the logic and the importance of hope. Like the exilic prophets before them, they were addressing the needs of a community faced with dire prospects, a community on the brink of being snuffed out by formidable obstacles, a community with no reason to expect anything but the worst. Thus the New Testament authors presented the suffering, death, and resurrection of Christ as a metaphor for the Christian life. Fortified by the gospel narrative, Christians could cope with anything—humiliation, defeat, ridicule, persecution, even fear of death—for they have the assurance of Christ's victory to show that the misfortunes of nature and history are never the end of the story for those who walk with God.

The action tendency associated with hope is to endure and to remain steadfast in a positive course of action. Hopeful Christians do not weaken and fold, they do not withdraw from commitments or abandon their companions. Rather, they stiffen their resolve, deepen their commitment to serving God's will, and offer encouragement and support to others.

As mentioned earlier, hope is a coping strategy, an attempt to manage an emotionally distressing encounter by reappraising the situation at hand. When an appraisal triggers fear, anxiety, or sadness, the pious Christian may thwart the negative emotional response by reappraising the encounter in light of the ultimate promise of salvation. The death of a loved one, for example, will result in grief and sadness, but these emotions themselves will be appraised by Christian piety as an occasion to cast one's thoughts beyond the grave and to dwell on the communion of saints in heaven. Indeed, any misfortune or defeat becomes an occasion to deepen one's commitment to serving God's will on the promise of eternal salvation.

FEAR

The promise of eternal salvation is balanced by the possibility of eternal damnation. Many Christians will insist that the saving grace of Christ puts an end to fear, and that anyone who associates fear with Christian piety just doesn't get the central point. This sounds good, but the fact remains that Christian tradition has exploited the fear response consistently and effectively. A central feature in the Christian myth is the final judgment to come at the end of history. Those who are found on the side of Christ will be saved forever, whereas the unfaithful will be damned to the eternal torments of hell.

Damnation remains a real possibility for everyone. Christians who have managed to live with complete assurance of eternal salvation have always been in the minority, whereas most Christians are easily moved to fear of hell. And there has been no shortage of inducements to fear. The New Testament envisions hell as a lake of fire and sulfur, where unrepentant sinners will be tortured day and night for eternity (Rev. 20:10). And Jesus says, in no less prominent a place than the Sermon on the Mount, that anyone who becomes angry or verbally abusive is "liable to the fires of hell." (Matt. 5:22) Who can remain assured in light of this? The action tendency associated with fear is avoidance, which may be achieved by running away or by submission and appeasement. Christianity has always insisted that the only realistic option is the latter. Fear, then, is a powerful motivator to repentance and self-reform.

LOVE

"I may have all knowledge and understand all secrets; I may have all the faith needed to move mountains—but if I have no love, I am nothing." (1 Cor. 13:2) Love is many things: a positive attitude toward persons or things, an intimate social relationship, a commitment to persons, projects or principles, and, of course, an emotional state. The appraisal pattern for conventional forms of love (romantic or companionate) always involves an assessment of the qualities of the potential beloved by the potential lover. The appraisal process asks, "Is there goal congruence here?"—which might mean one of several things, such as: Is he physically attractive? Does she like me? Has he done me a favor? Is she amusing? Does he share my values and interests? The threshold to a love response may be lowered by an affirmative appraisal on any of these items. If the answer is no on all counts, then there is no goal congruence and the love response is blocked. In other words, love may proceed only when the person encountered is appraised in some way as being lovable.

Christian love (*agape*) does not proceed according to this conventional appraisal pattern. Agape, we are told, is unconditional love: It does not depend on the lovability of the person encountered, and may proceed even when the conventional appraisal is entirely negative. Pious Christians are sincerely motivated to act on behalf of the needs of others regardless of their worthiness and regardless of the sacrifices involved in doing so. It should be noted that agape is not simply loving God, it is loving *like* God. Love for God unfolds by a conventional appraisal pattern—God loves us, provides for us, and saves us, so we reciprocate spontaneously in loving God, just as the genes have predisposed us to do. But loving *as* God loves—that is, loving the unlovable—requires more than the genes can deliver. To nurture agape, therefore, the Christian tradition has had to focus its attentions at the boundaries of the appraisal process, creating the conditions for lowering thresholds to love and enhancing the self-esteem linkages required for a love-biased coping process. For Christians to love as God does, they must be nurtured in their moods, attitudes, and goal hierarchies.

In part one it was observed that background feelings—moods for example—have a significant bearing on thresholds to emotional responses. Positive moods elevate thresholds to negative emotions while lowering thresholds to positive ones. The Christian tradition has consistently sought to enhance positive mood states by affirming God's certain and steadfast love for all persons. And as everybody knows, the ultimate mood enhancer is the sense of being loved. St. Paul provides the classic assurance of divine love:

> For I am certain that nothing can separate us from his love: neither death nor life, neither angels nor other heavenly rulers or powers, neither the present nor the future, neither the world above nor the world below—there is nothing in all creation that will ever be able to separate us from the love of God which is ours through Jesus Christ our Lord.

(Rom. 8:38–39)

The assurance of God's steadfast love also plays to the cognitive side of our nature by shaping our attitudes toward fellow creatures. As noted, Christians love God because God first loved them (1 John 4:19). This is a straightforward example of reciprocity: We are wired with a tendency to love those who love us. But Christian piety wants to nourish individuals to love even their enemies, to love indiscriminately without regard for the cost-benefit calculus scripted into the emotional systems. This may be accomplished by instigating the attitude that lovers of God must rightly love what God loves: "The command that Christ has given us is this: whoever loves God must love his brother also" (1 John 4:21). That is, anyone worthy of God's perfect love, which includes everyone, should ipso facto be appraised as worthy of a Christian's love. Thus it is expected that pious Christians will enter all their encounters with the prejudicial attitude that "any friend of God's is a friend of mine." If this positive attitude toward fellow human beings is consistently reinforced, then it will alter emotional thresholds and bias individuals in favor of charitable responses.

Even pious Christians are human, however, which means that their thresholds to positive emotions will not always be low enough to produce spontaneous charity. But even when encounters are appraised uncharitably, there remains the coping process, which enables a person to reappraise encounters in the course of an unfolding emotional event. This coping process requires the formulation of a self-esteem link to the performance of loving behaviors. Once such a link is formed an individual has the resources to short-circuit negative emotions in favor of genuine love. A Christian may find herself initially reacting in disgust to a street beggar because everything about the beggar screams goal-incongruence. But then she quickly appraises her own disgust as incongruent with her desire to be a loving person, whereupon her feelings toward the beggar begin to change.

Christian tradition nurtures this coping process by influencing goal hierarchies. Specifically, the tradition asks Christians to transform themselves completely by deleting their own goal hierarchies and replacing them with God's hierarchy. The language of the New Testament speaks of rebirth—that is, death to the old Adam and resurrection in the spirit of Christ, the new Adam. "When anyone is joined to Christ, he is a new being; the old is gone, the new has come." (1 Cor. 5:17) Here we are brought to the essence of Christian piety, which is to engage in the life of diligent self-monitoring. That is, pious Christians will appraise all their own thoughts, feelings, and actions in light of the image of Christ. The idea is that self-esteem becomes a monovalent process: The only good thing a Christian can do is serve God's will, as Jesus did. Anything else is sinful. Some

Christians even make a practice of wearing bracelets bearing the letters W.W.J.D. ("What Would Jesus Do?") to prevent lapses into un-Christlike thoughts and feelings. When negative feelings begin to emerge they ask what Jesus would do, and the answer is always the same: Jesus would do the loving thing.

ANCILLARY STRATEGIES

The social impact of the church's universal mission was nothing short of astonishing. At the end of the first century there were fewer than eight thousand Christians. A century later, there were nearly a quarter of a million, and by the end of the third century the church numbered over six million. The numbers surged following the conversion of the Emperor Constantine (313 C.E.), and by the end of the fourth century, when Christianity was declared the official religion of the empire, the number of Christians topped fifty million.[4] This remarkable pattern of growth followed the development of ancillary strategies for expressing, clarifying, defending, and transmitting the Christian proclamation.

INTELLECTUAL STRATEGIES

Christianity originated within Jewish Apocalyptic, a movement characterized by fervent proclamation and florid symbolism, but having little interest or expertise in systematic discourse. As Christianity grew more distant from Judaism, and as the appeal of apocalyptic began to wane, we observe the gradual emergence of the apologetic tradition of philosophical theology. The very earliest examples of Christian apologetic thinking are preserved in a few New Testament fragments,[5] but it was not until the second century that apologetics began in earnest, as Christians found a need to defend themselves against charges of godlessness and immorality.

Apologetic writers grasped the importance of adopting the language and thought forms of their accusers (usually Stoics, Skeptics, or Platonists) in order to formulate pertinent responses. As it turned out, however, the apologetic treatises were seldom read by the critics of Christianity but were assiduously studied by influential members of the Christian community itself. In practice, therefore, the disputations of the apologists played a formative role in educating the church to the rational discourse of philosophical theology, a mode of thought and expression that remains central to the church even today. It also turned out that the philosophical mode of thought was instrumental in sorting out disputes within the church, which was increasingly threatened by disunity as a result of rapid growth and intellectual diversity. The Christian theological tradition ripened to maturity with St. Augustine (354–430 C.E.), whose work was decisive in both answering external critics and extinguishing internal heresies. Augustine addressed virtually all the theological concerns of his age, leaving behind an ideological legacy that would remain normative in Christian culture for a thousand years.

The great strength of Christian theology has been the success of the tradition in demonstrating that its myth was systematically consonant with the best knowledge of the day. This would mean, of course, that mythic meanings would acquire all the cognitive force of genuine knowledge. A coherent theological vision can make nonrealism about the myth appear downright unreasonable. And even though Christian theologians have consistently insisted that faith is not knowledge, it remains true that an effective theology can sufficiently displace doubt so that faith is rendered neurobiologically indistinguishable from knowledge. This is the ultimate point of doing theology. When it comes down to the business of encounters and appraisals, an intellectually fortified faith is virtual knowledge.

The danger inherent in any theological tradition is the tendency to conflate theology and myth. That is, the cosmological and moral conventions utilized in theological formulations may become so inextricably associated with mythic images that the myth becomes dependent on the conventions. The trouble is that theological models and paradigms are routinely borrowed from contemporary sources and come laden with particulars that may not endure the tests of time.

For example, belief in miracles was commonplace during the first century when the myth of God incarnate was formulated. It is not surprising, therefore, that New Testament portrayals of Jesus would feature him as a miracle worker. But miracles—in the conventional sense of events that defy known laws of nature—do not resonate well with twenty-first century mentality. Contemporary men and women tend not to be realists about miracles any more than they are realists about the amazing exploits of Superman or Wonder Woman. So to the extent that the Christian myth is bound to the thought world of the first century, it may be judged by contemporary critics as having all the plausibility and relevance of a comic book.

I am suggesting here that the entire history of Christian theology can be read as a sustained effort to enhance realism about the entities, events, properties, and relations embedded in the myth of God incarnate. And this burden, we see, is both difficult and perpetual, requiring deconstructive as well as constructive objectives. As theological conventions pass out of fashion, it happens that nonrealism about the myth increases, leaving theologians to disabuse their audience of implausible and irrelevant constructions of the past to make way for new models and paradigms more palatable to the contemporary mind. Aquinas had to undo much of Augustine's Neoplatonic rendering of the myth in his efforts to offer a more credible Aristotelian interpretation. Enlightenment theologians had to critique medieval images of an inscrutable God in favor of mechanistic paradigms that were more compatible with the new science. Twentieth century theologians were left to combat mechanistic conventions in favor of existentialist analysis, and turn-of-the-century theology has been dominated by efforts to show that the Christian myth is compatible with the evolutionary cosmology of contemporary science.

And so it goes. For better or worse, Christian tradition has placed a lot of emphasis on belief as a decisive factor in the appraisal process. The idea is that wrong thinking about God's nature cannot result in righteous conformity with God's will. Faith must therefore seek proper understanding, for who can love what they do not know? The critical importance of the intellectual dimension was responsible for the foundation of countless monastery schools and universities throughout Christian culture. The impressive edifice of the Western educational establishment is firmly rooted in the Christian theological tradition.

EXPERIENTIAL STRATEGIES

I have maintained that religion is about nurturing people to think, feel, and act in ways that are judged to be good for us, both individually and collectively. Christianity makes essentially the same point, but does so in the language of theology: What is good for human beings, both individually and collectively, is to do God's will. The point of Christianity, therefore, is to nurture persons to abrogate their own private goal hierarchies in favor of God's goal hierarchy, to forego a self-centered existence for a God-centered existence.

For some individuals intellectual strategies may be sufficient to effect a reorientation of this sort, but for others a goal hierarchy transformation may require an extraordinary and poignant experience. From its inception the Christian tradition has stressed the importance of religious experience, but it has also been at least tacitly aware that the nature and value of religious experiences will vary according to personality differences. Thus Christianity has been tolerant of a wide range of spiritual adventures.

In his classic study of religious experience, William James refers to "once-born" and "twice-born" types of religiosity.[6] These categories may appear simplistic, but they do succeed in capturing two important ways in which Christian tradition has nurtured religious experience. The difference between once-born and twice-born Christians may be seen in the following autobiographical sketches.

> It is difficult to explain when I became consciously religious. It would be quite as easy for me to account for the fact that some years ago I was born weighing ten pounds and that I now weigh considerably more. This growth is not unmarked by incident, but the process, at least in retrospect, blurs into a continuum so completely that I cannot isolate the point at which I became "religious" any more than I can remember the point at which I emerged into conscious awareness of myself.[7]

> This experience occurred in the Fall of my fourteenth year. I had been working in a field plowing. Suddenly a storm seemed to approach, and as though everything around me stopped—I felt the presence of God. The horses had come to a complete stop; the inky black sky rumbled and I prayed. The storm passed on quickly, but it was at this moment—as I prayed—that I decided I would become a Christian and serve the Lord if that was the Lord's desire.[8]

The spiritual life of once-born Christians is marked by continuous growth, whereas the life of twice-born Christians is marked by dramatic change. Both spiritual types may have profound experiences—the once-born Christian life is not "unmarked by incident"—but the dynamics of these experiences are quite different.

We may say that the value of religious experience for the once-born Christian is to strengthen the individual's realism about the meanings of the myth. Despite the rigors of theological inquiry, the endeavors of the intellect to apprehend divine reality somehow always come up short, leaving a residue of doubt that might be erased by an extraordinary experience. Thus Christianity has encouraged prayer and various meditation techniques as ways of producing experiential apprehensions of God that are not mediated by discourse. The Christian who believes she has a personal acquaintance with God, or has beheld visions of Christ, achieves a deeper sense of realism about the myth than one whose beliefs are mediated entirely by theological conventions. Once-born Christians are encouraged to seek immediate experiences as subjective validations of the truth of the myth. The result of these experiences is to strengthen the biases of memory systems such that appraisals and coping events will be more likely to reflect the Christian pattern of piety.

The experience of the twice-born Christian seems to have much broader consequences for the individual's personality, for it involves a sudden and radical reorganization of a person's goal hierarchy. Twice-born Christians undergo a decisive change in personality so radical that they tend to adopt an attitude of contempt and rejection toward their previous life. The conversion experience of twice-born Christians is invariably marked by heightened conflict and anxiety, and usually by a sense of self-loathing or intense fear, as we see in the confessions of Henry Ward Beecher:

> There was the constant thought that I was an awful transgressor; every little fault seemed to make a dreadful sin; and I would say to myself, "There, I am probably one of the reprobate." So I used to live in perpetual fear and dread, and often wished myself dead.[9]

The pattern of such conversion experiences is to pass through a spiritual crisis during which one's entire life provokes disgust, but then suddenly the torment passes and the individual emerges with a completely new and exquisitely integrated perspective. The despair of the sinner gives way to the joy of the saved. The salient feature of such conversion experiences is that the newly integrated person adopts a completely new goal hierarchy.

St. Paul himself had an experience of spiritual transformation and made several attempts to articulate its theological significance. He relates the transformation experience to the crucifixion and resurrection of Christ. Just as Christ died for our sins and was raised for our salvation, so we must undergo a spiritual death and resurrection. The death of the Old Adam (read: the expunction of your self-centered goal hierarchy) is necessary for rebirth as the New Adam (read: the ingestion of God's goal hierarchy). The old ego with its self-serving perspective and agenda has vanished completely, and in its place now dwells the God-serving spirit of Christ: "I have been put to death with Christ on his cross, so that it is no longer I who live, but it is Christ who lives in me." (Gal. 2:19–20)

Paul's commanding influence has been sufficient to convince various movements and sub-traditions in the history of Christianity that his form of conversion experience is normative for Christian piety. These groups have typically insisted that true Christian faith requires one to be "born again" in the Holy Spirit. The techniques for provoking conversion experiences have been practiced for centuries, and while there have been many variations, the fundamental principle remains the same: disintegration (spiritual death) followed by reintegration (spiritual rebirth).

The disintegration of an organized personality might be accomplished by various means, including harsh personal derogation, extreme fear, intensifying a sense of conflict, guilt, or worthlessness, overstimulation of cognitive faculties, sensory disorientation, inducing confusion and uncertainty about what is true and false or right and wrong, and so on. The disintegration of personal organization drives the individual into a state of despair, exhaustion, and loss of control, eventually reaching the point of helpless surrender and receptivity to suggestion. Relief from despair is contingent on surrendering to a new self-understanding and direction in life: The goal hierarchy so integral to the former self is displaced by a new goal hierarchy. The relief is accompanied by feelings of joy and a well-focused sense of purpose. Religious conversion, like other extraordinary experiences, conditions the appraisal and coping process by playing on the bias of working memory to retrieve information marked with strong affective valence.

INSTITUTIONAL STRATEGIES

The institutional strategies of Christianity emerged in the complicated transition from the improvised and rustic community of apostolic times to the megalithic state church of the late Roman Empire. Primitive Christianity had all the appearances of a loosely organized revivalist movement. Among the early Christians there existed a sense of urgency (the end of history is at hand) as well as a sense of resignation (the Holy Spirit will guide us), precisely the conditions that would have made formal organization and deliberate policies look utterly pointless.

There was a modicum of polity emanating from the Jerusalem church, but after the destruction of the city in 70 C.E., there remained few constraints on how the proclamation would spread. And spread it certainly did, into most of the territories where the Roman Empire had a presence. It would be a mistake, however, to assume that the emergent Christian communities had much in common religiously. There may have been a slender core of shared beliefs—that Jesus was Lord; that he was raised from the dead; that he would return— but these were variously understood and freely mixed with local superstitions. The growth of Christianity was phenomenal but unruly, carried forward by a motley fraternity of itinerant evangelists, charismatics, freelance prophets, enthusiastic healers, and self-appointed proclaimers of the truth, not a few of whom were crude charlatans. Such was to be expected given the Pauline conception of a free-wheeling Spirit-led church. Authority was not invested in formal offices and regulated by policy, but was bestowed directly through gifts of the Holy Spirit. The Spirit elected individuals and led them on unique paths, their wonders to perform.

Eventually the church had to face the consequences of its own success. A laissez-faire mission was generating aberrant forms of Christian belief and practice, some bizarre and extreme enough to excite the hostility of the Roman establishment. The Romans, understandably, were left to form their general impressions about Christianity on the basis of its attention-getting extremists, with the result that all Christians, including responsible moderates, were presumed guilty of excesses by mere association. By mid-second century it was clearly time to harness the Holy Spirit by imposing order and discipline on the church, but it was not until the end of the third century that anything resembling orthodoxy prevailed.

The campaign for orthodoxy was led by the older churches claiming apostolic origins, especially the church in Rome, boasting the double authority of Peter and Paul, both reputedly martyred there. The campaign was premised on a fixed canon of authoritative scriptures, a growing body of doctrinal theology, and, especially, the establishment of a clerical hierarchy. The ascendancy of a clerical class put an end to the Pauline idea of the Spirit working through gifted individuals in favor of the notion that the Spirit worked through the ministrations of ecclesiastical offices. By the third century it was established that bishops held the exclusive power to remit sins—not because they were Spirit-led charismatics, but because they were bishops, authorized and anointed by apostolic succession. Apart from the office of the bishop there was no church, and apart from the church there was no salvation. The Spirit-led church became the bishop-led church, and henceforth Christianity would draw closer and closer to secular patterns of governance.

By the end of the third century the clerical hierarchy had all the appearances of a state within a state. The emperor Constantine perceived a common purpose between empire and church: Orthodoxy within the church would enhance stability within the empire. So Constantine allied himself with the church and enforced its campaign against heresy. An important feature of this alliance was Constantine's willingness to favor the clergy with legal and financial privileges, which considerably elevated the status and power of the clerical class. Now the divide between clergy and laity, between rulers and ruled, was finally pronounced.

In the Constantinian alliance the Empire became Christian and Christianity became imperial. But this partnership was not long to endure. Before the end of the fifth century the Roman Empire had collapsed completely, leaving the state church to reconsider its identity in the absence of a state.

The principal resource for the church's new self-understanding was St. Augustine's *City of God.* Writing after the sack of Rome (410 C.E.), Augustine contrasted the Earthly City with the Heavenly City, the former consisting of those who love and serve the self (even to the contempt of God), and the latter consisting of those who love and serve God (even to the contempt of self). These cities co-mingled in the temporal realm of history, but they will be separated absolutely in the last judgment. Augustine's vision is one of radical asymmetry between church and state. The church serves a divine and eternal agenda: God's will. The state, meanwhile, merely does what it can to contain the chaos and injustice inherent in a fallen humanity. True justice (that is, well-ordered love) is possible only within the Christian church. The eternal church, therefore, is in every way possible superior to the secular state. Augustine believed it was God's will that the Christian church become truly universal—it had already become imperial in scope and would eventually, according to divine plan, embrace the entire world.

Augustine envisioned a theocracy, a totalitarian community of saints in which the authority of the church is reflected in all facets of faith and life—not a state church, but an ecclesiastical state. God's grace and will were embodied in the church, and if God wills that the City of God become universal, then it was the calling of the church to transform the world. And if this required harsh discipline (such as punishment, intimidation, "constructive" torture), then so be it— eternal ends justified temporal means. As the church moved into the Dark Age, it faced the challenge of reconquering a world that had been largely de-civilized and re-paganized. But the church had now learned the arts of asserting absolute authority, and it was prepared to use these powers without hesitation or apology to "compel them to come in" (Luke 14:23).

The campaign against heresy remained a high priority and was pursued with disciplined efficiency. The idea that salvation was the exclusive property of the institutional church equipped the clerical class with a formidable emotional lever: fear. Those failing to conform to church authority were threatened with excommunication and severe punishment. And the meaning of "authority" was rendered unambiguous by a series of institutional principles. For example, the authority of scripture was reduced to the church's interpretation; the authority of individual bishops was trumped by councils of bishops; and the authority of the bishop of Rome took precedence over rival bishops.

These principles increased the efficiency of the church's campaign to enforce uniformity in all matters of doctrine, practice, and morality. In general, the institutional church used its formidable powers to foster conditions that would enable other ancillary strategies to do their work. For example, the intellectual dimension was helped by the campaign against heresy and the program to advance learning. In addition, the church promoted aesthetic and ritual strategies by sponsoring an ambitious program of building churches, where the arts and worship would flourish.

AESTHETIC STRATEGIES

The early church, as we have seen, adopted the language and thought forms of Greco-Roman philosophy and used these conventions to clarify, interpret, and defend the meanings of the Christian myth. The church also adopted Roman institutional strategies to develop a centralized and hierarchical power structure, which it used to expand the mission of the church and to impose constraints on its tendencies to diversify. In addition, the early church freely adopted aesthetic conventions from Jewish, pagan, and imperial sources, using these to embody the Christian myth and to evoke responses to it.

In chapter four I suggested that whereas theology attempts to narrow the scope of meaning to precise doctrinal formulations, the arts expand access to the myth by presenting multivalent images. Tension between exclusive intellectual strategies and inclusive aesthetic strategies was evident in the early church, especially during periods of rapid growth, giving force to the hypothesis that theology and the arts developed under different understandings of the church's ambiguous claim to universality.[10] Christianity is the one path to salvation for all humanity, but emphasis can be placed on either "one path to salvation" or "salvation for all humanity." It seems that whereas theology was guided by the exclusivist reading, the arts were guided by the inclusivist one.

While the theology of the early church sought ideological integrity, its art sought pragmatic results. That is, images were used if they were successful in engaging the sensibilities of a wide and diverse audience. Christian artists therefore helped themselves to a host of images and symbols that had previously worked in pagan settings, using these to evoke the range of emotions they felt were appropriate to the Christian context. An image having emotional appeal among pagans (the good shepherd, for example) was likely to have the same power when adapted to the Christian myth, thereby broadening the appeal of Christianity throughout the pagan world.

But let us not stray from the principal point of aesthetic strategies, which is to influence the emotional systems in such ways that appraisals will be less likely to default to intuitive morality and more likely to reflect the Christian pattern of piety. Thus we should examine a small selection of aesthetic expressions within the Christian tradition to see how they educate the emotions.

To begin, it is clear that much of Christian art, especially music, functions at the level of enhancing background feelings, especially mood states. The power of music to command our moods, and thus to bias our wills, is a well-known and heavily exploited phenomenon. We now know that musically induced mood states have a significant bearing on attention, perception, and memory. Play sad music to subjects and they will attend more closely to negative words, interpret neutral facial expressions as sad or angry, and demonstrate better recall for negative meanings. Play happy music and you will get the opposite results.[11] In short, music can influence what we look at, how we see things, and how they will be remembered. By altering our moods music can bias our appraisals.

None of these effects escaped notice by the early church. The question for Christians was not whether music was useful in the management of human nature, but rather to what ends music should be used. In Roman culture music was highly developed and used for diverse ceremonial, military, and entertainment purposes. Professional musicians were available to entertain guests at weddings and private parties. Large ensembles produced loud and elaborate concerts, instrumental competitions were common, and the virtuosi of the period enjoyed a celebrity status comparable to contemporary rock stars.[12]

For Christians, however, the utility of music had nothing to do with producing pleasure and everything to do with promoting piety. The point of music was to render the listener more receptive to the meanings of the Christian myth. The early church therefore eschewed instrumentation and showmanship in favor of a musical style that would evoke a solemn and contemplative mood. Simple, sweet, unadorned melodies would suffice—dignified enough to exalt the text, but not pleasing enough to compete with it.

It is not surprising that the early church derived its musical tastes from the Judaic chant tradition, which allowed melodic and rhythmic features to be defined by the grammatical structures of psalms, canticles, and hymns. Christian musicians experimented with many styles of chant until the Gregorian style came to dominate the liturgical practice of the church. No one should underestimate the contribution of Gregorian chant to the church's mission of restoring religious unity and political harmony to the de-civilized territories of the fallen empire. Music, no less than the gospel itself, was the avant-guard of cultural transformation. Here is St. Basil's testimony on the contribution of this medium to the achievement of personal wholeness and social coherence:

> A psalm is the tranquility of souls, the arbitrator of peace, restraining the disorder and turbulence of thoughts, for it softens the passion of the soul and moderates its unruliness. A psalm forms friendships, unites the divided, mediates between enemies. For who can still consider him an enemy with whom he has sent forth one voice to God? So that the singing of psalms brings love, the greatest of good things, contriving harmony like some bond of union and uniting the people in the symphony of a single choir.[13]

While much of Christian art and music was produced in the service of modulating moods and shaping attitudes, other practices appealed more intensely to emotional systems. It was not uncommon in medieval Christianity for images to become objects of veneration and legend, in the manner of holy relics. Many portraits of Christ, the Virgin Mary, and various saints were reputed by legend to have supernatural origins and to be responsible for miracles and conversion experiences.[14]

A fine example is the Mandylion, a Syrian cloth image of Christ. Legend has it that Christ washed his face with a cloth, leaving behind an impression of his features, and when the cloth was later beheld by King Abgar, he was converted and instantly cured of leprosy. The legend further relates how the "holy Mandylion" caused heathen idols to crash to the ground, and how oil flowing from the image burst into flames to thwart a Persian invasion.[15] Similar traditions surround the Veil of Veronica and the Shroud of Turin, as well as countless other images.

Cherished images were normally hidden from public view, to be displayed honorifically in ceremonial processions where they were greeted as royalty. Appearing suddenly from behind a curtain, the image would move from place to place like a living person. The practice of veneration was often shamelessly hyped, a strategy that heightened expectations and lowered thresholds to emotional responses, which were often intense.

Some of the earliest images of Christ picture him in a position of power and majesty: as emperor, as the sun god Apollo, or as the ruler of the cosmos. Such images were intended to inspire a response of awe and humility, as were the dozens of paintings showing kings and dignitaries kneeling before the infant Jesus and countless representations of miracles performed by Christ. Sympathy was provoked by the many portrayals of Christ's suffering: the humiliation before Pilate, the flagellation, the bearing of the cross, the crucifixion, and lamentations over the dead body. Deliverance scenes—the tale of Jonah, the parting of the Red Sea, the raising of Lazarus, Daniel in the den of lions, healing of the paralytic—were common in funerary art and were effective in eliciting both gratitude and hope. Renditions of the Adam and Eve story or the betrayal of Christ would instill guilt. Gratitude and affection were inspired by images of Christ as the good shepherd who lays down his life for the flock. Love was aroused by images of Madonna and Child, and by the scores of portraits projecting the serene beauty of Christ. The action tendencies associated with these emotional responses would vary somewhat, but they would all have a bearing on the willingness of subjects to reconfigure their goal hierarchies to conform to God's will.

The reign of Constantine brought with it an ambitious program of constructing churches and cathedrals as venues for Christian worship. But churches and cathedrals were also built to accommodate an aesthetic program that was meant to educate the emotions. The exterior appearance of these new churches was subordinate to the task of creating interior spaces where visual effects could be exploited. Brightly colored mosaics, paintings, and carved panels depicted biblical motifs on floors, walls, doors, and ceilings. "A stunning array of colors, textures and materials met the eye everywhere."[16] To step into a Constantinian church was to enter a transcendent realm where even light and sound behaved differently. The contrast between the grubby world without and the heavenly world within was dramatic, and was sure to have palpable effects on moods and emotions. The size and grandeur of Christian churches inspired an immediate sense of awe and humility, preparing individuals for the additional impact of particular images arranged throughout the building.

Any attempts to associate particular images with corresponding emotional responses can never be precise. The enduring power of religious images is bound up with their multivalence, such that many images have the power to appeal to a range of emotions. Consider that encounters with images of the passion of Christ may result in guilt, sympathy, gratitude, love, or hope. And the potential remains for a different response with each encounter. It must also be noted that images having strong appeal under one set of social circumstances might have weak appeal under different circumstances.

These considerations may confound attempts to describe the relations between art and feelings with precision, but they do not diminish the claim that Christian art has demonstrated its power to alter moods, shape attitudes, and arouse emotions in systematic ways.

The intended consequence of all these aesthetic effects remains exactly the same as that sought by intellectual, experiential, institutional, and ritual strategies, namely to influence the goal hierarchies that individuals bring to their everyday encounters.

RITUAL STRATEGIES

In the broad sense, a ritual is any repeatable unit of behavior, the performance of which engages people in the meanings of a myth for the sake of harmonizing the relationship between humanity and ultimate reality. Christian rituals regularly involve individuals in the world of the myth, thereby playing on the bias of working memory to supply the appraisal process with memories that are recent, frequent, or poignant.

The myth of God incarnate is dramatized each year in the ritual cycle of the church. The church year is divided in halves: The first half (sacred time) features events in the life of Christ, while the second half (ordinary time) is focused on Christ's work as it is carried out in the mission of the church. Sacred time has five seasons (Advent, Christmas, Epiphany, Lent, and Easter), and ordinary time is the twenty-four week season known as Pentecost. Each of these six seasons prescribes activities and preoccupations for the Christian life. Special prayers and scripture readings are assigned to each day of each season.

During the four weeks of Advent the pious Christian should be in a state of anticipation and preparation for the appearance of Christ. The mood of Advent, instilled by music, prayer, and scripture, is serious but happy. It is a time for getting things in order, for setting aside non-essentials and focusing on things that really matter. Most importantly, it is a time for attending closely to one's goal hierarchy.

The twelve-day Christmas season (Dec. 25–Jan. 5) is a period of joy and thanksgiving. The birth of Christ is regarded as God's greatest gift to the world, an event to be greeted with a sense of gratitude and its associated action tendency to repay by revitalizing one's commitment to do God's will.

The Epiphany season continues from January 6 until Lent. The focus of this season is on Christ's revelation to the Gentiles. Epiphany marks the arrival of the wise men to worship the infant Jesus, and is thus a time of worship, praise and adoration. The meaning of the Christmas event (the incarnation) deepens in the life of pious Christians during this season.

Lent is a solemn, month-long season of contemplation, self-examination, and penitence during which Christians are admonished to reflect on the sinful life emanating from a self-centered goal hierarchy. As Good Friday approaches, Christians identify closely with their humiliated lord and prepare to participate in his sacrifice by dying to the old self.

On Easter morning Christians perceive the entire creation as transformed. The dead Christ lives, and those who shared in his crucifixion are newly reconstituted by the goal hierarchy of discipleship. Easter celebrates God's decisive victory over the cosmic forces of evil, but it also celebrates baptism, the rite that enables Christians to share in the victory. The ascension of Christ is celebrated forty days after Easter, and then the Easter season concludes with Pentecost, marking the arrival of the Holy Spirit to guide the mission of the church.

The twenty-four weeks following Pentecost define the "ordinary time" of the church year, the season of discipleship when Christians are preoccupied with asking "What would Jesus do?" as they work out their commitments to lead Christ-centered lives. The church year is one long dramatic reenactment of the life and work of Christ, the ultimate point of which is to organize the neural systems of Christians. It is a script for educating the emotions, a measured program for reconfiguring and revitalizing the Christian's goal hierarchy to conform to a vision of God's will.

The Christian sacramental system is a set of ritual strategies overlaying the cycle of the church year. The earliest sacraments were baptism and the Eucharist (Lord's Supper), but five additional rites were given sacramental status during the Middle Ages. The Eastern Orthodox tradition recognizes more than seven sacraments, and protestant traditions have reduced the number to the original two.

Baptism is the rite of initiation into the Christian community. When this rite is administered to infants, it serves as an occasion for the community surrounding the child to revitalize their own identity with Christ, and to reaffirm their commitment to nurturing children toward a life of service to God's will. When the rite is administered to adults, emphasis is placed on the individual's repentance and conversion, a symbolic death and resurrection wherein the old self-centered goal hierarchy is displaced by a new Christ-centered goal hierarchy.

The Eucharist (literally "thanksgiving") concentrates the salient meanings of the myth in a single rite. The consecration of bread and wine (the body and blood of Christ) recreates the incarnation, and partaking in the Lord's Supper reenacts the crucifixion and resurrection. The ritual concludes with a prayer of thanksgiving, arousing a sense of profound gratitude that motivates worshippers to recommit themselves to a life of discipleship.

The Roman Catholic tradition grants sacramental status to

marriage, confirmation, ordination, last rites, and penance, all of which provide occasions for revitalizing one's goal hierarchy, but only the last of which is meant to be repeated. Penance, or the rite of reconciliation, enables individuals to ritualize the action tendency associated with guilt. There are four steps in the logic of penance: contrition (the awareness of guilt), confession (the verbal repentance of sin to a priest), absolution (whereby the penitent is released from the eternal consequences of sin), and acts of satisfaction (which remove the temporal consequences of sin). A priest may impose any of a variety of acts of satisfaction, including prayers, pilgrimages, acts of charity, payment of alms to the poor, or spiritual retreat—any and all of which may be expected to have consequences for goal hierarchies and memory retrieval.

PERSONAL WHOLENESS AND SOCIAL COHERENCE

George Bernard Shaw once quipped that Christianity appeared to be such a good idea that it should be tried sometime. One wonders what psychological and social consequences might follow if something close to a majority of self-professed Christians practiced their religion in the manner envisioned by its principal architects. Christianity, like any religious tradition, requires constant vigilance against the backsliding influence of our default mentality. Pious Christians must therefore combat spiritual entropy (or atrophy, if you like) with efforts to assure that working memory will be dominated by the meanings of the myth of God incarnate.

The ancillary strategies of Christian tradition were designed to enforce this process. If Christians seriously apply themselves to resolving intellectual challenges to the myth; if they seek personal validations of the myth in extraordinary experiences; if they saturate themselves with aesthetic expressions of the myth, making every effort to apprehend their multivalent meanings; if they participate in rituals that simulate the death and resurrection of Christ; if they pray without ceasing; if they organize their activities around the seasons of the church year; if they observe the sacraments; if they prompt themselves to ask what Jesus would do in the situations they face—if they do all these things with regularity and consistency, then we may expect that their neural systems will constrain them to think, feel, and act in harmony with the Christian vision of how things ultimately are and which things ultimately matter.

If the myth and ancillary strategies of Christian tradition work as they were intended to, then they will produce women and men who are, as Neusner put it, "intoxicated by faith." The goal is to prevent or override gut reactions to daily encounters in favor of Christlike reactions, and this may be achieved by applying a full-court press to behavior mediation systems. Moods will be altered to raise thresholds to anger, hatred, jealousy, fear, and anxiety, and to lower thresholds to sympathy, gratitude, and love. Attitudes will be forged to displace doubt and reinforce realism about the world of the myth. But most important of all, self-esteem links will be reconfigured so that individuals become wired to think badly of themselves when they fail to be Christlike, and to think well of themselves when they succeed. When self-worth is contingent on Christlike behavior, then we may say that a person's goal hierarchy has been displaced by God's goal hierarchy.

We may expect the kind of spiritual transformation described here to have far-reaching social consequences as well. Motivated by their obsession to serve God, Christians may be predisposed to initiate and join projects that serve the needs and interests of God's people—everyone—and they may eschew projects that reflect personal ambitions. In practical terms this involves a willingness to make personal sacrifices for the sake of social justice and the common good. "Do you love me?" Jesus asked, then replied: "Feed my sheep." But the idea, of course, is that prosocial behaviors would not involve sacrifice at all, they would instead provide opportunities for self-fulfillment. Self-esteem links to Christlike behavior is Christianity's distinctive answer to the perennial challenge of constructing overlaps of self-interest.

7

ISLAM

HISTORICAL CONTEXT

The prophet Muhammad was born into a world of intertribal conflict, social injustice, and religious confusion, a world primed for transformation. Islamic tradition looks back on this world as an age of barbaric ignorance, chaos, and violence.

Historically, the Arabian Peninsula was incapable of sustaining a population commensurate with its landmass. The land is dominated by barren volcanic steppes and desert wastes, making it unsuitable for agriculture except in places where irrigation is feasible. The southwest corner of the peninsula was one such place, where agricultural settlements gave rise to the earliest Arabian civilizations. But most of Arabia remained unsettled—in fact "Arab" literally means "nomad." In addition to their farming, southwest Arabs became active traders, taking advantage of busy trade routes up and down the western coast. Trade traffic also provided a livelihood for many Bedouin tribes whose caravans transported goods to and from the Mediterranean region. Thievery and extortion, too, became a way of life for many nomadic tribes that could not resist waylaying caravans for booty or safe passage fees. Thus, trading and raiding became important traditions in the economic and social life of pre-Islamic Arabs.[1]

Arabia had never been unified politically. It was all the Arabs could do just to keep their individual tribes intact against the relentless aggression of other tribes. The tribe was the ultimate unit of social organization and everything depended on its survival. Intertribal alliances, when they existed, were always pragmatic and temporary arrangements having a specific purpose such as common defense. But there was no true sense of solidarity or obligation beyond the tribe. If anything united the Arab people, it was a loosely shared ethos of strict tribal independence and sovereignty. It was a tribe-eat-tribe world. The ethos of tribal sovereignty is found in the concept of *muruwah,* a complex notion involving "courage in battle, patience and endurance in suffering, and a dedication to the chivalrous duties of avenging wrong done to the tribe, protecting its weaker members and defying the strong."[2] It was ultimately for the sake of tribal welfare that each individual was expected to uphold the values of muruwah. Personal fulfillment was important, of course, but it was utterly dependent on the resources of the group.

Arabian tribes did not maintain priests or shamans. Instead, they had poets, the most important bearers of tribal culture. The poets unified, glorified, and immortalized the tribe by retelling its story, celebrating its triumphs, and lamenting its defeats. They brought courage to warriors and vexation to their opponents, and not least of all they helped members of the tribe to accept and endure the twists of fate. The traditional religion of nomadic Arabia was an unsophisticated blend of animism, polytheism, and fatalism. The tribe was a sacred entity, and there was no expectation of personal immortality. The closest thing to life after death involved the howling ghosts of slain men, lingering in the air until their deaths were avenged.

Pagan Arabs were particularly reverent about sacred places, such as springs, wells, groves, and peculiar rock formations. These sanctuaries were often ancestral grave sites and were thought to provide refuge from danger and hostility. The most important sacred place was the Ka'bah in Mecca, an ancient building containing an unusual black stone together with more than three hundred idols, perhaps totems of the many tribes making the autumnal pilgrimage to the site. In the vicinity of Mecca were shrines to three major goddesses (al-Lat, Manat, and al-Uzza) said to be daughters of the high god Allah, and to whom pagans prayed and offered sacrifices for divine intercession.[3] But little was expected of these oblations. They were performed informally and infrequently—mere halfhearted long shots, really, offered up by simple folk for whom the world was fundamentally bewildering, unpredictable, and dangerous. The best one could do was take heart from the poets, do what was expedient for the tribe, and let fate take its course.

Tribal life may have been fraught with uncertainty and adversity, but it wasn't completely intolerable. Within the confines of the group there was a sense of belonging and a measure of security. These tribes were poor, but whatever wealth they had was evenly distributed.

When there was booty, everyone got their share, and when times were hard, everyone suffered equally. Easy come, easy go. And what was the use of material wealth anyway? Nomadic groups could only be encumbered by the acquisition of things. But this was all bound to change as opportunities for settlements increased.

Several cities matured along the western trade routes during the generations before Muhammad's birth. Mecca, birthplace of the prophet, became particularly important in the fifth century C.E. It was here that the Quraysh tribe gave up the nomadic life and established a permanent trading post. By the time of Muhammad's birth (570 C.E.), the Quraysh had come to dominate the commerce of western Arabia. The Ka'bah was a principal factor in the growth and prosperity of Mecca. Each year thousands of pilgrims came to Mecca to visit the shrine and to lose themselves in the trading and festivities of the annual fair. Concession holders at the fair were able to amass fortunes in a few short weeks.

The social consequences of urbanization and sudden prosperity among the Quraysh were profound. The solidarity and cooperation characteristic of the ancient tribal order declined as the old system was displaced by an emerging class system. The wealthiest tribes formed an oligarchy of rich merchants, much to the distress of the exploited "outskirt" tribes.[4] The elite tribes controlled all aspects of Meccan economic and social life, including the lucrative concessions associated with the Ka'bah.

The gradual breakdown of the tribal system deprived families and individuals of the social safety net intrinsic to the old order, and this at a time when victims of severe economic disparity were beginning to appear. Widows, orphans, the elderly, and the infirm would have been cared for under the tribal system, but the new urban order encouraged anonymity, individualism, and cutthroat fortune-seeking, leaving the less fortunate vulnerable to neglect and exploitation. It was not uncommon for greedy opportunists to swindle unsuspecting widows and orphans out of their rightful inheritance. Meanwhile, the elites of Meccan society were themselves falling victim to a lifestyle of hedonistic pursuits. Gambling became a favorite diversion, and drunken orgies punctuated by violence and bloodshed were commonplace.

Challenges to personal wholeness and social coherence of the sort just mentioned are precisely the issues one expects to be addressed by religious sentiments, but there was nothing in the religious outlook of the new urban rich to inspire anything but indifference. Apparently the practice of muruwah was too bound up with the nomadic life to have any relevance, and the sad truth was that the de facto religion of the new order was money grubbing.

Both Judaism and Christianity, of course, address the problems faced by the new Arabian order, and both had achieved a significant presence in the region well before the birth of Muhammad. The Arabs were aware that these monotheistic religions were far more sophisticated than their own provincial traditions, and more than a few of them had converted to Judaism and Christianity. For the most part, however, Arabs were suspicious of these religions because the great powers of Byzantium and Persia were known to use both faiths as instruments to advance their own imperial interests.[5] Many Arabs were attracted to the central ideas of Judaism and Christianity—especially their monotheism and the promise of immortality—but they remained wary of the potential for these religions to cost them their independence.

Into these circumstances entered Muhammad, whose remarkable achievement was to unify the Arab people under a common story that brought together elements from Judeo-Christian and traditional Arabian sources. And further, he laid the groundwork for advancing the new myth into vast territories beyond the Arabian Peninsula. Muhammad's work effected nothing short of a complete transformation of Arab identity and culture. About the only thing left unchanged by the rise of Islam was human nature.

Muhammad was born into one of the lesser branches of the Quraysh. He was orphaned at an early age and was raised by members of his extended family. As a boy he worked as a shepherd and was sent by his family to live for a spell among the desert Bedouin, from whom he learned the pure Arabic language as well as the traditional skills and values of nomadic life. In his twenties he entered the caravan business and eventually married his employer, with whom he had six children. As a young man frequenting the trade routes, Muhammad came into contact with many Jews and Christians, with whom he engaged in spirited conversations. From these conversation partners he learned the substance of Judeo-Christian tradition, and was able to apprehend how ethical monotheism could speak to the problems besetting Arab culture.

At the age of forty Muhammad received a visitation from what he judged to be an angel of God, telling him to bear a message of divine truth to humankind. This was the first of many such visitations spanning more than twenty years, up to the time of his death. Over the years these "recitations" (*qur'ans*) were recorded by scribes, and some time after Muhammad's death they were compiled in the Holy Qur'an.

THE MYTH OF FINAL PROPHECY

As wary as Arabs were about Judaism and Christianity, they had reason to be envious of the "People of the Book." Jews and Christians possessed something vital that was lacking in the lives of ordinary Arabs. It was a sense of belonging to some grand scheme of things, some sort of lofty cosmic enterprise. The familiar songs of the poets were beginning to sound parochial and shamefully out of step with the larger world that was descending on the peninsula. And the gods decorating the Ka'bah were beginning to appear rather smallish when compared to the God of Abraham, Moses, and Jesus. To make matters worse, Jews and Christians were not above taunting their Arab neighbors with the dig that they had been left out of God's plan.

These matters weighed heavily on Muhammad, driving him to make periodic retreats from Mecca where he could ponder them in solitude. It was during one of these retreats that he received his first vision. He became convinced that the Arab world had not been left out of God's plan. To the contrary, Arabs had been an important part of the plan from the very beginning. However, they had gone astray, as humans often do, and now needed to be reminded of their true origins and destiny. Muhammad knew that Abraham had fathered two sons: Ishmael (by Hagar) and Isaac (by Sarah). When Sarah bore Isaac, she insisted on the banishment of Hagar and Ishmael, but beyond this the Bible makes no further mention of Ishmael. Muhammad's visions informed him that Ishmael had in fact migrated to Mecca, where he and Abraham built the Ka'bah and dedicated it to Allah. The Arab people were, therefore, descendants of Abraham through Ishmael. And Allah, the high god of pagan tradition, was none other than the one true God worshipped by the People of the Book. Muhammad understood himself to be chosen by God as a messenger whose task it was to remind humankind of the truth. His message was a reminder not just to the Arabs, but to Jews and Christians as well, for they, too, had gone astray.

The first half of Muhammad's prophetic career was conducted in Mecca, and the second half in Medina. He commenced by preaching a mild form of ethical monotheism, the view that there is one God whose sovereign will directs the moral life of humans. He gradually stiffened his rhetoric until it became clear that he was offending the Meccan establishment. He focused on the idolatry practiced at the Ka'bah, the hedonism of Meccan society, and the unjust treatment of the poor and needy. These were, he insisted, affronts to God and must be stopped or there would be serious consequences to pay. Muhammad's warnings directly threatened the interests of Meccan elites, who became increasingly hostile to the prophet. Eventually Mecca became unsafe for Muhammad, so he withdrew to Medina (then called Yathrib) where the revolution began in earnest. The myth that sustained the new religious movement may be summarized as follows.

There exists one God only: the creator, sustainer, and judge of all things. Creation consists of three realms: heaven, hell, and earth. God created the universe in six days and then enthroned himself in the seventh heaven, surrounded by angels created to praise God and to execute his will. God created humans from the soil of the earth and ordained that they would have a unique place in creation. It was for their sake that God created the earth, and they were intended by God to be his special agents there. Humans are the supreme beings of the earth and will ultimately be judged according to how faithfully they serve the divine will. As a token of their exalted status God commanded the angels to prostrate themselves before Adam, the first human. One angel, Satan, refused to bow before Adam and was therewith banished from heaven for disobeying God's will. Satan would henceforth have no authority in creation except among those who, like himself, went astray. God allows Satan to remain on earth as a tempter until the Day of Judgment, when he and all those he leads astray will be banished to the eternal torments of hell.

Humans have access to God's will, which is essential to their purpose, but have no access to God's full nature. God's will is dispatched by his angels to specially chosen human prophets, and through them to the rest of humankind. Minor prophets bear messages intended for specific people or for a limited time, but major prophets (or messengers) bear universal truths, absolutely binding for all humans and all time. God ordained five messengers prior to Muhammad: Adam, Noah, Abraham, Moses, and Jesus. Each of these major prophets received truths from God, but in each case their messages were subsequently distorted, creating a need for the next messenger to bring a reminder. Muhammad is God's final and most decisive messenger. His recitation of the truth (the Qur'an) is an exact replica of the eternal and uncreated Qur'an that exists in heaven. Wherever the Qur'an disagrees with previous revelations (for example, the Torah and the Gospel), it must be assumed that these have been distorted.

The Qur'an is the final and perfect expression of God's will, the absolute standard to which every being will be held to account on Judgment Day. At the hour of God's choosing the dead will be resurrected and reunited with their souls to await the ordeal of judgment. On the Day of Judgment each individual will be presented with a Book of Deeds recording all the actions of their lifetime. If the good deeds outweigh the evil, then the person will be rewarded in paradise; but if the evil outweighs the good then the person will be punished in hell. Heavenly rewards and hellish punishments will be enjoyed or suffered with the body as well as the soul, and will vary according to the record of one's submission to God's will.

The fundamental aspiration of the Islamic tradition is given in the name "Islam," which means "submission." A "Muslim" is "one who submits." The point of Islam is therefore precisely the same as the point of Judaism and Christianity: to induce people to transform their goal hierarchies to reflect God's will exclusively. It is difficult to imagine that anyone might genuinely reorganize their personality— for this is what goal hierarchy transformation entails—around the meanings of the Islamic myth without being a realist about these meanings. If you believe, for example, that Muhammad suffered from epilepsy and that his visions were induced by seizures, then you will probably not regard his claims as absolutely binding on you, anymore than the claims of another epileptic proffering a different mythic vision. In other words, if you are a realist about Muhammad's epilepsy, then you are not very likely to be a realist about his claim to be a messenger from God. But if one is not a realist about certain meanings, then they will lack the cognitive status required for the emotional appeals of the myth to achieve their intended effects.

Regarding the matters about which Muslims are expected to be realists, we have no better source than the list provided by the Qur'an itself: "Any who denieth Allah, His angels, His books, His messengers, and the Day of Judgement, has gone far, far astray." (Surah 4:136)[6]

EMOTIONAL APPEALS

The expansion of Islam was such a remarkable phenomenon that it has continued to fascinate historians for 1,300 years. Within Muhammad's own lifetime virtually all of Arabia was united, and within a century after his death the Muslim empire had spread westward across North Africa and most of Spain; northward into Asia Minor and Central Asia; and eastward to the edge of India. Much of this success can be attributed to sheer military prowess: Muslim soldiers were experienced, brilliantly marshaled, and highly motivated. And the timing couldn't have been more fortuitous: The great empires of the age, Byzantium and Persia, had exhausted each other through generations of warfare and had little to offer in the way of resistance. Furthermore, many of the subjects of these empires bitterly hated their overlords, to the point of giving aid and encouragement to their Muslim liberators.

But there is more to the story than its military aspects: "As Islam penetrated new areas, people were offered three options: (1) conversion—that is, full membership in the Muslim community, with its rights and duties; (2) acceptance of Muslim rule as 'protected' people and the payment of a poll tax; (3) battle or the sword if neither the first nor the second option was accepted. The astonishing expansion of Islam resulted not only from armed conquest but also from these two peaceful options."[7]

For the most part, Muslims were gracious and tolerant victors. They left religious communities intact and allowed for a great deal of local autonomy. They were too confident of their own religion to impose it on others, offering incentives rather than ultimatums. Also, Muslim culture made it exceedingly simple to satisfy the universal human desire for social acceptance and belonging. To be a full member of the Muslim community one had only to confess that God is one and that Muhammad was his messenger. Do that and you are part of the ultimate in-group of those who are acceptable both to God and to the family of Islam. As converts were drawn further into the myth of final prophecy, they encountered the force of its emotional life. The emotional appeals of Islam tended to concentrate on capacities for awe, humility, gratitude, desire, and fear.

AWE AND HUMILITY

Muhammad is said to have declared that God has ninety-nine names, but he never specified the complete list, leaving it to Islamic tradition to outdo itself in producing superlatives that might express the reality of God. One gets a sense that there has been an ongoing competition among Muslims to come up with the most beautiful and exalted names. The Qur'an encourages this: "Say: Call upon Allah, or call upon Rahman. By whatever name ye call upon Him, it is well: for to Him belong the Most Beautiful Names." (Surah 17:110) The Qur'an itself takes delight in piling up superlatives: God is Totally Aware, Beautiful, All-Compelling, Creator, Evident, Subtle, Strong, Generous, Eternal, Merciful, Compassionate, Producer, Shaper of Forms, Absolute, Praiseworthy, Owner, Wise . . . and much more.

Nothing in Islamic tradition is hammered at more consistently than the oneness and wonder of God. This theme is most forcefully expressed in the idea that God is the only reality. Islam defines the cosmos as "everything other than God." Anything other than God is mere illusion, which means that nothing can be apprehended—or even said to exist—apart from its relation to God. This theme is relentless in Islam, the point of which is to trigger awe and humility in those who hear. And the point of triggering awe and humility is to affect the associated action tendency—that is, to cower and submit to the will of an overpowering reality. "Has not the time arrived for the believers that their hearts in all humility should engage in the remembrance of Allah and of the Truth which has been revealed to them?" (Surah 57:16)

GRATITUDE

A few more names: Helper, Most Bountiful, Most Kind, Beneficent, Sustainer, Giver of Life. Whenever the Qur'an speaks of the greatness of God (as it does incessantly), it does so to remind the hearer of the great things God has done for humankind. Here is one of countless reminders:

Or, who has created the heavens and the earth, and who sends you down rain from the sky?
. . . Or, who has made the earth firm to live in?
. . . Or, who listens to the soul distressed when it calls on Him, and who relieves its suffering and makes Mankind inheritors of the earth?
. . . Or, who guides you through the depths of darkness?
. . . Or, who originates Creation, then repeats it, and who gives you sustenance from heaven and earth? Can there be another god but Allah?
(Surah 27:60–64)

Before moving to the point of all this reminding, it will be useful to comment on a cluster of Arabic words. The Arabic word *iman* is usually translated as "faith," and *shukr* is translated as "gratitude." But interestingly, *kufr* is given as the opposite of both—that is, kufr appears to mean both "unfaith" and "ingratitude." The suggestion appears to be that iman (faith) is closely associated with shukr (gratitude): "Faith is nothing but a form of gratitude, and gratitude is a form of faith."[8] More can be said about kufr: The literal sense is to cover up, deny, or obscure something that is obviously and incontrovertibly true. When these meanings are brought together we get "ungrateful truth-concealing" as the best definition of kufr.[9]

This is way more exegesis than anyone needs, but it does bring us to the point behind Islam's constant reminders about the greatness of God: They are intended to provoke a response that is opposite to kufr—that is, a response that might be called "faithful gratitude." Faithful gratitude is a natural response to the recognition that humans owe their entire being to God. Failure to respond in this way is to dismiss the obvious, it is "the ugliest of human failings."[10] The action tendency associated with gratitude is to pay back the benefactor, or to do what the benefactor wants. And in this case the repayment must be radical indeed, for the debt includes everything one has. Nothing less than a complete dedication to serving God's will would be sufficient.

FEAR AND DESIRE

The Qur'an devotes more attention to the rewards of heaven and the punishments of hell than any other sacred literature. Pious Muslims remind themselves of the coming Judgment Day at least seventeen times each day in the course of their prayers. Clearly, the afterlife and its relation to this life play a fundamental role in Islamic piety, and this is where the myth of final prophecy makes its most potent appeal to the emotions. That is, fear of pain and desire for pleasure are the principal strings of human nature played upon by the Islamic myth.

All Muslims expect a cosmic day of reckoning at the end of time, when all the actions of their lifetime will be weighed in the balance of good and evil. The Qur'an takes pains to emphasize that Judgment Day is both unpredictable and inevitable: "When the Event Inevitable cometh to pass, then will no soul entertain falsehood concerning its coming." (Surah 56:1-2) Divine judgment will result in two groups: the Companions of the Right Hand, destined for paradise, and the Companions of the Left Hand, destined for hell. There is nothing vague or abstract about what these two groups might expect in the afterlife; the Qur'an is graphic about the sensuous nature of both heavenly delights and hellish torments.

On the matter of eternal rewards and punishments there is a marked difference between Christian and Muslim traditions. The New Testament contains a few graphic images of heaven and hell, but the majority of Christian tradition has maintained that these are symbols intended to accentuate the radical difference between this world and the next. Not so for Islamic tradition. For Muslims the mode of existence in the next life will be similar to life on earth, only much better (or much worse). In paradise there will be no pain or suffering—only joy and pleasure—whereas in hell there will be nothing but pain and suffering. Any differences will be quantitative—that is, experiences in the afterlife will be more real, more intense, and more enduring, but they will not be unlike the experiences we have known in this life. It is pretty obvious that any religious tradition that slips into abstractions and obscurities in its reward language might just as well not bother using reward language.

Islamic tradition has been far too astute about human nature to squander the possibilities lurking in our proclivities for seeking pleasure and avoiding pain. Instead, it exploits these features of human nature by offering up an afterlife worth thinking about. Here are some foretastes of hell:

> Those who reject Our signs, We shall soon cast them into the fire; as often as their skins are roasted through, We shall change them for fresh skins, that they may taste the penalty: for Allah is exalted in power, wise.
>
> (Surah 4:56)

> In front of such a one is Hell, and he is given for drink, boiling fetid water. In gulps will he sip it, but never will he be near swallowing it down his throat: Death will come to him from every quarter, yet he will not die: and in front of him will be a chastisement unrelenting.
>
> (Surah 14:16–17)

And a foretaste of heaven:

> They will be on thrones encrusted with gold and precious stones, reclining on them facing each other. Round about them will serve youths of perpetual freshness, with goblets, shining beakers, and cups filled out of clear-flowing fountains: no after-ache will they receive therefrom, nor will they suffer intoxication: and with fruits, any that they may select; and the flesh of fowls, any that they may desire. And there will be companions with beautiful, big, and lustrous eyes—like unto pearls well-guarded. A reward for the deeds of their past life.
>
> (Surah 56:15–24)

The psychological dynamics being played upon here are not very complicated: We are genetically predisposed to make sacrifices on the promise of avoiding pain or achieving pleasure. The action tendency for desire is to follow the necessary steps to possess the objects of desire, and the action tendency for fear is to follow the necessary steps to avoid danger. The necessary steps in the case of avoiding hell and securing heaven amount to reconstructing one's goal hierarchy for the sake of doing God's will.

ANCILLARY STRATEGIES

Encountering the word of God is one thing, but embodying it is something else. To embody a myth is to ingest its meanings and values into one's goal hierarchy, so that they might influence the outcome of all appraisal events. Islamic tradition has developed a range of strategies designed to achieve this embodiment.

RITUAL STRATEGIES

Islam specifies five essential practices that must be performed by any-one who desires to avoid hell and to become eligible for paradise. These "five pillars" of the faith are: bearing witness, prayer, almsgiving, fasting, and pilgrimage.

Every Muslim is required to make a public confession (*shahada*) to the following creed: "I bear witness that there is no god but God (Allah), and I bear witness that Muhammad is God's messenger." This public confession, made in the presence of two practicing Muslims, is the only formal requirement for one to become a Muslim. Theoretically, the Muslim oath is required only once, but in practice it is repeated as often as twenty times daily.

This confession is not taken lightly, for it is understood to have profound consequences for the individual. It is the first thing a new-born child hears from its parents, portending an absolute and lifelong commitment to sacrificing self-interest to God's will. So solemn is the taking of this oath that many Muslims will refuse to act as formal wit-nesses to it unless they are thoroughly convinced that the confessor's submission to God is genuine. The solemnity of the oath, and the fre-quency of its declaration in a typical Muslim's life, exploit the bias of human memory systems to favor information that is highly valenced or often repeated. A person who has, only minutes ago, declared an oath to obey God is unlikely to sin with abandon.

The second pillar of Islam, prayer (*salat*), also exploits the biases of human memory systems. Muslims are required to submit in formal prayer five times each day (dawn, noon, mid-afternoon, dusk, dark-ness). In Muslim cities the call of the minaret, summoning the faith-ful to prayer, is a familiar sound (as well as a familiar interruption to radio and TV programs). The prayers may be said anywhere, but gathering in mosques is preferred, especially on Fridays when there is an extended worship service.

Formal prayers are offered in the direction of Mecca—a gesture of solidarity with the rest of the Muslim world—and follow a tightly prescribed pattern. First, the worshipper must take care to remove all defilements from the body and from the foreground—a practice that heightens the importance of the act about to be performed. Then, standing erect with open hands raised, the worshipper utters a for-mula of disengagement from the world. Then the worshipper bows, stands erect, surrenders in prostration, kneels, prostrates again, re-turns to a standing position, and then repeats the whole cycle for a second time, all the while reciting verses selected from the Qur'an.

Muhammad taught that prayer was the centerpiece of Islam, the one act that God loves above all others. This teaching is certainly borne out by the level of investment it asks of Muslims. Consider a person who washes up five times a day and then scurries off to engage in a rigid display of bowing, kneeling, prostration, and recitation—any western secularist might think such a person borders on obsession. But of course that is precisely what Islam aims for: Men and women who are so obsessed with God that missing prayers would leave them in a deficit state. Obsession may be too strong a word, but one thing is sure: When the meanings of a myth are reinforced five times each day, the odds increase that these meanings will dramatically and systematically influence the appraisal process. Muslims who attend to their prayers dutifully will be biased by their brains to think, feel, and act differently than Muslims who don't.

The third pillar of Islam is almsgiving. In Muhammad's time destitute poverty grew in the wake of selfish fortune-seeking, provoking his bitter rants against the glaring injustices of Meccan society. It is therefore not surprising that the means for redressing social injustice—almsgiving—would find a central place in the new religion. Islam elevates charity to the status of ritual, frequently linking it to worship.[11] The principle of charity is *zakat,* which means "to purify or increase." The idea is that by giving alms, individuals may purify their own wealth—the more you give, the more God approves of what you keep. Almsgiving also increases the rewards one may expect in the afterlife—the more you give, the higher your station in heaven.

In the early centuries of Islam zakat was imposed as a graduated annual tax, ranging from 2.5 percent to 10 percent of each adult's accumulated wealth. The revenue from this tax was used for many prosocial purposes, including aid to the poor and orphaned, financing the building of schools and mosques, and ransoming Muslim war hostages. Nowadays, of course, taxation and welfare programs are mostly managed by nation-states, but zakat continues to provide self-esteem motivation—God still loves a cheerful giver—for supporting various philanthropic enterprises.

The fourth pillar of Islam, fasting, commemorates the gift of the Qur'an, which was first revealed to Muhammad during the month of Ramadan. Muslims are required to abstain from food, drink, sex, and other delights of the body from daybreak to nightfall for the entire month. Muhammad was familiar with the Jewish practices of spiritual discipline and fasting during the High Holy Days and adapted them to the ancient Arab observance of a holy month. Ascetic practices do not come naturally to Islam, which generally encourages Muslims to relish in God's gifts to the body. But Islam abhors the thought that these gifts might be taken for granted, or that they be considered the main point of human existence, as they were by the hedonists of pagan Mecca. The discipline of Ramadan reminds Muslims that all good things come from God, and that God is at liberty to withhold his bounty.

More broadly, Ramadan is a time for heightened moral and spiritual reflection, for considering what sort of Muslim one really is. Fasting during this period is a concrete way of testing one's resolve to put God's will before one's own. Breaking the fast betokens a spiritual deficiency, a tendency in one's character to be led astray. Ramadan is therefore a month-long exercise in devotion and self-discovery, a strategy intended to revitalize and deepen the Muslim's commitment to embodying God's goal hierarchy.

The fifth pillar of Islam is the pilgrimage to Mecca, a compelling testimony to Muhammad's sensitivity to the power of ritual and tradition. In one brilliant stroke Muhammad co-opted Judaic tradition and used it to reinterpret an ancient pagan custom of Arabian culture. It was Abraham, the Qur'an reveals, who originated the pilgrimage to the Ka'bah after he and Ishmael were commanded by God to build it (Surah 22:26–27). Pilgrimage to this sacred site is required at least once in the lifetime of every able-bodied Muslim who can afford it. So important is this event in the lives of pious Muslims that some even arrange for their bodies to be brought to Mecca after death.

The pilgrimage (*hajj*) involves an elaborate series of ritual performances spread over a ten-day period. As pilgrims enter the consecrated area near the Ka'bah, they clothe themselves in sandals and two simple white linen wraps. The series of ceremonies engages pilgrims in various reenactments of the past: They pray where Abraham prayed, they encircle the Ka'bah several times as Muhammad did, and they offer a blood sacrifice to commemorate Abraham's sacrifice of a ram in place of his beloved son (identified in the Qur'an as Ishmael, not Isaac).

The culminating event is the *wuquf*, a ceremony performed near Mt. Arafat, where Muhammad delivered his farewell sermon, and where the final part of the Qur'an was revealed. In this massive ceremony pilgrims stand erect before God and recite special prayers.

The pilgrimage experience revitalizes Muslim piety in several ways. Like other ritual obligations it tests the Muslim's commitment to embody the will of God. But further, it provides the most forceful affirmation of global Muslim solidarity one might imagine. During the climactic standing ceremony as many as three million Muslims of all races gather from every part of the world, all dressed simply to amplify their equality and humility before God. If there is an inherent human desire for inclusion and belonging, then this ceremony must represent the ultimate opportunity for its satisfaction.

Finally, the pilgrimage is regarded by Muslims as a transforming event, a form of rebirth, from which individuals gain a sense of deliverance from all their sins. When they return to their homes, pilgrims enjoy a new status as living saints. The range and the depth of emotional engagement one experiences during the pilgrimage are without compare in the life of any Muslim. Hearing secondhand reports of the event is itself enough to strengthen commitments to embodying God's will.

This may be the place to comment on the controversial concept of *jihad,* since there are some who insist that jihad constitutes a sixth pillar of Islam. Jihad literally means "struggle" or "exertion," but it is also used to translate the English "war." Jihad has become a media buzzword in recent years, often connoting (wrongly) that a "holy war" has been officially pronounced by some magisterium of the Muslim world. Distortions of this sort are troublesome. There is a story told about Muhammad that helps to clarify the Muslim understanding of jihad. Upon returning from a battle against the enemy, he said, "We have returned from the lesser jihad to the greater jihad." He was then asked what sort of struggle could be greater than armed struggle against the enemies of God. His reply was, "Struggling against the enemy in your own breast."[12]

The most important sense of jihad has to do with the struggles any human being faces when she tries to do what is right. Jihad is what happens when you attempt to resist a temptation. Muslims use the term to refer to the exertion it takes to live up to the demands of God's will. Performing the prayers, giving alms, fasting, making the pilgrimage—each of these engages the individual in a struggle against their own selfish tendencies, a struggle to override the values rooted in their intuitive goal hierarchies. The best full translation of jihad would be "struggle on the path of God." Equivalent secular locutions might be "working for a just cause" or "trying to do the right thing."

The concept of jihad is broad enough to apply to political affairs. When the Bush administration waged war against Iraq in the name of freedom and democracy, they were engaging in jihad in the sense of a holy war. Struggle for the sake of advancing or defending something of excellence is jihad. American politicians who declare that "God is on our side" are advocating a holy war, no less than the medieval Crusaders were. Jihad might be considered a pillar of Islam because all good Muslims are required to exert themselves on the path of God. And where Islam itself is faced with some clear and present danger, then struggling on the path of God might entail armed conflict. All of this is to suggest that there is nothing exclusively Islamic about the idea of a holy war. Indeed, the idea has been unmistakably implicit for centuries in the hereditary title of the English monarchs: Defender of the Faith.

INSTITUTIONAL STRATEGIES

When Muhammad died, his associates moved at once to institutionalize his role as leader of the Muslim community. But there was a problem: If Muhammad truly was the final prophet, then it would be impossible for anyone to replace him. Nevertheless, the Caliphate was established. Caliphs were "deputies" of the prophet, elected with temporal but not spiritual authority. The early caliphs tried to exercise both forms of authority, of course, but their efforts were doomed from the start. Under the caliphs the Muslim community expanded to an empire of truly awesome proportions, but then it began to decentralize and fragment politically, and as it did, the Caliphate was left with no powers, temporal or spiritual. The enduring institutions of Islam—those having centralizing and cohering effects—were those responsible for the development of legal institutions.

Islam is a global community united by a commitment to live, individually and collectively, as God intends for all humans to live. But how does God intend for humans to live? To know this would be to have precise answers to the following questions:

- What is required by God?
- What is encouraged by God?
- What is permitted by God?
- What is discouraged by God?
- What is forbidden by God?

If it could be determined, for any possible action, where it belongs among these categories, then humans would be in no doubt about how to live. It is precisely to make such determinations that Islamic law (*shari'ah*) took shape.

The legal scholars of Islam (called *ulama*) agree that there are four resources available to determine the righteousness of human actions. The first is the Qur'an: If the Qur'an requires certain actions, then the matter is closed. The same is true if the Qur'an encourages, permits, discourages, or forbids an act. If the Qur'an makes a determination, then the determination is made absolutely. The Qur'an does make many determinations—the prayers, for example, are required—but it does not cover every human action.

So there must be a second resource: *hadith*. Hadiths are written accounts of Muhammad's actions, judgments, habits, and extra-Qur'anic sayings. Together, these records project a reliable picture of the prophet's *sunna*, or way of living.

In addition to these written resources, Islamic legal traditions accept two procedural methods as reliable: analogical reasoning and consensus. Analogical reasoning, the third resource, allows jurists to determine what is righteous in a case by reference to parallel cases. The fourth resource, consensus, is justified by Muhammad's claim that the community of believers was incapable of agreeing on an error. Equipped with the categories of right behavior, and the resources to fill them, the legal schools of Islam were confident in pronouncing God's will regarding any conceivable human choice. Ignorance of God's will could never excuse disobedience.

The embodiment of God's will is to be achieved at every level: individual, family, neighborhood, city, state, and world. No conceivable human act falls outside the scope of religious significance. Thus we find laws governing worship, diet and hygiene, marriage and divorce, child rearing, inheritance, economic life, political life, and everything in between. The embodiment radiates from the individual soul outward to the social and political domains—that is, from the greater jihad to the lesser. It must be this way because individuals are ultimately the ones to be judged for the embodiment.

Guidance for this embodiment process is provided by the ulama, the ones most learned in the faith. If an individual is in question about the righteousness of an act, then it is her responsibility to inquire. If she is unable to find guidance from the Qur'an, she may ask a family member or seek advice from the *imam*, the prayer leader at her local mosque. If the imam cannot make a determination, then referral is made to someone more learned, and still more learned, until God's will is clarified. There is no rigid clerical hierarchy in Islam, no hereditary offices and no ordination; authority is vested in law, not office. There are, however, substantial differences among Muslims regarding their knowledge of the law. Most ordinary Muslims learn enough to lead satisfying and inoffensive lives, but others, the ulama, are drawn to make the study of law their life work.

The extent to which Muslims have pushed the embodiment of God's will into the domain of the lesser jihad has varied with time and circumstance. In theory, of course, Islam calls for a theocracy, as any form of ethical monotheism does, Judaism and Christianity notwithstanding. The reason is that God's will always trumps human will, implying that any customs or laws not conforming to God's will are misguided and should be put right. In practice, however, most Muslim leaders have been content to focus on the spiritual and to let the temporal realm follow its own agenda.

But the concession is not complete, for in predominantly Muslim countries the temporal rulers must appear to ground their legitimacy in Islamic principles, which effectively grants "soft law" status to sha-ri'ah. Any secular government willing to ignore this pragmatic principle runs a risk that activist ulama will ignite a countervailing movement. Muslims typically frame all political issues in religious terms, which means that the ideological case for activating the lesser jihad is never difficult to make.

The establishment of a uniform and coherent legal tradition provided the girders for a unified Islamic culture. But the legal tradition could never have flourished without effective educational institutions. Education has always been highly valued by Muslims: "It would be impossible in the Islamic context to discourage learning."[13] Muhammad himself declared that every Muslim was obligated to search for knowledge. The curriculum for traditional education was uniform from the beginning, centered on the Qur'an. Committing the Qur'an to memory was the primary goal of education. Teachers were not concerned whether their pupils actually understood what they were memorizing. There would be plenty of time for understanding to develop later on, but initially the important thing was to get the words into the brain—not just the letters, but the sounds as well. Learning the Qur'an was an affair of the whole body, as the hadith records:

> O God, I ask thee . . . that Thou givest me the provision of the Koran and knowledge, that Thou makest it blend with my flesh, my blood, my hearing, and my sight, and that Thou puttest my body to work through it.[14]

Embodiment of the Islamic myth begins by exploiting the memory systems. Muslim educators were sensitive to the fact that small children have an unusual capacity for rote learning—a gift from God— and that this capacity gradually declines. So there was a window of opportunity for instilling God's word in a manner that would make it a firm foundation for lifelong learning.

Islam prohibits translation of the Qur'an. This prohibition has been broken many times, of course, but there remains an insistence that the only true Qur'an is the Arabic Qur'an. Whatever the theological grounds for this prohibition may be, there remain good practical grounds for it. The Qur'an was meant to be recited; its rhythm, resonance, and poetic structure are intrinsic to its power to captivate and move the hearer. To translate the Qur'an would be to lose these essential qualities.

This has meant that the Arabic language would become and remain a fundamental unifying factor, the lingua franca, of Muslim culture. Wherever Islam went, Arabic went with it: "It was these two things, their language and their faith, which were the greatest contribution of the Arab invaders to the new and original civilization which developed under their aegis."[15] The process of Arabization was advanced by the creation of centers where the greatest works of Sanskrit, Greek, Latin, Syriac, Coptic, and Persian were translated. The vast learning of classical civilizations—literature, philosophy, science—was thereby assimilated, laying the foundations for an astonishing golden age of Islamic culture during the ninth and tenth centuries C.E.

INTELLECTUAL STRATEGIES

Religious traditions deploy intellectual strategies designed to support their myths by clarifying, interpreting, and defending cosmological and moral components. These strategies emerge naturally from disputes about various obscurities and logical bewilderments, or in response to critiques of the myth from external sources. As mentioned earlier, the intellectual energies in Judaic tradition were concentrated on moral issues, while Christian intellectuals found themselves forced into a preoccupation with metaphysical disputes.

Islam is more akin to Judaism in these matters. Most Muslim intellectuals found it more constructive to inquire into what God wants than to wrangle over what God is like. We have already discussed Islamic moral theology in the section on legal traditions, so the focus here will be on metaphysical and epistemological challenges to the myth.

Perhaps the earliest metaphysical dispute in Islamic history was one that the Qur'an itself addresses, the credibility of the afterlife. Pagan poets scoffed at Muhammad's confident promise of resurrection and eternal life: "And they used to say, 'What! When we die and become dust and bones, shall we then indeed be raised up again?'" (Surah 56:47) The Qur'an answers this skepticism with a simple and direct argument: Surely the power that created humans in the first place is sufficient to the task of restoring them. "He says, 'Who can give life to dry bones and decomposed ones?' Say, 'He will give them life who created them for the first time.'" (Surah 36:78–79)

Not very deep stuff, perhaps, but it didn't take long for Muslim thinkers to acquire both the taste and the resources for sophisticated theological reflection. These were acquired through regular contact with Christian thinkers and the translation of Greek philosophical texts, both of which helped to shape the vocabulary and the agenda for theological disputes.

Given the prominent and explicit nature of Islam's claim to revealed truth—"and Muhammad is his messenger"—it was inevitable that the faith-reason debate would erupt in Muslim circles. And so it did, with the appearance of the "Separatist" movement. Proponents of this movement, influenced by Greek rationalism, came to the view that reason should guide one's interpretation of the Qur'an. If passages in the Qur'an violate principles of reason, then they should be reinterpreted in such a way that inconsistencies vanish. The extreme form of this position holds that reason is the judge of revelation.

Separatists were vehemently opposed by the traditionalists, who regarded rationalism as a direct threat to the myth. Traditionalists insisted on using the Qur'an to interpret the Qur'an. They agreed that reason was a gift from God and was therefore a beautiful thing, but they opposed any doctrine that allowed the gift to judge the giver. They insisted that reason, like vision, had limits, and that God was not constrained by any limitations.

Various other standard theological problems came in for debate: the existence of God; realism about divine attributes (were these anthropomorphic symbols or were they accurate descriptions?); the problem of evil; the relative importance of works and faith; whether the universe was eternal or created; and so on. But the issue receiving most serious attention concerned human freedom and responsibility.

If everything in all creation derives from God, then certainly this must apply also to human choices. But if God is responsible for human choices, then what is implied about the coherence of the doctrine of divine judgment? Conversely, if humans are genuinely free agents, then there are some very consequential events in the universe that escape God's power (if not his judgment). Some parties in the debate saved the doctrine of free will at the expense of limiting divine power, while others emphasized God's sovereignty to the point of defending a fatalistic doctrine of predestination. The problem of free will versus predestination has proven to be just as intractable to Muslims as it has been to Christians, with the result that it has been virtually abandoned.

As important as metaphysical and epistemological questions were to some Muslims, they continued throughout the tradition to represent minority concerns. The central concerns remained with moral and ritual obligations. The dominant Muslim attitude toward philosophical inquiry is reflected in the differential treatment of two of Islam's greatest thinkers, Al-Ghazali (d. 1111) and Ibn Rushd (known to the West as Averroës; d. 1198). Al-Ghazali mastered the discipline of philosophy in order to demonstrate that it is ultimately detrimental to the life of faith. Ibn Rushd rejoined with an Aristotelian defense of the authority of reason as the measure of faith. In the end, Al-Ghazali was deeply revered and Ibn Rushd was exiled as a heretic.

The important thing to note, however, is that when challenges to cosmological elements of the Islamic myth did flare up, the tradition responded by developing the intellectual strategies for addressing them—thereby to safeguard and revitalize the myth. The irony of Islamic philosophy is that it had greater impact in the West than it did in the Muslim world. When European theologians rediscovered the wisdom of classical Greece, returned to the West through Muslim sources, they received an intellectual tradition that had been immeasurably enriched by Islam.

EXPERIENTIAL STRATEGIES

I have claimed earlier that religious traditions provide occasions for individuals to validate the meanings of a myth by seeking certain kinds of extraordinary experiences. Such experiences may have several effects. For one thing, they provide an effective doubt inhibitor, and for another, they create emotional markers to bias memory systems and to reinforce goal hierarchies, thereby conditioning the appraisal process. In other words, extraordinary religious experiences are likely to make a difference in the meanings we find in everyday events, and in the prosociality of our reactions to them (one of the twin teloi). But extraordinary experiences are also likely to be deeply satisfying in the most immediate subjective sense—that is, they have the potential for filling subjects with an overwhelming and enduring sense of personal fulfillment (the other telos).

The pilgrimage to Mecca is Islam's principal strategy for producing extraordinary religious experiences. Many Muslims would claim that the hajj experience is the apex of the religious life, the nearest one can draw to God in this life, a savory foretaste of heaven. The standing ceremony near Mt. Arafat—which, if you miss it, invalidates the hajj—leaves one with a sense of the sanctity of all creation. Here the command from God to submit unconditionally and gratefully to his majesty is fully realized; here the command from God to create a community of equity and justice is embodied. In this setting, like no other, there comes an intimation of the oneness, the love and the glory of God himself. The strategy works. Pilgrims do not return home from the hajj as skeptics or nonrealists about the myth of final prophecy; they return transformed, fulfilled, and with a deepened commitment to serve God.

For most Muslims the hajj experience prefigures the final destiny, the ultimate return to God. But intimations of God and foretastes of the return to God are not sufficient for everyone. For some, the path of return to God has to be more immediate and intense than ritual and moral righteousness, or reflective theological insight. The gardens of paradise are too symbolic, and the loving presence of God is too remote for some temperaments.

Islamic mysticism (Sufism) has its origins in a form of asceticism that was inspired by Christian hermits, but it gradually developed the meditative disciplines and theoretical footings characteristic of mature mystical traditions everywhere. Critical for this transition was the example of Rabia Al-Adawiya, an eighth-century slave girl who declined offers of marriage in favor of a lifelong spiritual journey. Rabia was not touched by the standard emotional appeals of Islam:

> O my Lord, if I worship Thee from fear of hell, burn me in hell, and if I worship Thee in hope of paradise, exclude me thence, but if I worship Thee for Thine own sake, then withhold not from me Thine Eternal Beauty.[16]

For Rabia the embodiment of Islam meant the embodiment of love. God is love, he saturates creation with his love, and the point of the spiritual life is to submit to God in love.

Subsequent Sufis became more explicit about mystical union with God, some claiming to be intoxicated with God, others claiming to be indwelt by God, and others claiming total annihilation in God's being. In many cases Sufi writings were dismissive of traditional Muslim preoccupations with legal and theological matters. Orthodox Muslims took exception to the claims of mystics, which often smacked of pantheism, antinomianism, and incarnationism, all severely heretical.

It was not until the career of Al-Ghazali in the eleventh century C.E. that Sufism was delivered from the opprobrium of heresy and made acceptable to the mainstream. Al-Ghazali was a well-known and highly respected legal scholar and theologian, but he was also a restless seeker. Having tried all the conventional paths to God, he turned in desperation to the mystical path, and there he found satisfaction. Al-Ghazali's extensive writings presented an interpretation of Islam that emphasized the domain of personal experience. Henceforth, orthodox Islam would be more open to spiritual variations and Sufism would be less inclined to unorthodox extremes.

Sufi practices called for abstinence, renunciation, repentance, and absolute trust in God as integral to spiritual discipline. These practices have the effect of relativizing the individual's goal hierarchy and creating the conditions for reaching the ultimate goal of Sufism, which is "rectification of character." The Sufi jihad is to displace one's own goal hierarchy in favor of God's goal hierarchy, to embody God's character:

> That is why . . . the Prophet used to pray, "O God, Thou hast made my creation beautiful, so make my character beautiful too." Without the possibility of the rectification of character, the whole idea of voluntary return to God loses its meaning.[17]

AESTHETIC STRATEGIES

In the Christian context many works of art—paintings, sculptures and relief carvings, mosaics, stained glass windows—were created with the express purpose of representing various elements of the myth, sometimes even rivaling the importance of the biblical texts. This practice of allowing images to substitute for the text has been assiduously avoided in Muslim culture. Islam is an iconoclastic tradition—that is, it imposes a theological ban on the creation of images. Interestingly, the source of Islam's iconoclasm is not the Qur'an, but rather the hadith tradition. Muhammad's early campaign against Arab paganism left him overly sensitive to the practice of idolatry, prompting him to apply to all aesthetic expressions the Qur'anic mandate that no being may be allowed parity with God.

Islam may be iconoclastic, but it is not anti-aesthetic. Far from it. In fact, of all the world's religions Islam is the most explicit in its regard for beauty. Beauty is divine: All of God's attributes are beautiful, as attested by the Beautiful Names tradition. One does not exaggerate in saying that beauty is just as central to Islam as love is to Christianity.[18] The principal arts of Islamic tradition are recitation, calligraphy, and architecture, all three of which are focused on glorifying the Qur'an. The Qur'an is the most beautiful gift in human possession, and it must be voiced, copied, and housed in forms befitting its beauty.

Traditional Muslims believe that the Qur'an is the literal word of God, revealed through the recitations of Muhammad. The practice of recitation is therefore taken very seriously—indeed, the Qur'an should always be heard, not read. There are two basic styles of Qur'anic recitation: one for private instruction and another for public performance. The instructive style is quiet and controlled, concentrated on the clarity of pronunciation. By contrast, the performance style is rhythmic and melodious, intended to elicit emotional responses from the worshipping audience. Most Muslims will tell you that when the Qur'an is recited by a trained voice, it is the most beautiful sound imaginable.

But this is by no means a simple beauty, for the power of Qur'anic recitation to command the full range of human feeling is legendary. It can mesmerize, enchant, and excite. Audiences experience everything from deep sorrow to ecstatic joy, and their responses include weeping, shrieking, swooning, and (according to Al-Ghazali) even dying. And the tradition continues to thrive in all parts of the Muslim world. In Egypt, Saudi Arabia, and Malaysia, stadiums and concert halls fill to capacity for recitation contests. But this is not merely entertainment, for the art of recitation is still focused on its original intent, which is to move listeners to revitalize their commitments to embody the myth of final prophecy.

Muslims are admonished to observe strict rules when handling the Qur'an: It must never be placed on the ground; it should never be placed beneath another book; the handler must not be ritually impure; it should be kissed or placed over the eyes before setting it down. In view of such customs of veneration it is not surprising that inordinate care would be taken when making copies of the book. Calligraphy and manuscript decoration became extremely important and highly developed art forms in Muslim culture. If the Qur'an is the most beautiful thing heard, then it should be the most beautiful thing seen as well.

Calligraphy, the only genuinely original Muslim art form, may have developed from attempts to render the melodic tones of the Qur'an in visible form. Early Muslims were probably inspired when they encountered the illuminated manuscripts of their Byzantine neighbors, and before long they were creating copies of the Qur'an that far surpassed the grandeur of the finest Christian works.

In addition to calligraphy Muslim artists embellished the Qur'an with intricate geometrical designs. The meticulous care and skill reflected in these decorative arts had the effect of exciting admiration and awe in those encountering them, thereby deepening their veneration for the word of God.

Muslim architecture finds its most impressive expressions in the mosque, a place of prayerful submission. The liturgical and ritual requirements of mosques are minimal (for example, no altar and no choir), the most important functional aspect being floor space to accommodate rows of worshippers. The two standard features of mosques are the *minaret* (the tower used to summon Muslims to prayer) and the *mihrab* (a small niche orienting the faithful toward Mecca).

The emphasis in mosque architecture is more focused on details of decoration than on structural features: "What the princes and the populace alike admired in the Muslim buildings were less the architecture as such than the mosaics or paintings on the walls, the decoration in abstract motifs, and the beautifully lettered inscriptions."[19] Inscriptions from the Qur'an are prominently displayed on both interior and exterior surfaces. In some respects the Ka'bah in Mecca, the most revered structure in all of Islam, best exemplifies the simple ideals of Muslim architecture. The Ka'bah is basically a huge cube draped with colorful fabric on which are emblazoned words from the Qur'an.

Another important feature of mosque architecture is color. Much of the physical environment across the Muslim world is barren, colorless, and devoid of features. As one author puts it, "The unrelieved monotony of both the surrounding landscape and the urban setting demanded at least psychological relief."[20] Mosques provided the needed relief. They are oases of brilliant colors and interesting patterns punctuating a mud-colored world. They provide visual excitement to cheer the weary. To approach a mosque is like discovering a box of jewels half-buried in the sand. All of these features conspire to stimulate impressions of refreshment and splendor. The mosque, like the call from the minaret, summons the attention and submission of the faithful.

PERSONAL WHOLENESS AND SOCIAL COHERENCE

The viability of our species depends upon a reciprocal relationship between personality and sociality. That is, stable social groups require individuals whose homeostatic systems—physiological, cognitive, emotional—are fully operative and maintained at or near equilibrium. Correlatively, maintaining homeostatic systems at or near equilibrium requires a social order that is predictable, cooperative, and well integrated.

A central part of my argument has been that these reciprocal dynamics are capable of generating very large social groups, but only on the condition that cognitive and emotional systems are uniformly modulated by symbolic learning—that is, by ingesting memes about how things are and which things matter. This is an overly complicated way of saying that our species cannot thrive in large groups unless we create effective symbolic conditions for the simultaneous achievement of personal wholeness and social coherence. Another central part of my argument has been that the provision of these symbolic conditions is fundamentally what religious traditions are all about. Islam is a good example.

In fact, Islam may be the best example of all, because in this case the entire process inherent in these principles unfolded in a manner so quick, so stepwise, so uncomplicated by misleading variables and extraneous factors that it has all the appearances of a controlled laboratory experiment. Take a random sample of human beings too psychologically and socially dysfunctional to count as a coherent group (in this case, pre-Islamic Arabs); then introduce an appealing and well-packaged myth (Muhammad's version of ethical monotheism); and presto! within a generation you have the predicted outcome: a coherent group of fulfilled individuals.

The secret of Islam's success is that its emotional appeals, and the ancillary strategies it deploys to keep them pressing on the attention and memory systems of individuals, create significant overlaps of self-interest. That is, it motivates individuals to enact the prosocial behaviors that are precisely and exhaustively specified in Muslim law. This happens if the individual assimilates the law into his goal hierarchy, thereby linking self-esteem to prosocial behavior. If self-esteem is linked to behaviors that serve the interests of others, then the individual performing those behaviors will not have a sense of making a sacrifice. If you have displaced your old selfish goal hierarchy with God's goal hierarchy, then your goals will be God's goals, and you will help the needy widow simply because you want the widow helped. You would feel badly about yourself if you failed to help her. It's all benefit and no cost, for the widow, too.

All three of the Abrahamic religions exploit the dynamics of self-esteem linkage in order to create overlaps of self-interest, thereby establishing the conditions for personal wholeness and social coherence. I am tempted to characterize the Abrahamic traditions as "new agenda" religions because they call on us to renounce the self-centeredness of our own agenda and to assimilate the other-centeredness of God's agenda. Under the old agenda the value of something is judged relative to the self, but under the new agenda value is judged relative to God's will. I want to say that the ultimate objective of the Abrahamic traditions is to educate the emotions toward this transformation from an old agenda to a new agenda.

I have made all of these points earlier, but I repeat them now for the sake of drawing attention to a striking irony at the heart of the Abrahamic traditions. It seems to me ironic that these traditions would use the language of rewards and punishments, because implicit in this language is an appeal to the old agenda of calculated self-interest. Which is it? Should I help the needy widow because I want the widow helped (new agenda), or should I help the widow because

I want the rewards of heaven (old agenda)? One might suppose that for a fervent believer any talk about rewards and punishments for the self would be irrelevant or even offensive. Yet the language is there.

It seems as if the Abrahamic traditions are speaking out of both sides of their mouths on the business of motivating prosocial behaviors. A God-centered goal hierarchy is in fact compromised by the language of rewards and punishments, but the ego-centered goal hierarchy is exploited and reinforced by such language. I often observe this fundamental tension in my students. I offer them the assertion that a real Jew, Christian, or Muslim doesn't care two figs about the afterlife, because the main point of these religions is to serve God without regard for oneself. They agree heartily. But when I ask them whether the afterlife is of central importance to them personally, they unequivocally say yes. And when I ask them whether the Abrahamic traditions would be viable without heaven and hell language, they unequivocally say no.

The point here is not criticism. I am not saying that this fundamental irony in the Abrahamic traditions is a sign of incoherence, and therefore weakness. To the contrary, I believe it represents a pragmatic strength, a willingness to create overlaps of self-interest by catching people where they are, and playing on all the strings of human nature. We see the same two-pronged approach during Public Broadcasting System pledge drives: In one sentence we hear all about the wonderful things PBS does for the common good, and in the next sentence we're told that if we send in our generous contributions now, they'll send us a special gift. Something for everyone. But why not? If the ultimate point of religion is to create overlaps of self-interest (and thereby to advance prospects for achieving personal wholeness and social coherence), then we should expect to hear from both sides of the mouth.

The thesis of this book is that religion is about human viability, not about God. But the material in these three chapters covering the Abrahamic traditions may give exactly the opposite impression—that religion is *exclusively* about God and what God wants, and has nothing to do with human goals. I certainly agree that when you examine these traditions closely, and when you talk to their practitioners, there can be no mistake that the focus of their attention and the object of their subjective experience is God. But it is often the case that what we see on the surface of human activity and experience is best understood as reflecting a deeper, more universal reality.

For Chandler, breakfast may be about spooning in a bowl of cornflakes and enjoying the taste, but for Joey breakfast may be about the preparation and delights of pancakes, and for Ross it might be about poached eggs. The focus of their attention and the quality of their experiences will vary somewhat, but everyone can agree that all three are engaged in an activity that transcends the details of their attention and experience. You may learn something important about Joey if you ask him about the meaning of breakfast and he tells you "pancakes," but you haven't learned much about breakfast. If you want to know what breakfast is really about, you would do well to speak to a nutritionist, who will tell you that breakfast is really about nourishment.

Maybe this is too obvious (and by now irritatingly familiar), but the point I hope remains after these three chapters is that Judaism, Christianity, and Islam are examples of particular traditions designed for the pursuit of the twin teloi of personal wholeness and social coherence.

HINDUISM

This part of the book is meant to show how a small selection of religious traditions bears out the features of the theory set forth in chapter four. If Islam is our best example, then Hinduism must be the most challenging. Whereas Islam burst onto the scene as a well-defined and unified mythic tradition, Hinduism has always defied definition and has never been unified. In fact, scholars constantly remind us that "Hinduism" is a Western construct that misleads us to expect more conformity of beliefs and practices than the facts can deliver. Nevertheless, it has become conventional to speak of "Classical Hinduism" as a recognizable and coherent mythic tradition. This will be the Hinduism of the present chapter.

HISTORICAL CONTEXT

Archaeologists have revealed the remains of a great Indus Valley civilization that flourished before 2500 B.C.E. What little is known about this Bronze Age culture indicates that they were farmers and traders who built large cities with sophisticated sewage disposal systems, street lighting, and capacious grain stores. There is evidence that the Indus Valley civilization exported agricultural products, including cotton, to Sumer, Egypt, and Crete. This was a highly developed and affluent society with a centralized government and a network of cities covering an area larger than the civilizations of Egypt and Mesopotamia.[1]

The language of the Indus Valley people remains undeciphered, and almost nothing is known of their religious life, although the presence of large public bathhouses invites speculation that they practiced some form of purification rituals. This remarkable civilization was already in a state of serious decline by 1500 B.C.E., when Aryan pastoral nomads began pressing into India over the mountain passes to the northwest. For the next thousand years the history of India would be one of gradual Aryanization.

The splendid isolation of India makes it ideal for empire building. With rugged mountains to the north and seacoast everywhere else, the subcontinent is a virtual island, full of natural potential for political unification. But the Aryans were not empire builders. Empire building presupposes more unity of purpose than these loosely connected tribal units were able to muster. They were just as likely to fight among themselves as with the indigenous groups they encountered as they penetrated eastward and then southward into the peninsula. The Aryans were nomadic cowherds, basically, who took their time in making the transition to a settled life. They built no cities for several centuries, preferring instead to dwell in tribal villages where they could pasture their herds and learn the arts of agriculture. The process of Aryanization was therefore slow and relatively bloodless, less characterized by brutal conquest than by compromise, assimilation, and hybridization with local populations that fell in the path of expansion.

By 500 B.C.E., Aryan social and political organization had stabilized into a system of minor kingdoms. Tribal villages were governed by chieftains or assemblies, giving way to larger territorial units ruled by kings (*rajas*) who administered taxation and justice through appointed bureaucrats. Aryan kings were elevated tribal chiefs who commanded a military force and performed important symbolic functions, but they were not feudal overlords and possessed no property rights. Property was owned by families, not individuals. Despite constant threats of intertribal warfare, tribal villages enjoyed a fair amount of stability and a high degree of autonomy.

The telltale marks of Aryanization were language, religion, and social hierarchy, but even these pillars of Aryan culture were cross-fertilized with local traditions along the way. The result was that Aryanized India became a highly pluralistic culture—not a large melting pot but many little pots, each with generous dollops of Aryan influence.

The Aryans were blessed with a gift for lyricism, and they had just the language for it. Sanskrit is a beautifully expressive instrument, whether in the service of Vedic hymns or the philosophic discourse of the Upanishads. Everywhere the Aryans went, they brought the Vedas, an oral tradition that was committed to text between 1500 and 600 B.C.E. The Vedas contain countless hymns addressed to many gods. They speak of creation stories, musings on the origins and nature of humankind, cosmic battles between gods, heroic exploits from the Aryan past, and much else.

An especially important aspect of Vedic tradition concerns the conduct of ritual sacrifices, the central element in Aryan religious practice. Early on, most Aryan-Indic religion was animistic—that is, it was preoccupied with forces in the world that were beyond human comprehension and control. These forces were rendered more intelligible and less worrisome by personifying them. And the good will of these gods might be influenced by ritual sacrifice, which became almost an obsession in Aryan-Indic culture. There were private sacrifices relative to every conceivable uncertainty in domestic life, and public ones to honor the king, thereby to win divine support in the persistent wars between tribes. Some of the sacrifices could be performed by novices, but rituals were increasingly left to the expertise of priests who were able to perform them according to detailed instructions provided in the Vedas.

The origins and development of India's caste system is a hotly disputed subject. It is not known whether the Aryans possessed a system of social stratification upon their arrival in India, but it is clear that Aryan-Indic communities almost universally reflected four distinct social classes: seers-priests, warriors-rulers, artisans-producers, and followers-servants. This class system was important enough to be given a place in the sacred Vedic literature:

> From his mouth came the priestly class
> From his arms, the rulers.
> The producers came from his legs;
> From his feet came the servant class.
> (*Rig Veda* 10.90.12)

It is possible that the first three classes originally represented simple divisions of labor, and that the fourth class emerged as an expression of Aryan chauvinism toward their Indic neighbors. Whatever the origins, in time these class distinctions hardened into castes, eventually acquiring hereditary dimensions as well as prohibitions against marrying out of one's caste. Various abuses crept into this system in the late pre-classical period, but one in particular deserves further attention here.

The priestly caste, called Brahmins, rose to a position of inordinate power and authority in Aryan culture, enabling them to benefit from the naive insecurities of the prosperous artisan caste. The primary instrument of Brahminic exploitation was the system of magic reflected in the *Atharva Veda,* the last of the four Vedas. This system presumes the power of ritual sacrifices to affect outcomes in the natural order. The *Atharva Veda* contains over five hundred magical charms and incantations meant to control nature in specific ways. For example, there were hymns for "protection from snakes, good luck in gambling, release from demons or witchcraft, riddance of a cough, successful childbirth, success against enemies, success in love affairs, recovery of virility, or destruction of the virility of a rival." [2] If the rituals were performed improperly, they were likely to produce adverse effects, so it was essential to procure the services of a Brahmin (for a hefty fee), since priests were the only ones who held the requisite magical powers.

It is easy to see how this exploitative aspect of Aryan-Indic religious life might erode the conditions for personal wholeness and social coherence, even to the point of provoking a reaction. The sacrificial system encouraged psychological dependency and fatalism. It undermined any enduring sense of personal efficacy: If nothing in my life is left up to me, then I cannot feel in control of anything. In the end it left individuals without a sense that their circumstances in life might improve. How fulfilled can one feel when trapped in an endless and futile cycle of escaping one peril after another? The system undermined conditions for social coherence as well. For one thing, it squandered time, energy, and material resources that might otherwise have been used for more constructive social goals. But more importantly, it did nothing to generate overlaps of self-interest. Just the opposite: It became an oppressive form of social parasitism.

It is also easy to see how the sacrificial system would eventually appear downright implausible to attentive minds. What was one to think when the rains came before the priests had a chance to perform the necessary sacrifices? Or when the sacrifices failed to bring the rain? Or when the ritually treated cough deepened and eventually killed? How long would it take for even the most casual observer to realize that the priestly record for producing results was no better than blind chance? It was inevitable that the limitations of the rituals would become widely recognized, and that the whole fabric of Vedic religion would be called in for serious examination. New directions in reflective thought were already being mapped out before the seventh century B.C.E., and by the sixth century reactions against the religious past began to materialize. Two of these, Jainism and Buddhism, rejected the Vedic world view completely, while a third, the Upanishadic tradition, sought to reinterpret its meaning.

The Upanishads, written during the sixth and fifth centuries

B.C.E., were commentaries on the Vedas. These texts moved from specific concerns about the power of rituals to more general and speculative questions. What is the nature of the universe? The nature of the self? What is the relation between reality and appearance? Between unity and multiplicity? Between permanence and change? Between experience and objects? Attempts to resolve such perplexities prepared the way toward Classical Hinduism.

THE MYTH OF MANY PATHS

One may rightly say that Hinduism contains many myths—that is, there are many different stories having particular cosmological and moral elements, each with potential to inform human thinking, feeling, and acting. If we were to focus on these many particular stories, then we would never discover Hinduism, but rather a confusing array of individual religious orientations. To describe this diversity as confusing is an understatement. There are literally thousands of gods and myths falling within the family of Hindu traditions. So diverse is this family that one may be said to qualify as a Hindu simply by not explicitly rejecting the label.

In order to speak of Hinduism, therefore, we must focus on a meta-myth, a more comprehensive story of how things are and which things matter, which provides a unifying framework for the many particular stories. The principal accomplishment of the Upanishads was to formulate this meta-myth, which is summarized in the following.

> The absolute and final truth about reality is that Reality is One, not many. When we speak of "many" real things (including millions of gods), we are not speaking of Reality as it is "in itself," we are rather speaking of Reality "from itself," that is, Oneness manifested in manyness. The One True Reality (*Brahman*) is primary and original, while the many real things are derivative from Brahman. The realm of the many manifested is a lawful, orderly realm, strictly governed by *dharma*, the law that sustains the derivative world. Dharma is the ultimate order, "how things are," in the derivative realm. It is known both as natural law (*rita*) and as moral law (*karma*). All change, all action, all causation in the manifested realm is conditioned by dharma. What happens to the human body is governed by rita, the laws of nature, and what happens to the human self (*atman*) is governed by karma, the laws of action. The self may try to ignore or rebel against dharma, but this is futile, for the consequences of actions are inexorably enforced by karma.
>
> The absolute and final truth about the self is that it is ultimately none other than Reality: Atman IS Brahman. The self is Brahman in a manifested state, a dharmic state. As long as the self is held in a dharmic state it is subject

to *samsara,* a cyclical pattern of successive rebirths, a pattern that follows the ironclad dictates of karma. For the self to leave its dharmic state and achieve its true, unmanifested state of unity with Brahman (known as the state of *moksha*), it is necessary for the self to follow a path of liberation from dharma to moksha. Each person is a unique combination of abilities, limitations, temperaments, and experiences. Each person has a unique set of obstacles to achieving moksha, and the path of liberation is therefore somewhat different in details for each individual. The right way for you is your personal dharma, your personal law. Dharma, the principle of order underlying both natural and moral law, is also recognized as conditioning each individual self. The pathways to moksha are therefore as numerous as human selves. But just as humans can be grouped in terms of personality types, so the paths of liberation may be grouped in three types of disciplines: the paths of insightful knowledge (*jnana yoga*), the paths of dutiful action (*karma yoga*), and the paths of loving devotion (*bhakti yoga*). To follow any one (or some combination) of these pathways is sufficient for liberating the self from the karma-driven cycle of rebirth.

Many have observed that Eastern and Western cultures reflect fundamentally different concerns. The critical question for the Western mind is "What is the truth?" but for the Eastern mind the question is "What is the way?" It is clear, however, that Hinduism concerns itself with both questions. Hindu tradition recognizes that metaphysical claims and assumptions are inherent in every religious orientation, and the meta-myth of the Upanishads is an attempt to make these explicit. But the tradition has also recognized that metaphysical truths—even the meta-myth—cannot take us very far toward the goal of human existence, which is to transform the self. The anatomy of Hinduism may be reflected in the meta-myth, but its physiology is found in the many particular paths to liberation of the self. I will have more to say about these paths shortly, but for the moment there is more to say about the meta-myth.

First, the concept of dharma is the conceptual device that integrates cosmology and morality into narrative form. That is, dharma does for the Hindu myth what the metaphor of God as person does for the Abrahamic myths. Dharma is the ultimate explanation for events in the natural order as well as the ultimate justification for actions in the moral order.

The meta-myth assumes realism about dharma—that is, one must accept dharma as more than a useful way to think. It is a useful way to think because it is an accurate account of how things are. One must also be a realist about the blissful transcendent state of moksha as a possibility for the self. Without realism about the extra-mental status of *dharma-moksha* duality, there would be no source of motivation to follow any of Hinduism's many paths. Implicit in the realistic stance toward *dharma-moksha* is realism about samsara, the cease-less cycle of rebirths endured by the self, as well as realism about karma, the law of consequences that consigns the self to successive births. Skepticism or nonrealism about these fundamental principles will effectively neutralize the emotional appeal inherent in Hindu myths. But if one maintains an attitude of realism about these fea-tures of the meta-myth, then the cognitive stage is set for the individ-ual to appreciate the stakes in adhering to a disciplined path.

Modern theories of learning emphasize fundamental differences in learning styles. Some people are visual learners, while others may be auditory or kinesthetic learners. Educators are encouraged to honor these differences by varying the presentation of material in their classrooms. Similarly, Hinduism honors human diversity in the religious life by encouraging individuals to seek paths of piety that are most suitable to their own dispositions and temperaments. Intellec-tuals who favor the contemplative life of study and reflection will be drawn to the disciplines of jnana yoga; pragmatic take-charge types who cannot sit still or withdraw from the world will find the path of karma yoga more agreeable; and passionate, sensitive, and emotion-ally expressive individuals will find bhakti yoga more satisfying.

JNANA YOGA

Jnana yoga, the path of insightful knowledge, may be approached by contrast with the philosophy of Immanuel Kant. Kant made a funda-mental distinction between *noumena* (reality as it *really* is, in itself, apart from any experience of it), and *phenomena* (reality as it appears in the experience of sentient beings). This is similar to the Upani-shadic distinction between Brahman and the derivative world of many manifestations. Kant insisted on complete agnosticism with re-spect to the noumena—we cannot know what reality is like in and of itself. All knowledge is conditioned by contributions from the side of the knower, and is consequently relative to the knower's instruments of reason and sense. For Kant, therefore, metaphysical claims such as "the essence of Reality is Oneness" or "Atman is Brahman" cannot be known. They may be asserted as practical postulates to quiet the urge to know ultimate truths, but they are not known.

Hinduism disagrees sharply and vehemently. We *can* know the noumena, we *can* know the absolute and final truth about Reality as it is in itself. Granted, this is not the conventional knowledge of rea-son and sense, but it is much deeper and more certain than conven-tional knowledge ever could be. The knowledge of direct insight into

Reality is made possible by rigorous intellectual disciplines designed to abrogate and deconstruct all the influences of the instruments of conventional knowledge. The trouble with conventional knowledge is that it is never disinterested, it is always *someone's* knowledge, inevitably reflecting the knower's self-interests, stakes, and desires—the knower's goal hierarchy. The secret of insightful knowledge is systematically to remove the goal hierarchy, to get the autobiographical self out of the picture completely. Then, presumably, one no longer apprehends "what is happening to *me,*" but rather apprehends "what is *really* happening." And what is really happening is a disinterested experience of pure undifferentiated being. To enter into this state of knowing is to destroy the effects of karma and thereby to achieve liberation from the endless cycle of rebirths.

The disciplines leading to insightful knowledge may vary somewhat, but typically involve the practice of moral purity, instruction from a liberated teacher (*guru*), reflective study, and lots of strenuous meditation.[3] Having directly experienced the absolute unity of Being, the individual returns to the phenomenal world in a state of complete emotional detachment. Cruising blithely toward a birthless death, the liberated self has no agenda, no goal hierarchy, and no longer makes stake appraisals.

KARMA YOGA

Karma yoga, the path of dutiful action, is bound up with the caste system and with the Hindu concept of "life stages." We have already noted that Aryan-Indic society reflected four castes (priests, rulers, producers, servants). What was considered righteous or dutiful varied with one's caste. Righteousness and duty also varied with respect to one's stage in life. The four stages in life were student, householder, recluse, and wandering ascetic. Thus one's station in society as well as one's stage in life were variables having a bearing on one's duties. Each combination of variables, together with temperamental differences, determined a unique dharma for each person. One's personal dharma (duties) was assumed to be just recompense for the actions of previous incarnations. A person should never make the mistake of serving the duties appropriate to the circumstances of another person, for to do so would only worsen the accumulation of karmic effects (who would deign to take another person's medication?). It is better to do your own duties poorly than to do someone else's duties well.

In the second century B.C.E. a class of texts, called *Dharma Sastras,* began to appear for the purpose of spelling out both general and specific social duties appropriate to the different combinations of caste and life stage variables. Students, for example, had duties to remain celibate and ritually pure, to study hard, beg for their food, and do the bidding of their teachers. Veneration for teachers was especially important, as this passage from *The Laws of Manu* shows:

> By censuring the teacher, though justly, he will become in his next birth an ass, by falsely defaming him, a dog; he who lives on his teacher's substance will become a worm, and he who is envious of his merit, an insect.[4]

Householders were duty bound to produce and to reproduce—that is, to marry, have many children, and apply themselves to the demands of an honorable vocation for the support of their families and communities. They were also expected to study scripture and offer sacrifices, as well as practice the social virtues (honesty, kindness, and respect) expected of everyone else.

Performing the duties of one's personal dharma produces good karmic effects, earning for the self better circumstances in the next incarnation. If one fails to perform duties well, then the next life will be less dignified and more unpleasant. The path of karma yoga is one of measured improvement: If you are a dutiful producer, then in the next life you may be born into a higher caste with a new set of duties. If this progress continues, then one will eventually be reborn into the priestly caste where the duties are focused on spiritual disciplines leading directly to moksha. Consistently good karma will, in time, liberate the self.

The path of karma yoga could, however, be shortened if one combined it with the disciplines of jnana yoga. If an individual performs personal dharma in a mode of self-detachment, then karmic effects will be avoided. Ordinarily we take actions for the sake of desired outcomes. This action, we say, will have better consequences than that one. The normal mode of action has an agenda at work, whether it is motivated by selfishness or by the desire to help others. Either way, where there's a desired outcome, there's a goal hierarchy. But it is precisely the attachment of the self to a preferred outcome that gives traction to the laws of karma. If, however, the self is indifferent about outcomes, if the self is not in any way invested in its actions, then there is nothing to attract karmic effects. Such an agenda-free self is liberated:

> He who regards pain and pleasure alike, who dwells in his own self, who looks upon a clod, a stone, a piece of gold as of equal worth, who remains the same amidst the pleasant and the un-

pleasant things, who is firm of mind, who regards both blame and praise as one; he who is the same in honor and dishonor and the same to friends and foes, and who has given up all initiative of action, he is said to have risen above the modes.[5]

BHAKTI YOGA

Bhakti yoga, the path of loving devotion, has been by far the most popular form of piety within the Hindu tradition. The way of insightful knowledge is too rigorous and impersonal for many Hindus, and the way of dutiful action is too rigid and legalistic. By contrast, the path of loving devotion is more narrative, more personally engaging, and much more satisfying emotionally.

The origins of bhakti yoga are obscure, but it appears to have its roots in a flourish of cultural creativity following the reign of King Asoka (268–233 B.C.E.). By this time the influences of Buddhism and Jainism were stimulating all sorts of things, not least an interest in popular stories and epic narratives, in which the gods appeared more interesting, more personal, and more accessible. Localized forms of theism were expanding, offering a different way to think about divine realities than one finds in the world of the Upanishads. Popular theism is more emotional than conceptual, more outward-reaching than inward-searching, more dualistic than monistic, more worshipful than reflective. It is characterized by a subject-object relation, and is felt by the worshipper as personal intimacy with God, not metaphysical identity. As the sagely Sri Ramakrishna put it: "I want to *taste* sugar, not *be* sugar." Bhakti yoga would thrive on such sentiments.

The earliest unambiguous reference to bhakti yoga is found in the *Bhagavad-Gita,* probably composed sometime in the first century B.C.E. One of the marvels of this most beloved book is that it achieves a sweeping synthesis of many of the most important ideas in Hindu tradition. In this text Krishna, one of Vishnu's divine manifestations, instructs Arjuna (read: "every self") on the nature of divinity, humanity, and the paths of liberation. Krishna teaches Arjuna various paths and explains the importance of striving for insight and righteous action. Important as these paths are, however, at the end of the day loving devotion is sufficient for salvation:

> Listen again to my supreme word, the most secret of all. Well beloved art thou of me, therefore I shall tell thee what is good for thee. Fix thy mind on me; be devoted to me; sacrifice to me; prostrate thyself before me; so shalt thou come to me. Abandoning all duties, come to me alone for shelter. Be not grieved, for I shall release thee from all evils.[6]

Popular theism continued to develop throughout the first millennium C.E., finding expression in the *Puranas*, a genre of literature used as devotional material for the growing bhakti yoga movement. Devotees of the many gods of the Puranas typically choose a deity that manifests some personally attractive or ideal qualities, and then attach themselves to the deity for life. They strive to make God an intrinsic part of their personality, seeking to embody the divine qualities, seeking to be simultaneously emptied into God and filled up with God. Various strategies support the effort, including worship in temples and household shrines, hymns, mantras, dancing, and so on. The point of it all is to achieve an intensely intimate personal relationship, a state of absolute love, devotion, and obedience to God.

It is becoming evident that Hinduism reflects both similarities and differences with respect to the Abrahamic traditions. I have claimed that the Abrahamic traditions are "new agenda" religions. That is, the point of the religious life in these traditions is to displace the old self-centered goal hierarchy with a new goal hierarchy centered on God's will. This same dynamic is at work in the bhakti yoga path, where the point is to give oneself in loving devotion and obedience to God, to identify so intimately with the object of devotion that God's will becomes one's own. The karma yoga path, too, displaces the selfish agenda when one undertakes the pursuit of dharma. But the path of jnana yoga (and the shortened path of karma yoga) is different. Here the objective is to have "no agenda," that is, to deconstruct all goal hierarchies completely, thereby to enter a state of emotional detachment where the self no longer engages in appraisals of meaning. Despite these differences, there remains a fundamental agreement that the good for both individuals and society depends on strategies for goal hierarchy management.

EMOTIONAL APPEALS

The Hindu meta-myth is a formal conceptual framework, a sort of philosophical display case featuring the plethora of particular mythic traditions that comprise Hinduism. As such, the meta-myth plays only indirectly on the strings of human nature, leaving the most concentrated emotional appeals to the imagery of particular mythic paths, of which there are far too many to consider here.

FEAR AND DESIRE

The meta-myth does, however, appeal to human capacities for fear and desire insofar as it articulates the common system of rewards and punishments that operates in the various sub-traditions. Rewards and punishments in all variants of Hinduism are reckoned within the system of karma and rebirth. This system is Hinduism's version of the universal religious insight that justice will prevail; that sooner or later we must all face the consequences of our behavior. Whether one's personal dharma is the rigorous path of jnana yoga or the more popular devotional path of bhakti yoga makes no difference. The basic understanding is the same: If one's personal dharma path is followed diligently, then good karma will accrue and one may expect a better future existence. If, however, one fails to perform personal dharma, then negative karma will inevitably result in dire consequences:

> Accordingly, those who are of pleasant conduct here—the prospect is, indeed, that they will enter a pleasant womb, either the womb of a Brahman, or the womb of a Kshatriya, or the womb of a Vaisya. But those who are of stinking conduct here—the prospect is, indeed, that they will enter a stinking womb, either the womb of a dog, or the womb of a swine, or the womb or an outcast.[7]

It is difficult to see how a realist about this system of justice might encounter such admonitions and not be moved either by fear or desire toward a deeper commitment to following a path of "pleasant conduct."

To see how Hinduism plays more fully on the strings of human nature, it will be necessary to focus the remainder of this chapter on a particular path within the broad expanse of Hindu tradition. For this purpose a choice from devotional Hinduism seems most appropriate, since it is the most common type of piety. The tradition of Krishna devotion should do nicely, though it should be remembered that this is only one of very many particular mythic traditions within Hinduism. The choice of Krishnaism is almost completely arbitrary.

LOVE IN THE KRISHNA-BHAKTI TRADITION

Krishnaism is not mostly about love, it is *all* about love. There are, of course, many forms and expressions of love, and Krishnaism seeks to exploit all of them. The point, it seems, is to open every portal of the human heart so that the love of Krishna gushes in and out without ceasing. But first, a bit of background.

Legends and lore about the enigmatic Lord Krishna have a fascinating literary history. In the epic literature Krishna appears as an adult: an avatar of Vishnu, a warrior chief of the Yadava tribe, and as charioteer and teacher of Arjuna in the *Bhagavad-Gita*. Later, in the

Puranas, we encounter stories of Krishna as a remarkable and mischievous child, a fun-loving cowherd boy and slayer of demons, and then as an irresistible seducer of milkmaids. These sources were lavishly embellished by later accretions in poetry and song. The basic narrative goes as follows:

In a time when the earth was oppressed by demon-kings, Vishnu announced that he would enter the world as the eighth child of Vasudeva, both to rid the earth of demon-kings and to establish a cult of devotees. The demon-king Kamsa hears a prophecy to this effect and resolves to kill all of Vasudeva's children. Kamsa's plan is foiled when Krishna and his brother Balarama are secreted to a foster home in a nearby cowherd settlement.

Krishna's divine powers were evident from the beginning. On one occasion a demoness was sent to nurse the infant with her poisoned breast, whereupon Krishna, immune to the poison, sucked the life out of her. On another occasion he was put down to nap beneath a loaded wagon and when he awakened hungry he kicked over the wagon and shattered it to pieces. On yet another occasion Krishna's mother tried to restrain him by tethering him to a heavy stone, only to discover him uprooting trees as he crawled through the woods dragging the stone behind. Once, Krishna's mother suspected him of eating clay, so she peered into his mouth and there she beheld all the splendors of the entire cosmos.

The foster family took such incidents as omens and moved to a new location, called Vrindavana, where Krishna and Balarama were raised as cowherds. There they spent their time enchanting the forests with their flutes and songs, and protecting the world by slaying demons and serpents. As a youth Krishna was fond of teasing the milkmaids who came to bathe in the forest streams. He would steal their clothes and make them come one by one to retrieve them beneath the tree where he was hiding, and when they did he promised to marry each of them. One summer, after the rains had lifted, Krishna was strolling in a forest glen playing his flute with such allure that all the wives and daughters in the village fell instantly in love and rushed away from their duties to join the charming and beautiful youth. As they arrived Krishna multiplied his form to accommodate each of them, and there, for two months, they all enjoyed the delights of music, dance, conversation, and lovemaking. Then Krishna quickly disappeared, leaving the maidens lovesick and forlorn, as he was called to vanquish the demon-king Kamsa and to release his parents from bondage. This he accomplished with dispatch and then enthroned a Yadava ruler before going on to further world-protecting exploits.

In the course of events Krishna multiplied his form and took several wives, treating each one with tenderness and devotion. On one occasion he married 16,000 brides to free them from bondage to a demon-king. The narrative concludes when Krishna is fatally wounded by a hunter's arrow. As he lay in agony he gave final instructions for the care of his family, and then as praises sounded from the heavens, the Lord Krishna disappeared from earth in a flash of lightning.

The source material about the life of Krishna includes hundreds of very diverse stories, presenting a vastly complex and interesting set of images. Sometimes Krishna is a naughty little boy, sometimes a randy youth, sometimes a courageous hero, a wise advisor, a tender lover, a radiant god, and so on. But never do we find him boring. At the heart of the great popularity of Krishnaism is a prodigiously attractive personage. Human beings are naturally drawn to individuals who are unusual or complex. Anyone who is exceptional in power, wisdom, intelligence, creativity, courage, or even cruelty makes promising material for a best-selling biography. Krishnaism plays on this predisposition in every way it can, with the result that Lord Krishna is indeed a figure who offers something for everyone to love.

The childhood stories are especially effective. We have already seen how Christianity appeals to the human propensity to relate affectionately with children. The baby Krishna is no less adorable than the baby Jesus:

> With your tiny little face, and your tiny arms and legs, with a tiny bit of
> butter on your tiny little hand;
> You talk a tiny bit, and you fret a tiny bit, and you're a tiny bit delighted
> at the tiniest little thing.
> Your tiny little cheeks and your tiny little teeth form a tiny little smile
> that captivates us all.
> Just a tiny, oh a tiny little closer, Lord, let Sur approach;
> And with a tiny bit of grace, grant me a tiny bit of shelter.[8]

Even Krishna's naughty pranks are occasions for endearment:

> "Mama, I didn't eat the butter!"
> "It was all just a trick the other boys played: *they* smeared it on my face!
> Go look for yourself! The butter jar's hanging way high on a hook;
> tell me how I could reach it with my tiny little hands?"
> So said Krishna as he wiped the butter from his face, and tried to hide the
> plate behind his back.
> But Yasada just dropped her stick, and smiled, and took him into her arms.
> So the Lord once again showed the power of his *bhakti,* as his childish
> antics charmed his mother's mind.[9]

The stories of Krishna as seducer make powerful appeals to human sexual desires. These stories have an impressive history of eliciting erotic responses from devotees, as the following example attests:

> How can I describe his relentless flute,
> which pulls virtuous women from their homes

and drags them by the hair to Shyam
as thirst and hunger pull the doe to the snare?
Chaste ladies forget their lords,
wise men forget their wisdom,
and clinging vines shake loose from their trees,
hearing that music.
Then how shall a simple dairymaid withstand its call.[10]

Many Krishna bhaktas believe that the most intense feeling of love is the heartsick longing for the departed Lord. He who has satisfied all desires has now withdrawn, and who can bear the loss?

It was in bitter maytime my lord
renounced the world, and shaved his head,
and took to the roads with only a staff and a begging bowl.
My heart sickens, tears sting my eyes.
The hope of my life went with him.
How long will my days drag on without him, my Gaura?
The springtime, when the world brims over
with joy, comes round again, bitter to me.
My old love for my lord aches in my heart,
all I remember makes my life a noose
tightening around my throat.[11]

The lovesick longing for God is a burning in the heart that further heightens the devotee's desire for reunion. This is the highest form of bhakti love, for unlike sexual desire, which may be relieved temporarily, this desire can be satisfied only by the permanent, transcendent release of salvation, expressed here as mingling with God in ways the body cannot:

Let the earth of my body be mixed with the earth my beloved walks on.
Let the fire of my body be the brightness in the mirror
that reflects his face.
Let the water of my body join the waters of the lotus pool he bathes in.
Let the breath of my body be air lapping his tired limbs.
Let me be sky, and moving through me
that cloud-dark Shyama, my beloved.[12]

Just as there is something for everyone to love in the Lord Krishna, so there are many ways to love him. Here we have seen bhakti expressions of love in the modes of parent/child and lover/beloved. But one may also love Krishna as a servant loves the master, as a student loves the teacher, as a friend loves a friend, or as a grateful admirer loves a protective hero. Bhakti yoga encourages devotees to experience all of these forms. The objective is for the devotee to love without ceasing, to be caught up in one or another mode of love perpetually. Perpetual love is the personal dharma of everyone who follows the path of Krishna-bhakti, irrespective of caste or stage in life.

The impulse associated with love is to approach and maintain contact with the beloved. In the case of Krishna devotion the particular responses may vary according to the type of love experienced. One who loves Krishna as a parent will want to hold and protect the child; one who imagines Krishna as a master will want to bow down before him and serve his will; one who beholds him as a teacher will want to draw closer and express humility; and one who regards him as a lover will languish in his absence and obsess over the prospects for reunion. Each of these responses predisposes individuals to do whatever is possible to please the deity.

ANCILLARY STRATEGIES

We turn now to consider various strategies designed by Krishnaism to create the conditions for perpetual love.

INTELLECTUAL STRATEGIES

Devotional movements do not typically distinguish themselves in the disciplines of philosophy and theology, and the Krishna-bhakti movement has been no exception. Still, in a culture as deeply shaped by philosophical speculation as India, some attempt to ground religious movements in philosophical principles would appear almost obligatory. Again, the Krishna-bhakti movement was no exception.

In the sixteenth century the Krishna movement enjoyed a phenomenal growth in popularity under the spiritual guidance of Mahaprabhu Caitanya. By the time of Caitanya's death the movement had come to dominate the greater part of eastern India.[13] Caitanya was no philosopher himself, but he understood the importance of making a philosophical case for the movement, so he commissioned a school of philosophy for this purpose. The philosophers of this "Bengali school" set out to show that Krishna-bhakti was both legitimate and true. This was established by their claim that revelation alone, not rational proof, was the ultimate test of scriptural authenticity, and the most authentic texts were the Vedas and the Puranas. And further, the greatest of the authentic texts was the *Bhagavata Purana,* sourcebook of Krishna-bhakti.[14] Bengali philosophers busied themselves with elaborating and classifying the doctrines embedded in this text.

The Bengali school drew on the work of earlier philosophers Ramanuja, Madhva, and Nimbarka, all dualistic thinkers who had broken with orthodox Upanishadic tradition to distinguish themselves as apologists for bhakti yoga. These philosophers insisted that love would be impossible if the lover and beloved were essentially identical, or if either one were essentially non-personal. Thus they argued that the personal nature of God is an essential quality, not a derivative manifestation, and that the distinction between the self and God is real, not illusory. The version of dualism settled on by the Bengali school has been characterized as "simultaneous difference and non-difference," which means that the self is a part of God and shares in God's infinite qualities, but that a decisive quantitative difference remains between the self and God. Even though this doctrine was asserted by the theologians, they did not pretend that it was fully comprehensible to the intellect.[15] The Bengali school also put forward the doctrine that Krishna was the true essence of divine reality, and not merely an avatar of Vishnu.

The moral teachings of the Bengali school were virtue based, not utilitarian or legalistic. The *gopis* (milkmaids) were the paragons of morality: They abandoned their duties for the sake of loving God. They cared only for Krishna and his happiness, not for their own reputations or for the welfare of society. It was Krishna's will that his devotees would forsake all and embody the life of perpetual love. This life, however, entails definite virtues:

> The true bhakti is merciful, and harms no one; he is truthful and dispassionate; he is pure and "free from sin"; he is charitable, gentle, lowly, and humble, he is a benefactor to all; he is calm; he has Krishna as his only refuge and is free from desire; he is moderate about his food, nor does he indulge in anything too much; he honors all things other than himself; he is humble, grave, compassionate, poetic, friendly, skillful and silent. These are the signs of the true bhakti.[16]

The philosophers of the Bengali school were themselves first-rate intellectuals, and they tried to emphasize the importance of intellectual discipline as well as devotion. Their endorsement of intellectual pursuits, however, did not produce a robust theological tradition, and most subsequent "Krishnalogy" has amounted to little more than regurgitation of Bengali doctrine. Once the obligatory nod to philosophy was made, it seems, the movement felt free to concentrate on the more serious practice of devotion, with the assurance that all possible intellectual doubts and uncertainties had been put to rest.

INSTITUTIONAL STRATEGIES

Nothing about Hinduism lends itself to easy description, perhaps least of all its institutional aspects. Some forms of Hinduism are more fully and formally institutionalized than others, but it is difficult to say much in general terms about pan-Hindu institutions. A Western observer might conclude from the abundance of priests and temples that there exists a widely institutionalized practice of temple worship, but this has never been the case. The ancient Vedic literature does not mention temple worship, and many devout Hindus have never worshipped in a temple. Many do, of course, but like so many things in Hinduism, temple worship is a matter of personal or family preference.

If the principal role of religious institutions is to transmit a mythic tradition from one generation to another, then a much better candidate for a pan-Hindu institution would be the guru, or spiritual master. The guru as transmitter of religious tradition has been a remarkably stable institution, going all the way back to the Upanishads.

India is awash with *sannyasis,* "holy men" who have taken up the life of wandering ascetics. Some of these sannyasis are merely unfortunate old men who wander from place to place begging for food, but many are wise and learned teachers, gurus who have themselves reached the ultimate goal of liberation and are prepared to offer guidance along the path to moksha. What makes the role of the guru an institution is not some centralized administration, but rather a carefully conserved tradition of unbroken pedagogical succession through generations of masters and disciples that can be traced directly to ancient authorities. Gurus in the Krishna-bhakti tradition, for example, may claim spiritual pedigrees that go back to the Lord Krishna himself. The etiquette governing this institution commits disciples to being faithful conduits who will transmit the wisdom of the path without distortion or innovation.

There are also rules of etiquette governing the master-disciple relationship. All Hindus are free to choose their own spiritual path, including the deities they find personally appealing. Normally, the next step is to find an appropriate guru who will accept the disciple. The master-disciple relationship is formally initiated when the guru imparts a mantra and a set of instructions to the disciple. This relationship represents what was earlier called an "apprenticeship in thinking." In addition to instructions on how to worship, the guru is likely to offer guidance in proper attitudes, proper emotions, and even proper hygiene. The guru amounts to a personal trainer for the purpose of goal hierarchy reconstruction. An important aspect of this spiritual training program involves instruction for maintaining a Krishna shrine in the devotee's home. The real center of Hindu piety, regardless of one's path, is the household shrine and the various rites associated with it.

The Krishna-bhakti path has always been flexible and adaptive to local circumstances, with the result that most of its institutional strategies have been spontaneous and ad hoc. This feature is nicely illustrated by Milton Singer's description of self-organizing devotional groups that became fashionable in twentieth-century Madras.[17] In the mid-1940s neighborhood groups of Krishna devotees began meeting in private homes to celebrate their love for Krishna in prayer and song. The weekly meetings were small (ten to twelve people) and informal, with leadership responsibility falling to individuals who were especially devout. Eventually, monthly meetings were organized, attended by as many as forty devotees, some from other groups in the city. The monthly meetings were more formal, but still led by whomever was willing and able. Within a few years the activities and organization of the groups had developed into a rather complex institution featuring daily, weekly, monthly, and yearly occasions for group worship. A city-wide association was formed, replete with dues, officers, secretaries, outreach and charity programs, an annual conference, fund-raising campaigns, publicity outlets, organized pilgrimages and, of course, an annual audit.

The development of a "church-like" institution in Madras was not necessarily in conflict with the institution of the guru. Indeed, the phenomenon could not have developed without the approval and participation of local gurus. Nor did the development of formal worship groups interfere with the maintenance of household shrines. The Madras experience illustrates a general pragmatic element found in most religious traditions—an openness to using whatever institutional strategies are deemed conducive to the perpetuation and vitality of the myth.

EXPERIENTIAL STRATEGIES

All religious traditions encourage their devotees to achieve some sort of extraordinary experience to validate the myth. For some traditions the value of religious experience is found in its utility as a doubt inhibitor, but for Hinduism extraordinary experience is absolutely essential to the religious life. It should not be supposed, however, that extraordinary experience is itself the ultimate goal of the religious life. In all forms of Hinduism the ultimate goal for the self is moksha, or liberation from the binding effects of karma. Extraordinary experience is important mostly because it is an essential mark of moksha. One might have all sorts of extraordinary experiences without achieving moksha, but it is inconceivable that one might achieve moksha without extraordinary experience. Thus Hindus are well able to understand the ebullience of charismatic Christians who report feelings of ecstasy at the moment of salvation, but are mystified by Christians who claim salvation without the ecstasy.

Many observers have noted that Krishna devotees have distinguished themselves, even among Hindu mystics, for the range and intensity of their religious experiences:

> [Krishna-bhakti] manifests itself in the soft melting of the heart and expresses itself in tears, inarticulate utterances of speech, laughter, songs and dances, such as can only be possible through a mad intoxication of love . . . They [the bhaktas] come to experience such intense happiness that all their limbs and senses become saturated therewith and their minds swim, as it were, in a lake of such supreme bliss that even the bliss of ultimate liberation loses its charm. . . . Such a person is beside himself with this love of God. He sings, laughs, dances and weeps. He is no longer a person of this world.[18]

It is easy to understand how devotees who have endured such transforming experiences might find them personally fulfilling, and might want to have them again and again. And presumably they can, for they are assured by spiritual masters that however wonderful the experience was this time, there are many more and much better to come. The literature produced by Krishna swamis is vast, including commentaries on commentaries, which go to great lengths attempting to classify the countless stages of perpetual love.[19] The strategies for provoking these extraordinary states of love are also numerous and varied, many of which are bound up with ritual practices.

RITUAL STRATEGIES

Much devotional worship is centered on images of deities, little statues of gods or goddesses kept in household and temple shrines. These objects of worship are treated with deep reverence, as one might treat a royal guest. Image worship, called *puja,* involves a variety of rituals and ceremonies for seeing, touching, praising, parading, and caring for the deity. The household deity, for example, will be greeted each morning and offered food. Periodically the deity will be bathed, dressed, and decorated.

The treatment of deities is far more lavish in temples, where they are attended by priests. In a temple dedicated to Krishna there will be found a highly decorated shrine where the image is kept under a canopy and behind a curtain. The priests wake the deity each morning, offer him food, entertain him, dress him in clothes suitable to the season, take him for exercise, put him down for occasional naps, and even fan him when the heat is extreme. When devotees come to worship the temple deity, they will bring food or flowers to accompany their praise, in exchange for which they will be given a glimpse of the deity and perhaps a small token of his blessings. Detailed protocols for honoring the deity are found in ancient scriptures and have been further elaborated by layers of commentary.

To outsiders the activities of puja may resemble childish play with dolls. On the surface there may appear little difference between a priest dressing up Krishna for a festival and a little girl dressing up her Barbie for a date with Ken. The difference shows up beneath the surface where, for Krishna-bhaktas, the concept of *murti* is at work to condition the imagination. Murti means "image" in the sense of a real manifestation. A murti of Krishna is a forceful manifestation of Krishna himself—less than a full flesh-and-blood incarnation, of course, but considerably more powerful than a mere toy. Murtis of Krishna are believed to embody certain aspects of the divine essence or power, perhaps in the same sense that relics of Christ impressed medieval Christians with their holiness. Even skeptics can appreciate the effect: A twenty-first-century historian standing in the presence of the Rosetta Stone will have an appreciably different experience than her colleague who sits at home viewing its photograph in a textbook. The difference may be faintly analogous to the way the concept of murti prepares the imagination. It supplies the cognitive conditions for emotional flights, so that when the devotee receives a glance from the temple deity, she *feels* blessed, or when she touches the deity's feet, she *feels* cleansed of her sins.

In the end, of course, it may all be nothing more than conditioned imagination, but then, in the end, imagination is nothing less than experience. Hindus tend to be more comfortable with the interplay between reality and illusion than their Western counterparts. Krishna devotees, for example, are not bothered by the morality of Krishna's nocturnal trysts with the wives of cowherds because he took the precaution of inducing in their husbands the delusion that their wives remained at home.

Krishna devotees carry their conditioned imaginations into a range of ritual contexts where they are predisposed to strong emotional responses. In addition to daily worship at household and temple shrines, there are open-air shrines where small groups of devotees may gather informally to dance and sing praises to Krishna. There are also occasions for all-night *kirtana* sessions of chanting, instrument playing, and dancing, all intended to recreate the atmosphere in which Krishna dallied with the gopis. During these sessions devotees drive each other to paroxysms of laughter, praise, and joy, occasionally spilling out into the streets at impolite hours of the morning.

Numerous festivals throughout each year provide Krishna devotees with opportunities to congregate for longer periods and in more formal settings to celebrate their love. Krishna's birthday and his marriage to his favorite gopi, Radha, are favorites. Annual festivals vary substantially according to local traditions, some of them lasting for more than two weeks. A typical feature of these festivals is the staging of *lilas,* dramatic reenactments of well-known and much-loved stories about Krishna's childhood pranks, his love sports with the gopis, and his heroic achievements. Many devotees will prepare themselves for the festivals by reading poems about Krishna's separation from the gopis, thereby to approach the festival in a heightened state of longing and anticipation for the moment when Krishna appears before them on stage or in a temple (the actors, while not deities, are murtis of Krishna). The plays are long—several days in many cases—and full of inspiring poetry, song, and dance, invariably deepening the bhakta's devotion.

The pilgrimage is an important ritual for many Hindu devotional sects, including Krishnaism. Most religious traditions regard certain places and times to be sacred, and for Krishnavites the most sacred place is Vrindavana, where Krishna was raised as a cowherd and where he sported with the gopis. To visit Vrindavana is to draw closer to the presence of Krishna, partly because he is thought to have lived in the area, but also because the temple deities there are believed to have more power. Krishna loves all those who come to him in devotion, but there are special blessings in store for the ones who undertake hardships and expenses in order to make a pilgrimage. For many devotees a pilgrimage represents a lifelong dream, similar to the hajj for Muslims.

Thus it is not surprising that pilgrims typically set out with expectations for transforming experiences. The theme of separation and reunion is especially strong in the Krishna myth, and the pilgrimage ritual is the most dramatic way to embody it. Many pilgrims prepare for the reunion by shaving their heads and making special clothes. Vrindavana never disappoints, for it remains in a state of festivity year-round, with lilas performed in auditoriums each day of the year, and many informal lilas enacted in the streets.

Anyone attempting to perform all the rituals prescribed in Krishna-bhakti teachings would be engaged in devotional service non-stop, leaving absolutely no time for anything else. But this seems to be the point of perpetual love, to be completely consumed by loving thoughts of Krishna and nothing else—ever. Household worship, temple worship, kirtana sessions, prayers, festivals, scripture reading, pilgrimages—these are all part of the full-court press to keep one's life centered on Krishna.

Another important element is the familiar *maha-mantra:* "Hare Krishna, Hare Krishna, Krishna Krishna, Hare Hare, Hare Rama, Hare Rama, Rama Rama, Hare Hare." Krishnavites believe that uttering this mantra of God's names will neutralize the effects of karma. More than that, the mantra itself is a sort of murti—that is, repeating the mantra puts one directly in the presence of Krishna. As the *Padma Purana* puts it: "There is no difference between the holy name of the Lord and the Lord himself." [20] The important thing about the mantra is its availability to the devotee at any time, any place. Whatever the circumstances, the mantra ritual provides instant access to God.

AESTHETIC STRATEGIES

One of the most interesting aspects of Hindu aesthetic traditions is that the creative act of the artist is itself considered to be a religious event. Artists are seen to engage in a kind of yogic discipline involving spiritual techniques and rituals. [21] Thus we should not be surprised to find that virtually all the art of India's past (and most of its present art) is focused on explicit religious themes. In Hindu culture art is naturally religious and religion is naturally aesthetic. Hindu art, generally speaking, prefers refinement over realism, elaboration and embellishment over imitation. This aesthetic ideal is consistent with the Hindu ethos that the central task of human existence is to refine or purify the self. Nature is to be perfected by culture, life is to be corrected by discipline. A second general characteristic of Hindu art is its preoccupation with narrative themes, with the epic stories of heroes, gods, and goddesses drawn from classical texts. These general characteristics are amply illustrated in the aesthetic traditions of Krishnaism.

Most of the art associated with the Krishna myth has concentrated on the youthful figure, the child prankster and irresistible lover. These stories have inspired an exceptionally rich and enduring poetic tradition. One of the great masterpieces of Sanskrit literature is the *Gita Govinda,* a twelfth-century poem celebrating the love of Krishna and Radha, written by the Bengali court poet Jayadeva. The

Gita Govinda represents a complex genre that combines elements of lyric drama, folk pageant, and dance. The immense popularity of the poem is a tribute to its multivalence—that is, its power to evoke responses on many levels. On one level the poem creates miniature word-pictures that transform nature by infusing it with passion and romance:

> The scarlet blossoms of *palasa,* which flame red like the golden finger-nails of Kama, tear the hearts with amorous frenzy. The *ketaki* pierces the heart with the keen-edged aroma of its spears. The orange tree laden with white blossoms shivers at the pain of separated lovers. Even the sleeping mango tree, clasped by the *madhavi* creeper, grows conscious with love, quivers with joy, and bursts into tender purple shoots.[22]

The *Gita Govinda* works equally well on both the level of romantic love and the level of mystical devotion. It plays on themes of longing, hope, disappointment, anguish, and reconciliation. Thus the love between Krishna and Radha gives way gracefully to the love between God and humankind. Radha, the ideal of beauty and goodness, is saddened when the man she loves squanders his affections on trivial pursuits with the gopis. When Krishna realizes his folly, he implores Radha to forgive him, which she generously does. The subsequent reunion of the lovers is both physical and spiritual.

This poem remains one of the most popular and influential texts in all of India, rivaling the *Bhagavad Gita.* So revered is the *Gita Govinda* that it has spawned its own legends; for example, it is said that when Jayadeva struggled with difficult verses, Krishna himself finished them. And on one occasion when a temple deity heard a young woman singing verses from the poem, he left the temple and followed her through the countryside. There is even one account of a Mughal prince (a Muslim) who was so moved by Jayadeva's verses that he fell into an ecstasy of communion with Krishna.[23] In addition to popular legends, the *Gita Govinda* has inspired a wide range of artistic expression in painting, drama, dance, sculpture, and more poetry.

The *Gita Govinda* was given its most notable visual presentation in a series of paintings by Manaku, an eighteenth-century painter of the Kangra school. These paintings are valued by many for capturing perfectly the idyllic ambience of the poem. Like the poem, the paintings are simultaneously erotic and devotional. The pastoral scenes present nature as it should be: pulsating with life, yet perfected by divine love. Birds and dragonflies flutter among blossoming trees, and colorful wildflowers embroider the banks of pristine streams—all

delicately framing the embraces of Krishna and Radha. It's all perfection, a visual satisfaction of every devotee's longing for the dreamy
world of Vrindavana.

The combination of poetic and pictorial arts has been an important element for educating the emotions of Krishnavites. It creates
sensuous mental images of a perfect world in which devotees are intimately bound to Krishna. To the extent that individuals inhabit this
imaginary world, their appraisals of events in everyday life will embody the qualities of divinity.

Even more important as an aesthetic strategy for reinforcing the
devotion of Krishnavites has been the tradition of the dance-drama
(*rasalilas*). The dance-drama, not exclusive to Krishnaism, is found all
over India, from the urban Victorian theaters to the open-air stages
of more than 700,000 rural villages. Wherever it is found, the dance-
drama combines elements of poetry, dance, drama, music, and elaborate costuming. In the seventeenth century the dance-drama was elevated by royal patronage to become an esoteric art form, of which the
kathkali tradition is one manifestation.

Kathkali is a highly sophisticated and refined genre, full of subtleties of convention and technique. There are, for example, only nine
facial expressions and twenty-four hand gestures allowed in a kathkali production.[24] The training for actor-dancers in the kathkali tradition is extremely rigorous, demanding years of intense discipline.
The emphasis in this esoteric art form is more on aesthetic achievement before an audience of connoisseurs than on bringing a narrative
to life before an audience of eager devotees.

In their folk-art settings the dance-dramas have more the character of religious rituals than theatrical performances. In the region
of Bhaj Bhoomi, where Vrindavana is located, popular productions of
Krishna rasalilas combine some of the formal elements of the kathkali tradition with the informality of audience participation. The production develops in two stages: first the dance, then the play. At critical moments during the dance the audience will throw flowers, clap,
and shout. At the conclusion of the dance phase someone, usually a
priest, begins chanting and ringing bells, a cue for the audience to clap
in time with the chanting. As the tempo increases the priest becomes
more and more animated, until finally the entire audience is swaying
to-and-fro and waving their arms, chanting in unison and clapping.
Those who do not participate can expect to be born dumb or crippled
in their next life.[25] At the climax of the chanting a curtain is drawn to
reveal a poster of Krishna and Radha, a longed-for moment of sheer
ecstasy for Krishna devotees. Then follows the play, a dramatization
of one of the many episodes in Krishna's life.

The rasalila event, like the poetic and pictorial treatments of Krishna's life, has the effect of reinforcing a world of mythic meanings to inform memory and imagination, thereby to condition the appraisal of everyday events. The more frequent and meaningful these experiences, the more completely transformed the goal hierarchy becomes.

PERSONAL WHOLENESS AND SOCIAL COHERENCE

Krishnaism, it appears, is a totalitarian proposition, pressing the devotee into an obsession with God:

> An unalloyed devotee who has developed ecstatic love for Krishna is always engaging his words in reciting prayers to the Lord. Within the mind he is always thinking of Krishna, and with his body he is either offering obeisance by bowing down before the deity or engaging in some other service . . . In this way his whole life is engaged in the service of the Lord, with not a moment wasted on any other engagement.[26]

Accordingly, if you did absolutely nothing for the rest of your life except stay alone in your room reciting the Hare Krishna mantra continuously, then you would be the perfect devotee, for there would be no moments when your mind was "wasted on any other engagement." If Krishnaism is seriously proposing this sort of behavior, then it clearly represents a challenge to the thesis of this book. I have been arguing that religious traditions properly function to create the conditions for personal wholeness and social coherence. But in what sense might these twin teloi be even remotely served by obsessive and reclusive behavior? Krishnaism, it may appear, drives us toward personal dysfunction and social incoherence.

In all fairness, this reading must be seen as a distortion. One might produce a similar distortion of Christianity by exaggerating the injunction to "pray without ceasing." Religious traditions may be expected to overstate their teachings in response to what they see around them, and what they commonly see is behavior that is motivated by self-centered goal hierarchies. The suggestion that devotees should become totally obsessed with Krishna should be read as an antidote to the condition of being totally obsessed with oneself. It is a way to remind us that selfishness is a persistent default state, and that unselfishness calls for excessive vigilance.

At the end of the day all the teachings of Krishnaism emerge from a simple insight: There is no way for human beings to be too loving. Thus Krishnaism should be seen as a collection of strategies designed to provoke any kind of love it can, whenever and wherever it is possible. Perhaps we can be brought to love God through loving others: "Let me become a sincere servant of the devotees because by serving them one can achieve unalloyed devotional service unto the lotus feet of the Lord."[27] Or perhaps we can be brought to love others through loving God: "If we can learn how to love Krishna, then it is very easy to immediately and simultaneously love every living thing."[28] But either way, whether love originates or culminates in God, the essential point is the one succinctly captured by John Lennon and Paul McCartney (both of whom flirted with Krishna): "All you need is love."

Krishna-bhakti provokes one to love without ceasing, to be caught up in some form of love perpetually. The relevance of perpetual love for the simultaneous achievement of personal wholeness and social coherence should be obvious. When humans are in a state of loving and being loved, they feel personally fulfilled. But further, when in a state of love one's moods brighten and thresholds to negative emotions (anger, sadness, resentment) are significantly elevated, while thresholds to positive emotions (happiness, sympathy, gratitude) are significantly lowered, all of which predisposes individuals to make prosocial sacrifices resulting in overlaps of self-interest. Persons in a state of perpetual love are more forgiving, more tolerant, more generous, and more cooperative. Those who are moved to love Krishna will seek to embody his divine qualities, especially his devotion to those who love him. They will take seriously the words of the *Padma Purana:* "Even higher than the worship of the Lord is the worship of the Lord's devotees."[29]

Hinduism has often been dismissed as a life-negating tradition that offers an ideology and a set of strategies designed to remove the self from an unsatisfying and intolerable world. But if Krishnaism is in any way relevant to the matter, then it should rather be said that Hinduism is a life-affirming tradition offering an ideology and a set of strategies for educating the emotions, thereby to render the world more satisfying and tolerable.

9

BUDDHISM

HISTORICAL CONTEXT

The deep histories of Buddhism and classical Hinduism are the same, for both of these traditions emerged in response to a failure of the ancient Vedic religion to create conditions for personal wholeness and social coherence. The rise of Buddhism appears to have been largely an urban phenomenon.[1] During the seventh and sixth centuries B.C.E., the central region of the Indian subcontinent underwent a political transformation as many small kingdoms were absorbed into a few large ones. The economy of the region flourished as commercial trading and banking industries developed. A new money economy emerged, concentrating political and economic power in growing cities. Social structures began shifting as a new class of merchants and bureaucrats gained influence in the cities. Tensions grew between the newly wealthy urban merchants and the traditionally wealthy landowners, adding to the existing resentment of Brahminical hegemony in religious matters.

There has been much speculation about the conditions of urban life that made it especially receptive to the teachings of Buddhism. Perhaps the accelerated rate of change in urban settings opened ears to the Buddhist doctrine of impermanence; perhaps the increased freedom and social mobility of city life was conducive to experimenting with new forms of spiritual development; perhaps the urban priesthood was increasingly drawn into the public sphere to the neglect of families and individuals; perhaps the anonymity of the city was less conducive to older patterns of social cooperation; or perhaps it was a combination of these factors that produced in the cities a growing sense of anomie.[2]

Richard Gombrich offers the interesting thesis that the new Indian cities were rife with squalor, parasitism, disease, and high mortality rates, which may have increased receptivity to the Buddhist doctrine of *dukkha,* the idea that human life is essentially a vale of tears.[3] Whatever the reasons, Buddhist teachings caught on, especially in urban districts among the well-to-do, and by the third century B.C.E., under the patronage of King Asoka, Buddhism had become a major spiritual force in India.

Little can be said with certainty about the life of Siddhartha Gautama, the historical Buddha. The claims of tradition that he was born about 560 B.C.E. to an aristocratic family in what is now southern Nepal, and that he attracted followers to his teachings until he died at the age of eighty are probably reliable. Beyond these basic facts the line between historical truth and legend becomes obscure. Nevertheless, the traditional narrative of Siddhartha's life has been sufficiently important throughout Buddhist history to justify repeating it here as a prelude to the Buddhist myth.

Siddhartha is said to have led a privileged but somewhat confined childhood. His father, a king, had political ambitions for his son and conspired to protect him from the sorrows of life, lest a burst of compassion deter the young prince on a lengthy quest for spiritual insight. While touring the royal park, however, Siddhartha encountered four men whose circumstances would change his life forever. The first man suffered from sickness, the second was weakened by old age, and the third, a dead man, was being prepared for cremation. The fourth man was a monk whose serenity and detachment from the world stood in sharp contrast to the first three encounters. The prince, then twenty-nine years old, returned to the palace to reflect on the meaning of these encounters. That very night he renounced his life of privilege and embarked on a spiritual journey, determined to discover ultimate truth.

He began his journey by submitting to the instruction of teachers, first a wise sage and then an accomplished yogi. Both men were helpful, but neither could bring him to the enlightenment he sought, so he joined a small band of fellow seekers who were practicing a life of austere asceticism. For six years he disciplined his body and mind with meditation and fasting to the point of exhaustion, but still could not fathom the meaning of sickness, old age, and death.

Having explored the limits of the ascetic life, he resumed a healthful diet and sat down beneath a large fig tree where he resolved to stay until he achieved enlightenment (*nirvana*). When the moment finally arrived, the Buddha entered a blissful state of complete and undefiled awareness of the Four Noble Truths, an absolute vision of how things ultimately are in the cosmos and which things ultimately matter for human fulfillment:

> But as soon as the absolutely true knowledge and insight as regards these Four Noble Truths had become perfectly clear to me, there arose in me the assurance, that I had won to that supreme Enlightenment unsurpassed.[4]

The experience of direct insight into these ultimate truths had the effect of delivering the Buddha forever from the bonds of samsara (rebirth) and into the state of nirvana (absolute liberation). Assured that he had reached the plane of a *Tathagata* (a "thus-gone one"), he located the band of fellow seekers and instructed them in the Four Noble Truths, after which he ordained them to carry the teachings to others. For the next forty-five years the Buddha traveled and taught, finally dying at the age of eighty from eating spoiled food.

When the Buddha's disciples inquired about his successor, he replied that after his death they would be led by the *Dhamma,* the doctrines he had imparted to them. But when the Buddha died, the inevitable happened: The many disciples fell into dispute over the many doctrines. Within a few centuries original Buddhism had fractured into many distinctive schools of interpretation and practice. The process of diversification continued as missionaries carried the movement beyond the borders of India, where it was variously reshaped by accommodation to local traditions. The basic story is that the relatively simple and practical teachings of the Buddha became the diverse and highly sophisticated philosophies of Buddhism.

Eventually there were as many ways to be a Buddhist as there were to be a Hindu, leaving us with a problem similar to the one faced in the previous chapter on Hinduism—namely, which Buddhism shall carry the discussion forward? The basic option is between the Theravada tradition dominant in Sri Lanka and Southeast Asia, and one of the many Mahayana traditions that flourished in China, Korea, Tibet, and Japan. For our purposes the older and more conservative Theravada tradition will suffice.

THE MYTH OF SELF-EMANCIPATION

Picture a pedophobe, a man who suffers from a persistent fear of children. When someone mentions children, he grows anxious and quickly changes the subject. Whenever he encounters a child, he panics and runs off to hide. He awakens from sleep in a cold sweat when he dreams of children. And so on. What do we say of such a case? We might classify his condition together with all the other phobias and say the poor fellow is burdened with a groundless, irrational fear response. We might say he is deluded about children, that he thinks wrongly about them, and that his wrong thinking is responsible for misappraisals of what is at stake when he encounters children. This unfortunate man, we might say, suffers needlessly because in reality children present no danger to him. If we have sympathy for the pedophobe, then we might take measures to reduce his suffering by helping to uncover the psychological conditions that produce these debilitating misappraisals of meaning.

I use this example to introduce the Buddhist myth because it takes us straight to the heart of the matter. Buddhism, in my view, makes the quite radical claim that there is nothing terribly unusual about the pedophobe. His condition is universal: Every human being suffers needlessly because we all make misappraisals of meaning. In fact, Buddhism asserts that *all appraisals are misappraisals,* originating in delusions about reality and resulting in unnecessary suffering. The problem is that we persist in making appraisals at all. If we could abort the appraisal process altogether, then we would put an end to the evils of human suffering. This, I believe, is an accurate, demythologized version of the Buddhist gospel. Our business here, however, is with the mythological version, which is summarized in the following narrative.

> The absolute and final truth about reality is that reality is process, not substance, not form. Reality is a process of ceaseless change and impermanence. There exist no static enduring entities and no eternal changeless forms. What we suppose to be enduring things and characteristics are in truth passing moments of apparent duration in a field of flux. Whole universes arise and pass away in an endless stream of causation. Identity is a fiction, nothing is itself.
>
> *Dharma* is the ultimate principle of reality. At the bottom of everything, the point at which all inquiry and analysis stops, there are the dharmas, the ultimate constituents, an infinite number of discrete forces or ephemeral elements arising and passing away continuously. One might think of the dharmas as momentary packets of energy having their effects in a vast fluctuating sea of dharmas. Each one exists interdependently with all the others, and the combined impact they have during their fleeting existence is ultimately responsible for everything that is and happens. Every object, event, property, and relation observed in human experience can in principle be traced to the causal influence of these instantaneous bursts of reality. All the "realities" of

the universe(s) are merely composite states of dharmic process—they are what they are, not by virtue of essential qualities adhering to permanent substances, but merely because of previous states in the process.

The self is a fiction, too, a contingent passing state originating from previous states. You may believe that you are the same enduring self that existed minutes, hours, days, and years ago, but this is a fundamental error. Your belief in a self is merely a belief state arising from previous belief states about an enduring entity that has belief states. Belief states do happen, but there are no enduring selves to which they happen. Impermanence is the ultimate fact, and to live in harmony with the process of ceaseless "arisings and passings" is the ultimate wisdom.

Our universe arose eons ago and, like other universes, it passes through phases. In some phases wisdom (i.e., knowledge of reality and how to live in harmony with it) may arise and may also pass away. In the current phase wisdom has declined, and as wisdom declines suffering states (*dukkha*) arise, for suffering is a consequence of ignorance about how things really are. Wisdom may be declining in our era, but it is not forever lost. It has been rediscovered by the enlightened Buddha and presented in the four absolute and final Noble Truths about human existence.

- The first truth is that life inevitably leads to suffering states. There are many suffering states, such as birth, sickness, old age, death, pain, frustration, grief, anxiety, stress, and so on. Even the joyous states are sources of suffering, for they, too, soon pass away, giving rise to states of loss.
- The second truth is that suffering states are caused by other states. Suffering states arise because of states of desire, longing, thirst, craving, greed, and so on. Basically, suffering is inevitable wherever there exists an invested ego, a preference for some particular thing or event over others—that is, wherever there exists a hierarchy of desired outcomes.
- The third truth is that the self-perpetuating series of suffering states can be extinguished by preventing the desires and attachments that give rise to them. If the causes are eliminated, then the effects will not arise. The key is to find a way to disinvest the ego, to dismantle the hierarchy of desires.
- The fourth truth is that the causes of suffering states may be eliminated by following the Middle Way between self-indulgence and self-destruction. This is the wisdom the Buddha discovered and articulated in the Eightfold Path. The Eightfold Path prevents the ignorance that gives rise to the cascade of states leading to rebirth and perpetual suffering. The Path involves right views, right intentions, right speech, right conduct, right livelihood, right effort, right mindfulness, and right concentration.

Anyone taking refuge in the Buddha has found the right exemplar and teacher. Anyone taking refuge in the Buddha's doctrines (Dhamma) has found

the right source of wisdom. And anyone taking refuge in the community of
followers (*sangha*) has found the right context in which to learn and to practice the path of self-emancipation from all evil.

I have claimed repeatedly that a mythic vision integrates ideas
about how things are (cosmology) with ideas about which things
matter (morality), and that the integration amounts to a fusion of
facts and values. I have also claimed that an integrated mythic vision
is made plausible by some sort of central concept or root metaphor.
In the case of Abrahamic myths the metaphor of God as person fuses
cosmology and morality, while in the case of Hinduism the concept
of dharma achieves a comparable result. The Buddhist myth is integrated by the concept of dharma, too, but Buddhist traditions have
found distinctive ways to understand it.

In Buddhist thought dharma is both the ultimate principle of reality and the ultimate principle of value. It is "that which holds," the
absolute and final truth about reality from the momentary microcosmic constituents to the macrocosmic arising and passing of universes.
It is also that which holds true of human reality—that is, it is the ultimate expression of wisdom. As the central concept of the Buddhist
myth, dharma is the ultimate metaphysical explanation as well as the
ultimate moral justification. To avoid confusion in the discussion
ahead I will use three forms to refer to the different meanings of this
complex principle. "Dharma" will mean the universal principle of reality and value—the Ultimate Truth. The Pali word "Dhamma" will
refer to the doctrines taught by the Buddha. When referring to the elementary constituents of phenomena, I will use the lower case "dharmas," or simply "the elements."

It is important when trying to understand any religious tradition
to be clear on the things its myth presumes realism about, for these
presumptions determine how the tradition will shape its appeal to elements of human nature. Here I will venture a claim that the Buddhist
myth presumes realism about both the Dharma and the Dhamma.
That is, it presumes realist attitudes toward the Buddha's teachings as
precisely accurate accounts of ultimate reality and value. Realism
about the Dhamma holds that the Buddha did not merely invent these
teachings, he discovered them while in an undefiled state of immu-
nity from the ignorant delusions of ordinary experience. A realist
about the Dhamma will be confident that these teachings will work
for anyone because they are based on a perfect understanding of how
things really are (Dharma).

A medical analogy is pertinent: A patient undergoing surgery
will have confidence in the procedure only if he believes that the sur-

geon's understanding of how the body works is based on impeccable science. The Dhamma, of course, is open to a range of different interpretations, but it is difficult to imagine that the Buddhist myth would have much purchasing power among nonrealists who think it amounts to nothing more than a tissue of fancy.

Buddhist tradition has codified and qualified its doctrines in a dizzying array of numbered lists, the compilation of which has been something of an obsession from the beginning:

> Thus we have the Three Characteristics, the Four Aryan Truths, the Five Groups, the Six Bases, the Seven Factors of Enlightenment, the Aryan Eightfold Path, the Nine Transcendental States, the Ten Perfections, the Twelve Links, the Thirty-seven Wings of Enlightenment, and hundreds more.[5]

In what remains of this section I will attempt to amplify the myth of self-emancipation by focusing on the most salient features of the Dhamma. The discussion will cover karma, dependent origination, the Eightfold Path, the no-self doctrine, and nirvana.

KARMA

Buddhism started out by rejecting much of ancient Hindu tradition, including the privilege of the priestly caste, the religious ideal of devotion to the gods, and the notion that salvation might be effected by either ritual sacrifices or metaphysical speculation. But the doctrines of samsara (rebirth) and karma remained intact at the heart of the Buddhist moral perspective. It is obvious that human choices escape into the world beyond the will where they have consequences of various sorts, and every moral perspective must find a constructive way to think about these consequences.

Buddhist moral thought is less attentive to the social and political consequences of human choices than it is to the consequences for the actor's character. If I steal money from my neighbor, a Buddhist moralist will first call attention to how the act impoverishes me, not how it impoverishes my neighbor. The doctrines of samsara and karma attempt to organize our thinking about consequences in terms of if-then schemas. Karma (literally "action" or "volition") is the lawful connection between the "if" and the "then." If I sow this act, then I must expect to reap that consequence. According to the law of karma, consequences of volition may be experienced either in this life or in a future life. The prospects for future lives are defined by three realms of positive rebirth (you may be reborn as a human, a demigod, or a god) and three realms of negative rebirth (you may be reborn in one of several hells, as a hungry ghost, or in the form of some animal). The lifelong accumulation of meritorious karma may be expected to result in a better future life, for example, to be reborn with better health or to be reborn in heaven. Negative karma will result in a less fortunate future life: one might be reborn sickly or ugly, or in hell, or in some animal state.

The point is that karma keeps samsara going, and as long as samsara continues, there will be suffering (dukkha). It may be better to be reborn as a god than as a paralytic pedophobe, but the difference is trivial, one of degree only. If, however, the point of existence is to end suffering, then one must somehow prevent consequences from attaching to volition.

DEPENDENT ORIGINATION

The law of karma says that one's present existence is a consequence of past existence, and that one's future existence will be a consequence of present existence. Every thought, word, and action carries its effects forward to condition subsequent states in a continuous chain of causation. But the if-then lawfulness of karma is not simply a moral principle. It represents the specifically moral aspects of a general, cosmological process known in Buddhist thought as the doctrine of dependent origination: "Behold, this exists when that exists, this originates from the origination of the other; this does not exist when that does not exist, this ceases from the cessation of the other."[6]

Nothing about the ordinary world of human experience (including experience!) is unconditioned. All things, events, properties, and relations—all thoughts, words, and deeds—arise as they are conditioned by previous states, just as they will in turn condition future states. The doctrine of dependent origination is often pictured as a Wheel of Life, showing how one thing leads to another, and then to another, and so on. At the hub of the wheel are ignorance and desire. The central point of the image is that humankind is caught in a self-perpetuating cycle of rebirth, suffering, old age, death, rebirth, suffering, old age . . . ad nauseam.

If we consider the Five Aggregates (*skandhas*), we shall see how Buddhist thinking grounds the Wheel of Life in the cosmology of the elementary dharmas. Much effort has been expended to analyze and classify the various attributes of the elementary dharmas that give rise, in lawful and orderly patterns, to the different states of the universe. There arise, for example, rocks, trees, wind, stars, water, and so on, and these many states are presumed to result from different types of elementary dharmas.[7] It also happens that the elementary dharmas give rise to individual human beings with physical properties, modes of sensation and perception, emotional tendencies, habits of mind, and so on. All these effects are produced by dharma types, grouped together as the skandhas, or the Five Aggregates:

- *Form.* One class of dharmas gives rise to bodily forms and functions, or rather, our anatomical and physiological characteristics.
- *Feelings.* Another class of dharmas gives rise to feelings, such as pleasure, pain, and various moods.
- *Perceptions.* A third class of dharmas gives rise to the various senses such as sight, smell, taste, hearing, and touch. In addition, this group gives rise to awareness.
- *Impulses.* A fourth class of dharmas gives rise to a range of psychological phenomena, including various emotions (desire, hate, fear), virtues (courage, modesty, mindfulness), and vices (greed, deceit, sloth).
- *Consciousness.* A fifth class of dharmas gives rise to various states of consciousness. The general states are pure (having positive karmic effects), impure (having negative karmic effects), and neutral (having no karmic effects).

Here we have the five categories of elementary dharmas that account for everything one can say about a person. A person just is the activity of these dharmas, nothing more. If the dharmas ceased producing these effects, then there would be nothing left to talk about. There is no substantial ego or enduring "self" over and above these effects, no thing that the dharmas hang upon or happen to, no thing that can be said to have stakes in outcomes. To suppose there is such a thing is the fundamental error that gives rise to attachments, desires, and stake appraisals, the very effects that perpetuate the Wheel of Life.

But here's the problem: Once the Wheel of Life gets going, how can it be stopped? Once ignorance gains a systematic foothold in the streams of dependent origination, it is difficult to see how its influence might be extinguished. The attachments, desires, and stake appraisals giving rise to suffering states will condition the dharmas to give rise to further attachments, desires, and stake appraisals. And the wheel rolls on. The problem seems to be one of breaking into a self-perpetuating closed system.

THE EIGHTFOLD PATH

The Four Noble Truths tell us the Wheel of Life is *not* a closed system, for the effects of ignorance can be precluded by wisdom, by following the Eightfold Path. One might view the Eightfold Path as a strategy for using Dhamma to fight the elementary dharmas. That is, the Dhamma formulated in the Eightfold Path has no less the force of universal law than do the dharmas that condition the five skandhas. The Eightfold Path deploys causes against causes, states against states. If certain states can create the conditions for subsequent states, then why cannot some states subvert the conditions for these states? It is, after all, wrong views, wrong intentions, wrong speech, wrong conduct, wrong livelihood, wrong effort, wrong mindfulness, and wrong concentration that give rise to suffering states, so it stands to reason that getting these elements right would abort the process and put an end to suffering.

The first two elements of the Eightfold Path are concerned with wisdom, the next three address matters of moral conduct, and the final three are focused on discipline.

- *Right Views.* Wisdom involves a correct understanding of how things really are. The right views are given in the Four Noble Truths and in the doctrine of dependent origination. To cultivate the right views one should reflect on these elements of the Dhamma seriously and frequently.
- *Right Intentions.* Knowing the truths about how things really are is one thing, but resolving to live in harmony with these truths is quite another. Wisdom requires knowledge, but it also requires a deliberate commitment to embody the right views.
- *Right Speech.* Wisdom is connected to action through speech. It is therefore important that one's speech acts are well managed so as to minimize suffering. This means no lying, no slander, no rudeness, no boasting, no idle gossip, no inflammatory speech, and so on.
- *Right Conduct.* Proper moral conduct is essential for the pursuit of enlightenment, for one cannot make spiritual progress while leading a self-indulgent life. It is better to practice kindness than violence and killing; better to practice charity than to steal; better to practice self-control than to commit adultery; better to be honest than to cheat; and better to abstain than to weaken the mind with alcohol or drugs.
- *Right Livelihood.* One should practice a vocation that is consistent with right conduct. A life of service is better than one of profit and acquisition. One should avoid vocations involving warfare, prostitution, or butchering animals. One's profession should promote peace and well-being for all.
- *Right Effort.* One should discipline the mind to focus on spiritual ideals, and to purge the mind of self-interested states. One must strive to expunge unwholesome states of mind and to prevent new ones from arising. One must also maintain wholesome states of mind and encourage the arising of new ones.

- *Right Mindfulness.* The seventh step on the path is a pivotal one, preparing access to the insights of the eighth step. Right mindfulness involves disciplined attention to and keen awareness of the activities of one's body and one's mind. It brings clearly into focus the bodily activities that are normally left to habit (for example, our daily routines), and the psychological activities that are normally not reflected upon (moods, attitudes, opinions, emotions). The idea is to observe the rising and passing of physical and psychological phenomena in the mode of a disinterested spectator.
- *Right Concentration.* The final step on the path to nirvana and cessation of suffering has both active and passive aspects. Having prepared the mind by right mindfulness, one must sit quietly but alert, keeping the mind focused on its own essence. As passing thoughts arise, they should be dismissed as irrelevant. In the passive phase of rapture all dualities vanish as the mind is completely rinsed in waves of compassion, joy, peace, and equanimity.

NO-SELF

Imagine this: You are gazing across a meadow on a warm summer evening and a hoard of fireflies assembles itself into a pattern, and then suddenly they all light up simultaneously, creating the visual effect of a giant cow. Was that really a cow you saw, or was it the likeness of a cow created by special effects? The latter, you say. Fair enough, but are "real" cows any more substantial than special-effect cows? According to Buddhist thought, "real" cows are no more than the special effects of countless dharmas, conditioned to "happen" in a manner not unlike the assemblage of flashing fireflies. Real persons, too, are created by the special effects of the Five Aggregates of elementary dharmas. If the dharmas are the ultimate constituents of the universe, then it follows that the reality of persons and cows is a conditioned or derivative reality. That is, persons and cows are effects.

But does that mean we cannot refer to them as real? Surely they are more real than persons and cows effected by fireflies and novelists. Well, are they? In the case of the firefly cow we might want to say it is not real because there is nothing "behind" the firefly effect— when the fireflies are done flashing, there is nothing left. But this is precisely the case with "real" cows too. When the dharmas no longer have their effects, there is nothing left of them either. Maybe it would help to say that real cows have a kind of identity lacking in firefly cows. We can name them ("Erma" perhaps) and we can return after a week and Erma will still be there. But this strategy doesn't get us very far, since we can name firefly cows too—this one's "Florence," there's "Phyllis." Firefly cows generally don't hang around long enough for naming them to be practical, but why should that make any difference? Real cows eventually vanish too. What makes temporal duration an official mark of reality?

Let's look more closely at persons. The Buddhist view is that persons just are what dharmas do, nothing more. Are persons real? Yes, of course, just as real as you like. But their reality must be kept in the proper perspective—that is, we should say no more about them than is prudent. Persons eventually vanish, too, just as soon as the dharmas quit effecting them, and when the dharmas stop, there is nothing left to talk about. From the Buddhist perspective it is imprudent to say that persons have a soul or self that endures after the dharmas stop. This is a fundamental error, on the order of positing something "behind" the firefly cow. There is nothing behind (or beyond, or in addition to) the effects of the skandhas.

To assume the existence of a self that is something more than a temporary collection of dharmic effects is imprudent because the assumption gives rise to a whole stream of agendas designed to serve the self, protect the self, advance the self, fulfill the self, and so on. To assume the existence of a self is to generate the perspective of self-interest, and with it the absurd utilitarian practice of appraising every encounter with the environment relative to self-interest. Appraising everything relative to what is at stake for the self presumes that the self is some sort of absolute, independent, sovereign reality.

But this is to go straight against the facts, for the self is thoroughly and undeniably conditioned by many factors. Any attempt to deny this basic truth puts one out of harmony with reality. Not to see the conditioned nature of the self is to persist in making appraisals and constructing agendas that are ultimately futile—one cannot preserve the impermanent, one cannot safeguard the vulnerable, one cannot satisfy the insatiable.

The desire for absolute freedom is both a cause and an effect of the illusion of selfhood—indeed, the illusion of selfhood and the illusion of free will are indistinguishable. The illusion of a free self engenders illusions of entitlement and creates blindness to limitations, giving rise to endless agendas designed to acquire and to avoid, all of which are bound to result in disappointment, frustration, resentment, and failure. The issue of free will, so important in the Western perspective, is given a brisk dismissal by Buddhism:

> If the whole of existence is relative, conditioned and interdependent, how can will alone be free? Will, like any other thought, is conditioned . . . If Free Will implies a will independent of con-

ditions, independent of cause and effect, such a thing does not exist.[8]

It may be objected that the attribution of an enduring self is both highly practical and ultimately impossible to prevent. To these objections the Buddhist will answer, first, that there is very little of practical value in a deluded perspective that results in suffering. How practical is pedophobia? And second, the Buddhist would insist that the illusion of selfhood is possible to avoid. The Buddha achieved liberation from the illusion, and so can anyone else who follows the Dhamma. Buddhism does not offer the prospect of self-emancipation as an item of faith, to be accepted on the authority of the Buddha's example. Quite the opposite: The truth of the Dhamma is empirically testable by anyone.

The process of self-emancipation has a double meaning for Theravada Buddhism. In one sense it means liberation from the illusion of self and its insufferable effects. And in another sense it points to the idea that salvation is the sole responsibility of the one who suffers, not something that can be granted by some external savior or agent of grace. Some forms of Buddhism have adopted the view that prayers to transcendent Buddha figures will solicit their help in achieving salvation. But for Theravadins there are no saviors, no deus ex machina to relieve one of the hard work of self-emancipation.

NIRVANA

Most religious traditions get themselves into difficulties whenever they attempt to describe the ultimate goal of the religious life. The goal of the religious life is usually described in terms of extraordinary experiences of a transcendent realm of being, beyond the normal state of things where ordinary language and thought forms apply. Buddhism is certainly no exception to this general problem. Whenever it attempts to speak of nirvana, the goal of religious practice, it tends to generate incoherence and confusion. One recurring difficulty is this: If there is no self, then who or what achieves nirvana? Another is this: How can one expect to achieve nirvana without first wanting to achieve it? But if nirvana is the extinction of desire, then wanting it renders it impossible to achieve. In the language of goal hierarchy management, the difficulty amounts to this: If nirvana is a goal, then, like other goals, it will generate stake appraisals—but stake appraising is precisely what prevents achieving the goal of nirvana.

Buddhists allow themselves to speak about nirvana, but they always throw in the caveat that it transcends all the conditions of speech. Nirvana is described as the "unconditioned," but language is itself a thoroughly conditioned instrument for talking about thoroughly conditioned objects, events, properties, and relations. Language cannot, therefore, apply to nirvana. These difficulties help to explain why the Buddha is reputed to have fallen silent on many occasions when he was asked about nirvana, but they do not explain why, on several occasions, he did not fall silent.

Nor do they explain the extensive verbiage of the tradition. Nirvana is often described in positive terms: as a goal, a destination, an event, a state, an experience. It is the final validation of Buddhist doctrines, the ultimate truth, a state of pure consciousness, peace, and equanimity, a state transcending all the dualisms of ordinary experience, a state that may be entered into during this life, and one that continues after death.[9]

More commonly, however, nirvana is described in negative terms: as the cessation of desire and suffering, a state without conditions, liberation from ignorance and selfhood. Literally, nirvana means "to blow out"—that is, a state of being wherein the fires of desire and aversion have been extinguished. It does *not* mean extinction of the self, for if there is no self, there is nothing to be extinguished. It is sometimes described as the extinction of the illusion of selfhood, the extinction of a self-interested perspective. Nirvana does not entail a loss of consciousness, but rather a de-conditioning or a purification of consciousness.

Buddhists may discourse about the nature of nirvana, but they normally do so with the understanding that there is nothing much, if anything, to be gained by talk. "No attempt is ever made to establish the existence of the Unconditioned by argumentation."[10] Nirvana is what it is when one gets there. It is whatever is left when one does away with the causes of suffering. It is whatever consciousness and understanding can be when the ordinary conditions for consciousness and understanding have been quieted. It is what remains when the autobiographical self has been deconstructed. It is the meaning that prevails in the absence of self-interested appraisals of meaning. It is, most fundamentally, what one gets to by following the Eightfold Path. This, it seems, is the real substance of Buddhism: To get on the path of spiritual discipline and follow it diligently, letting the rest take care of itself.

EMOTIONAL APPEALS

It is important to recognize a fundamental distinction in Theravada Buddhism between "esoteric" and "popular" forms of piety. Esoteric piety is the form of spiritual life practiced by monks who devote their lives full-time to achieving the goal of nirvana. The foregoing discussion of Buddhist doctrine may have left the impression that all Theravadins actively seek nirvana, but this is not the case. In fact, most lay Theravadins expect not to achieve the ultimate goal in their current lives. In some future life, yes, but not in this life. Most lay Theravadins are concerned about improving their present circumstances (better health or greater wealth) and accumulating karma sufficient for a better rebirth.

Esoteric piety and popular piety therefore seek different goals: The monks are fixed on the ultimate goal of salvation, the final cessation of suffering, while the laity are fixed on the proximate goal of merit-making, by which they may achieve a reduction of suffering.[11] These different goals are advanced by radically different practices. The salvation seekers engage in rigorous meditation exercises, while the merit makers engage in ceremonies, ritual practices, devotion to the Buddha, and charitable service to the monastic community (sangha).

The pronounced differences between the esoteric piety of the monks and the popular piety of the laity suggest that in Theravada Buddhism we are faced with two completely different religious systems.[12] This thesis has some validity because the original teachings of the Buddha appear almost disdainful of popular piety. The reality, however, is that esoteric and popular forms of piety are extensively interactive, much like complementary components within a unified system. The monks sustain the laity by providing spiritual guidance, personal counsel, and education. Meanwhile, the laity sustain the monks by providing food and material support for maintaining monasteries. The symbiosis gives monks a livelihood and freedom to pursue a spiritually demanding path, and it gives the laity guidance and ample opportunities for merit-making. We shall return to this system of institutionalized reciprocity in the next section, but for the moment we are left to consider how the emotional appeals of the Buddhist myth reflect the bifurcation between esoteric and popular piety.

FEAR AND DESIRE

Implicit in the myth of self-emancipation is a system of rewards and punishments that exploits human proclivities to fear and desire. Noncompliance with the Dhamma results in the accumulation of negative karma, which leads to increased adversity either in one's present life or in a future life. Positive karma, meanwhile, leads to a reduction of suffering and to good fortune in this life or in future lives. Merit-making Theravadins take this system of rewards and punishments very seriously, and are constantly reminded that their present circum-stances, good or bad, are the fruits of their own actions:

> . . . it is through a difference in their karma that men are not all alike, but some long-lived and some short-lived, some healthy and some sickly, some handsome and some ugly, some power-ful and some weak, some rich and some poor, some of high de-gree and some of low degree, some wise and some foolish.[13]

The education of emotion in Theravada societies owes much more to popular literature than to the canonical texts so favored by Western interpreters of Buddhism. Most Theravadins learn their Buddhism in the context of rituals and ceremonies, but especially by listening to stories from the *Jatakas,* a collection of 550 folktales re-counting previous lives of the Buddha.[14] These stories reinforce the reward-punishment system of samsara and karma by dwelling on the consequences of behavior for future lives. For example, in the *Nimi-Jataka* the charioteer Matali takes King Nimi on a tour of the hells and heavens, where he encounters tortures and delights to rival Qur'anic imagery:

> "Yon women, bent and broken, stretching their arms
> And wailing, wretched, smeared with stains of blood,
> Like cattle in the shambles, stand waist-deep
> Buried in earth, the upper trunk ablaze!
> Fear seizes on me: tell me Matali,
> What sin has been committed by those women,
> That now they stand all buried in the earth
> Waist-deep, the upper trunk a mass of flame?"
> Then answered Matali the charioteer,
> Describing how sin ripens and bears fruit:
> "They were of noble birth when in the world,
> Lived lives unclean, did deeds of wickedness,
> Were traitors, left their husbands, and besides
> Did other things to satisfy their lust;
> They spent their lives in dalliance; therefore now
> Stand blazing, waist-deep buried in the earth."[15]

And then to heaven:

> "Yon mansion built of jewels, shining bright,
> Symmetrical, proportioned, a fair sight,
> Where in divinest melody around,
> Songs, dances, drums and tabours do resound:
> I never have beheld a sight so fair,

Nor sounds so sweet have ever heard, I swear!
Joy seizes on me: tell me Matali,
What good these mortals did, that now I see
Happy in this heavenly mansion of delight?"
Then answered Matali the charioteer,
Describing how good ripens and bears fruit:
"These were lay Brethren in the world of men:
Provided parks and wells, or water drew
In the well-shed, and tranquil saints did feed,
Found clothes, food, drink and bedding, every need,
Contented with these men of life upright,
Who kept the Sabbath day, and each fortnight
The eighth, the fourteenth and the fifteenth days;
Generous, controlled, they walked in holy ways,
And now dwell in this mansion of delight." [16]

The impulse associated with fear of negative consequences is to act in ways that will avoid them, while the impulse associated with desire for positive consequences is to devise behaviors for realizing them. Theravadins moved to fear or desire by the Jataka stories will therefore be motivated to reconstruct their goal hierarchies to conform to the system of merits and demerits. Fear and desire, however, are not considered in Buddhism to be noble emotional virtues. True, one may be moved by desire to act in ways that reduce suffering in the world, but the positive merits of reducing suffering in such cases might be neutralized by the negative karma accrued by acting out of self-interest.

COMPASSION

The cardinal emotional virtue in all forms of Buddhism is compassion, a prosocial emotion that is vigorously exploited by the tradition. Suffering is the central problem addressed by Buddhism—the First Noble Truth is concerned with this—and at every turn one encounters reminders that one must be sensitive to the suffering of all sentient beings. Here again, the Jataka stories are a principal resource for educating the emotions. In one story, for example, the Buddha discovers a starving tiger with her cubs in a forest. Noticing that the distressed mother is about to eat her own young, the Buddha tears off his clothes and offers himself to her in their place. The tiger devours the prince and she and her cubs are spared. One after another, these stories present the Buddha as a compassionate and selfless figure, a model for righteous living in a world full of suffering:

. . . as a Deer King, the Buddha risked his own life to free all creatures from danger; as a monkey he saved all the frightened beasts
from their own fears; as a parrot he flew selflessly through flames
to save all those trapped in a burning forest; as an elephant he offered his life so that starving men might live; as a king he offered
his own flesh to save a dove; as a prince he gave his life so that a
starving tigress and her cubs might live.[17]

These stories excite several prosocial emotions (gratitude, affection, respect), but compassion dominates. Every story is a lesson in
meritorious sentiments and behavior, moving the listener to identify
with the Buddha and to emulate his emotional virtues.

Compassion is a complex emotion in Buddhist tradition (on the
order of agape in Christian tradition). Here I will offer the view that
compassion comprises love, sympathetic suffering, sympathetic joy,
and evenmindedness. All Buddhists are encouraged to practice these
Four Sublime States regularly, to the point where they become spontaneous and habitual responses in the course of one's daily encounters (this, I take it, is the point of Right Effort, step six on the Eightfold Path).

Love is practiced by projecting the sentiment outward, from
one's self to others. You first acknowledge that you love yourself, and
then you extend the range of your affection to include your family,
friends, and associates. You continue to dwell on these "lovable" ones
until no boundaries remain between them—that is, until you sense
that it is the same love embracing each of them. Gradually you extend
your affections even further, to casual acquaintances and strangers.
Then you allow yourself to feel a sense of love for the unlovable ones,
for those who are unattractive or even offensive to you. And eventually you admit your enemies to the expanding circle of loving regard.
Ultimately you will find that it is not only possible but even liberating
to extend your affections to include all living things.

The same principle of expanding sentiment is practiced with respect to sympathetic suffering. Consider how sensitive you are to your
own discomforts and concerns. But other beings suffer too, and it is
possible to participate in their suffering. Your sympathy for the suffering of your loved ones may be expanded to include the sufferings
of those you hardly know, of those you will never know, and eventually the domain of your sympathy may be expanded to include the suffering of all sentient beings.

Next, one expands joy. Just as it is possible to suffer in sympathy
with others, it is likewise possible to rejoice in the happiness of others. Oddly enough, sympathetic joy seems more difficult to practice
than sympathetic suffering. We are more easily moved to feel badly

about a stranger's suffering than we are to feel gladly about a stranger's joy. There is too often a tendency to ignore the happiness of others or, worse, to feel resentment or jealousy when others are rejoicing. But sympathetic joy can be practiced as surely as affection and sympathetic suffering can be. We know we are making spiritual progress when we can reflect on the happiness of our enemies with a sense of joy for their sake.

Finally, one may develop the capacity for compassion by practicing evenmindedness with respect to all the vicissitudes of life. Synonyms for this state might include "impartiality," "equanimity," "informed indifference," or even "objectivity." The mental operation here is not to expand but to contract the scope of one's identification with the outside world. An important part of this practice is to become indifferent to both criticism and praise. As proficiency in evenmindedness increases, one's compassion becomes more spontaneous and less discriminating—that is, it becomes more genuine.

These exercises have the potential for effecting deep changes in the brain's memory systems, biasing them toward the emotional states that are practiced with regularity. The practice of loving regard, sympathetic suffering, sympathetic joy, and evenmindedness will increase the likelihood that these responses will arise in everyday appraisals of meaning. Yet these are not demanding exercises—one can practice the Four Sublime States effectively in daily half-hour sessions, as many Theravadins do.

In my own view, the education of emotion is fundamentally different for esoteric piety than it is for popular piety. Merit makers seek good fortune and a better rebirth by means of generating positive karma, whereas nirvana seekers strive for detachment and no rebirth by means of eradicating karma. Merit makers are "new agenda" pietists—that is, they are challenged to reconstruct their goal hierarchies in harmony with the law of karma. But salvation seekers are "no agenda" pietists who use techniques designed to deconstruct all vestiges of a goal hierarchy. For esoteric piety the ultimate mode of existence is "compassionate detachment."

COMPASSIONATE DETACHMENT

The mode of compassionate detachment strikes common sense as an absurdity. One may be either compassionate or detached, we say, but how could one be both at once? Compassion presupposes caring, but detachment precludes it. The paradox is that compassionate detachment sounds like an appraisal-free emotion. But how can such a state exist? Compassionate detachment is understood to be a state of perfection—the Buddha's last forty-five years exemplifies this—a state that transcends conventional understanding.

So perhaps our question should not be "how is such a state possible?" but rather "how does one make progress toward such a state?" To this question Theravada Buddhism has a clear answer: To make progress toward perfection one must practice the "mental culture" of meditation: Right Mindfulness and Right Concentration, the final steps on the Eightfold Path.

The details of Theravada meditation practices are exceedingly complex and varied, but the general strategies are easily grasped. There are two basic techniques: the practice of calm and the practice of insight. The calming techniques involve a progressive withdrawal from sensory input and a renunciation of interest in anything that can excite the senses. Breathing exercises are important for the practice of calming, and these are typically followed by contemplating various objects or topics, depending on the needs and temperaments of individuals. Buddhaghosa, a fifth century commentator, lists forty objects of contemplation that may be used in the process of calming the mind.[18] If an individual is particularly attached to the pleasures of the body, then he might be advised to dwell upon different states of corpses: a swollen corpse, a purple corpse, a hacked-up corpse, a dog-chewn corpse, and so on. The idea is to loosen one's attachment to the body by reflecting on its impermanence and death. During the mind-calming exercises the meditator may enter into various trance states, but these are to be dismissed as distractions while the mind continues in the effort of shut itself down.

When the mind is completely quiet, the meditator turns to the techniques for insight. These techniques involve exposure to sensory input again, but now the input is analyzed by a completely quieted mind. The ordinary ego-involved mode of experience has been transcended, and the still mind is free to apprehend reality as it is in itself, absolute and unconditioned. Awareness remains, responsiveness remains, joy and compassion remain, but these are no longer relative to any goal hierarchy. They are spontaneous, boundless, and uncalculated. In the state of nirvana one sees directly the impermanence and interrelatedness of all things and experiences a loss of the sense of selfhood:

> This, verily, is the Peace, this is the Highest, namely the end of all formations, the forsaking of every substratum of rebirth, the fading away of craving, detachment, extinction, Nibbana. And in this state he reaches the cessation of passions . . .
>
> And his heart becomes free from sensual passion, free from the passion for existence, free from the passion of ignorance.

Freed am I! . . . Exhausted is rebirth, fulfilled the Holy Life; what was to be done, has been done; naught remains more for this world to do.[19]

It is difficult to construe Theravada meditation techniques as making an appeal to human emotional systems in the way that Jataka narratives and Gospel narratives do. Indeed, the goal of these techniques appears to be one of extinguishing normal emotional reactions rather than exploiting them. But if this is the case, then how can we fit esoteric piety into the thesis that religion seeks to play on the strings of human nature? What strings are played on in the case of Theravada mystical practices? Thus, it seems, we are brought back to the impossible question: How is such a state possible? No definitive answer to this question is available, but there is at least one promising hypothesis proposed by the late Eugene d'Aquili and his associate, Andrew Newberg.

The d'Aquili-Newberg hypothesis amounts to a neuropsychological model of mystical states.[20] The model centers on a phenomenon called "deafferentation," which basically means cutting off neural input to a brain structure. Deafferentation can be caused either by physical lesions or by the inhibitory function of another neural structure. When deafferentation occurs in a neural structure, the cells are likely to activate according to the "internal logic" of that structure, resulting in spontaneous impulses being sent to other parts of the brain.

When, therefore, a meditator undertakes breathing exercises and resolves to remove all thoughts from the mind, impulses from the prefrontal cortex inhibit the transmission of information from the inferior parietal lobe to the posterior superior parietal lobule (PSPL), preventing words and concepts from being processed by the PSPL. To completely master this technique of deafferenting the PSPL may take years of discipline. Deafferentation of the right hemisphere PSPL initiates a cascade of spontaneous impulses going from the right PSPL to the right hippocampus to the right amygdala to the hypothalamus, culminating in stimulation of the peripheral parasympathetic system. The resulting subjective experience is one of relaxation and progressive quietude.

If there is continued input from the prefrontal cortex (that is, if the meditator persists), then eventually the parasympathetic system will reach a maximum level of stimulation and there will occur a "spillover" stimulation of the sympathetic nervous system. When spillover occurs, the left hemisphere structures are recruited into a reverberating circuit, which results in a deafferentation of the left

PSPL. If the meditator gets everything right, then both the right and left PSPL will be simultaneously deafferented, with spectacular results:

> We propose that the total deafferentation of the left PSPL results in the obliteration of the self-other dichotomy at precisely the same moment that the deafferentation of the right PSPL generates a sense of absolute transcendent wholeness . . .
>
> We believe that this results in the subject's attainment of a state of rapturous transcendence and absolute wholeness which carries such overwhelming power and strength with it that the subject has the sense of experiencing absolute reality. This is the state of Absolute Unitary Being. Indeed, so ineffable is this state, that even the memory of it carries a sense of greater reality than the reality of our everyday world.[21]

An interesting feature of the d'Aquili-Newberg model is that the state of Absolute Unitary Being (nirvana) may be reached either by the practice of calm (initiated from right prefrontal cortex), or by the practice of insight (initiated from left prefrontal cortex), or by some combination of the two. This would help to explain why there is so much complex variation in Theravada meditation practices. In any event—and at the risk of overloading the metaphor—I will suggest that the techniques of esoteric piety represent a strategy for goading some of the strings of human nature (or rather, brain structures and circuits) into playing on themselves.

ANCILLARY STRATEGIES

EXPERIENTIAL STRATEGIES

It would be an understatement to say that Buddhism encourages everyone to seek and achieve extraordinary experiences. Even if popular pietists do not actively seek nirvana, they are no less convinced than esoteric pietists that it represents the ultimate goal of the religious life. But these points have already been made, and I will forego further discussion of experiential strategies on the assumption that enough has already been said to serve the arguments of this book.

INTELLECTUAL STRATEGIES

In theory Buddhism has an ambiguous relationship with intellectual endeavors. On one hand, it claims that all evils result from ignorance, suggesting that everything depends on the success of intellectual pursuits. On the other hand, it claims that the Buddha came to relieve suffering, not to answer abstruse philosophical questions. The Buddha was nothing like the physician who refuses to remove an arrow from a wounded patient without first knowing what the arrow is made of, where and by whom it was made, who shot it, and so on. In practice, however, Buddhists have shown little ambivalence about the intellectual domain and have always embraced it seriously and enthusiastically. In effect, there may have been no choice in the matter, since Buddhist doctrines had to be clarified and defended against both external and external critics.

As one might guess, the major intellectual challenges had to do with the central doctrines: no-self, impermanence, and causality. The no-self doctrine was challenged on the grounds that it was incoherent. For example, if there is no enduring self, then the phenomena of memory make no sense. Also, if there is no self, then all subject-predicate assertions ("He saw the moon," for example) are absurd. And if there is no self, then the doctrine of rebirth is absurd. Furthermore, if there is no self, then nothing exists to endure suffering, in which case compassion would have no object and would therefore be absurd.

The doctrine of impermanence had problems too. For example, if the elementary dharmas (the only ultimately real things) exist only in the moment, then the past and the future are not real. But if the past is without reality, then it is without efficacy, which raises doubts about the law of karma. And what about those elementary dharmas? How long do they subsist? An instant? Just how long is that? And what causes dharmas to decay and cease having effects?

Concerns about causality were raised too. How can it be said, for example, that one's desire is the cause of karmic accrual when in fact everything is declared to be interrelated in a vast causal network? It seems arbitrary to select desire as the sole cause when it is merely one of countless conditions. How might I be consumed with desire for that pair of shoes in the shop window if the shop had not been there at the moment I passed by? Or if the shopkeeper had selected a different pair for display? Or if the shoemaker's parents had not met one another on an ocean cruise? [22]

It would take us far out of our way to consider the subtleties of Theravada responses to these intellectual challenges. Besides, the point we're after is not how the challenges were addressed, but *that* they were addressed by developing intellectual strategies for interpreting, clarifying, and defending the metaphysical aspects of the Buddhist myth.

Theravada metaphysics has tended to be dogmatic and

commentarial, rather than dialectic and speculative. That is, it fore-goes reasoned arguments in favor of citing proof texts from earlier writings and constructing endless lists to classify elementary dharma types and their characteristics. Theravada moral theory, too, has fol-lowed the same dogmatic style. Basically, it amounts to normative ethics, much of it given to rigid classifications of mental phenomena according to whether their karmic effects are good, bad, or neutral. The rationale behind the authoritative approach to metaphysical and moral problems is entirely pragmatic. The lists and classifications are put out there for the monks to learn, without the unnecessary bother of tedious arguments to justify them. If there is any doubt about the doctrines, then there remains only one sure way to validate them—to enter the state of nirvana, whereby argumentation is rendered com-pletely superfluous. In the end, therefore, theory gives way to prac-tice, precisely because logic is no substitute for insight.

INSTITUTIONAL STRATEGIES

One formally becomes a Buddhist by taking refuge in the Three Gems:

> I take my refuge in the Buddha
> I take my refuge in the Dhamma
> I take my refuge in the Sangha

The sangha is the community of the faithful, literally, the "assem-bly" of those who identify with the Buddha and his teachings. The term has been defined in various ways: as a monastic order (*bhiksu-sangha*), as a comprehensive institution including both monks and laity (*maha-sangha*), and as a spiritual elite (*arya-sangha*).

Originally, the sangha was composed of Buddha groupies, those who undertook the life of homeless wanderers in order to follow the Buddha and embody his teachings. After the Buddha's death, and as the assembly grew and diversified, there was a need to formalize a code of discipline to maintain the solidarity of the sangha. The intro-duction of a monastic code (*vinaya*), together with standards of or-dination, established a distinction between monks and laity.

At this point "sangha" denoted the assembly of ordained men and women who formally accepted the monastic rule. Eventually the itinerant monks settled down in permanent communities where they became more involved in maintaining the faith than proselytizing it. As this happened, the term "sangha" could be used in a broader sense, to denote a local monastery together with the lay congregation associating with it. The term has also been used in a strictly spiritual sense, referring to "an aristocracy of transcendental attainment."[23] In this sense the sangha is an invisible assembly of those who have distinguished themselves in terms of spiritual endeavor, whether ordained or not: "Monasticism counts for nothing; householding counts for nothing. Purifying the mind counts for everything."[24]

Institutional strategies are designed to manage the interpretation, transmission, and embodiment of a mythic tradition. In Theravada societies these functions are served in the context of the relationship between monks and laity, a relationship based on reciprocal generosity. The relationship is an ancient one, the terms of which have been attributed to the Buddha in the *Sigalaka Sutta:*

> Signifying [monks] whom their disciples should uphold in five ways: (1) by acting with loving-kindness; (2) by speaking with loving-kindness; (3) by thinking thoughts with loving-kindness: (4) by always welcoming them into their homes; (5) by providing them with material requisites.
>
> [Monks], having been upheld in these ways, should then help their followers (1) by helping them to avoid evil; (2) by encouraging them to do the good; (3) by helping them with a compassionate mind; (4) by teaching them what they do not know; and (5) clarifying for them what they might already know.[25]

Monks see it as their duty to "keep up the faith of the laity" by providing guidance and support. They do this by advising the laity on the details of merit-making, and by serving as personal counselors, family therapists, and teachers (in both religious and secular subjects). Monks also support the laity in the role of legal and political advocates. Theravada monks may be fixed on the ultimate goal of nirvana, but they do not regard this as inconsistent with their dedication to serve the laity. The two fit together rather well: "Temporarily and provisionally it is possible and necessary to allay much suffering; but permanent and really effective alleviation is possible only for each individual in himself."[26]

Theravada laity are fixed on the proximate goal of merit-making: developing the character traits and performing the actions that will bring positive karmic effects. For Theravadins the monastic community has always been the "field of merit par excellence."[27] Any offering made to ordained members of the sangha will bear more karmic fruit than offerings made elsewhere. The monastic community is associated in the minds of most laity with occult powers, and the monks are perceived as means of protection and blessing. Therefore, the more closely one associates with the monastery, and the more one supports and venerates the monks, the better off one can expect to be both in this life and in future lives. Lay members of the sangha regularly provide food, clothing, bedding, and other material support to the monastic community. In addition, they treat the monks with the kind of reverence and devotion most cultures reserve for royalty. In Theravada societies it is common practice that no one should sit in the presence of a standing monk, and one should always use honorific speech when addressing monks or referring to them.[28]

What really drives the institutionalized reciprocity of the sangha? Do monks serve the laity from a sense of duty or out of genuine compassion for them? Do the laity support the monks from calculated self-interest or out of a genuine sense of gratitude? Hard to say. The truth is probably that all these motives are involved. The important thing, however, is that the system works effectively to generate broad overlaps of self-interest.

RITUAL STRATEGIES

Buddhism, as previously mentioned, originated partly in reaction to the ritualism of Brahminical religion. The Buddha was severely critical of ritual practices because they had become an instrument of exploitation, but also because he thought that ritual sacrifices were irrelevant and ineffective means to self-emancipation. It is somewhat surprising, therefore, to discover that ritual strategies abound in all parts of the Buddhist world.

As Buddhism moved into new territories it had no elaborate rituals to impose. That is, it had no initiation rites, no puberty rites, no marriage or atonement rituals, nor many of the other formal practices commonly found in religious traditions. But neither did Buddhism take measures to expunge the existing ritual practices of new converts. The result has been that ancient local rituals have tended to persist alongside the practice of Buddhism, often with little or no attempt to legitimate them on Buddhist principles. Thus, for example, it is not uncommon for devout Buddhists to be married in non-Buddhist (Hindu, even) ceremonies without a hint of incongruity.

The closest thing to a purely Buddhist ritual (apart from meditation itself) is probably the ceremony of ordination into a monastic community. The formal ceremony begins with a procession to the monastery compound where the ordinand bows before his preceptor and "takes refuge" in the Buddha, the Dhamma, and the sangha. After the ordinand requests permission to enter the priesthood, the preceptor delivers an instructive sermon on Buddhist doctrines. The ordinand then commits to the Ten Precepts of monastic life and is given a Pali name together with a begging bowl and monastic robes. He is then questioned to determine whether he is free of impediments, and if so the ordinand is presented to the sangha with a proposal for admission. The power of admission falls to the monastic community as a whole, which acknowledges consent by collective silence.[29]

There are various monkish rituals associated with the Rainy Season Retreat, a three-month period when monks follow a special regimen of intensive study and meditation. At the conclusion of the retreat the monks gather in solemn assembly and each monk in turn asks the others to reproach him publicly for any offensive behavior he has committed as a member of the sangha. This ritual is intended to foster harmony and solidarity among the ordained.[30]

Virtually all Theravada rituals involving the laity (of which there are many) amount to opportunities for merit-making. The most important of these are the *Kathina* ceremony and the Buddha Day festival. The Kathina (literally "robes") ceremony takes place in early November after the Rainy Season Retreat. This is an auspicious time for merit-making because the monks, having just completed an intense period of discipline, are believed to be in an enhanced state of spiritual potency. The ceremony is focused on a sort of transaction between the laity and the monks: The laity present new robes and other material goods to the monks, whereupon the monks bestow a blessing of protection, power, and happiness on the laity.[31] The ceremony may last up to three days and involves nearly every family in the district.

The formal ceremony begins with a procession of laity to the monastery where they are greeted by the monks. The laity then "take refuge" in the Three Gems, and repeat after the monks the Five Precepts of Buddhist lay life: not to kill, steal, lie, misbehave sexually, or drink alcohol. After the formal presentation and blessing, the entire community joins in the festivities of food, music, and traditional dance.

According to Theravada tradition, the Buddha's birth, enlightenment, and death share a common anniversary, celebrated as Buddha Day at the full moon of April/May. Buddha Day (*vesak*) is an ancient festival, presumably inaugurated by King Asoka in the third century B.C.E. The threefold nature of this celebration of the Buddha's life makes it a prime occasion for merit-making. The colorful festival features an evening procession to the monastery through streets decorated with lanterns. At the monastery flowers are heaped on the altar and the narrative of Buddha's life is retold at length by the monks. Lay members of the sangha accrue merit by participating in the events, listening to the sermon, and doing their part to feed the large crowd.

Many of the ceremonies and festivals in Theravada societies have their origins in pre-Buddhist times, having since acquired merit-making status by superimposing Buddhist elements. This is accomplished by persuading monks to participate by providing *paritta,* blessings chanted from sacred texts. This practice has merit-making value because many of the verses touch upon moral virtues, but also because the sound alone is conducive to purifying the mind. Monks provide paritta for a variety of domestic rituals in addition to public ceremonies. The most significant domestic ritual is the funeral, observed either in homes or temples and officiated by monks. Funeral rites include chants, mantras, and a sermon that typically dwells on impermanence, karma, and nirvana. It is supposed that all the merit generated in the funeral service will benefit the deceased.

By now it should go without saying that frequent and emotionally engaging rituals such as these will have significant impact on the brain's memory systems and will bias them toward appraisals consistent with Buddhist ideals. Such appraisals themselves—if one appraises an event with compassion or detachment, for example—are understood to bring their own merits.

AESTHETIC STRATEGIES

Much is distinctive in Buddhist aesthetic traditions, but nothing more so than images of the Buddha and representations of events in his life. For the first five centuries no Buddha images were produced, not because Buddhism was iconoclastic but presumably because there was a bias against attempts to reduce a "thus gone one" to any conditioned forms. Instead of the Buddha, therefore, one encountered mere suggestions of him: a footprint, an empty seat, animal forms from Jataka stories, the Wheel of Life, or a Bodhi tree. By the first century C.E., however, all hesitations had vanished and Buddha images began to appear in abundance.

Buddha imagery reflects a complicated set of conventions. First, the range of body postures is very narrow, limited to standing, sitting in the lotus position, sitting with legs dangling, or reclining in repose. The Buddha's hand gestures conform to an aesthetic vocabulary sharing several features in common with Hindu and Jain art. Each gesture conveys a special meaning. Many statues and relief carvings feature the Buddha seated in the lotus position with his left hand resting palm-up on his lap while his right hand droops over his knee, allowing the finger tips to touch the earth. This gesture alludes to the episode when the tempter, Mara, demanded to know by what authority the Buddha claimed to be enlightened. The Buddha's response was to call the earth to testify on his behalf. In another distinctively Buddhist gesture the Buddha forms two circles by joining forefingers to thumbs, signifying the turning of the wheel of Dhamma. A right hand raised with palm outward conveys a blessing of protection, while a lowered hand with palm outward signifies compassion and generosity.

When the infant Gautama was examined by a Brahmin priest, he was found to possess thirty-two major birthmarks (*lakshanas*) and eighty minor ones. Lakshanas are associated in Indian tradition with great power, and on the strength of these birthmarks it was predicted that Gautama would become either a world emperor or a Buddha. The thirty-two marks feature prominently in Indian art to denote power, authority, or heroism in a subject. In Buddhist imagery the marks are associated with the Dharma, thus calling attention to the Buddha's significance as a "wheel turner" or cosmic authority.

Some of the marks listed in Theravada texts include a prominent node atop the head (a sign of superior wisdom), a hairy tuft on the forehead, forty teeth, rounded smooth ankles, protruding heels, straight torso, chest like a lion, eyelashes like a cow, flat feet with wheel prints, webbed fingers and toes, smooth and lustrous skin, and a long tongue. Every Buddha image is presumed to possess all the marks, even though they may not be discernable.[32] Buddha images typically include some traditional marks of distinction (halo, elongated earlobes) that do not appear in sacred texts. The hand gestures and lakshanas reinforce memory and stimulate reflection on the central meanings of the myth of self-emancipation.

Many Theravadins will insist that they are not devotionalists or image worshippers in the manner of Krishnavites, for example. That is, they do not regard the Buddha image as the real living presence of an object of devotion, but merely as a reminder of the great teacher and his doctrine.[33] This may well be the case for esoteric pietists, but the "Eye Opening" ceremony of popular pietism presents a very different picture. In Theravada settings Buddha images must be consecrated by a ritual intended to imbue the image with "the Buddha essence."[34] In Thailand, Laos, and Cambodia the monks will hold a sacred cord attached to the Buddha image, which transfers the power of their chants into the image. The image is also made a recipient of flowers, candles, and incense. The Eye Opening ceremony typically begins at dusk and ends at dawn the next day, when the eyes are uncovered, bringing the image to life and releasing its power.[35]

Jataka stories occupy an important place in the aesthetic strategies of Theravada Buddhism. Scenes from the Jatakas are represented in relief carvings in caves, on monuments and fences, and in temple statuary. The earliest surviving examples of Buddhist painting feature Jataka scenes, and countless monastery walls throughout the Theravada world are festooned with murals depicting favorite Jataka stories. They are also reenacted in traditional dance dramas performed during festivals. These stories provide constant reminders that the Buddha's long path through many successive lives to nirvana was paved by a compassionate nature.

The art of Theravada culture, no less than its many rituals and festivals, is tied into the system of merit-making. Donors providing resources for the production of artworks both give and receive merit, as do artists who produce the work. And not least of all, those who encounter the images acquire merit by virtue of the power of images to purify the mind. The merit-making system motivates individuals to attend carefully and regularly to Buddhist imagery, a practice that effectively revitalizes the meanings of the myth. The impact of Buddhist imagery on human emotional systems can be quite profound, as the following encounters attest:

> The thing about [these Buddha figures] is that there is no puzzle, no problem, and really no "mystery." All problems are resolved and everything is clear, simply because what matters is clear. The rock, all matter, all life, is charged with dharmakaya . . . everything is emptiness and everything is compassion. I don't know when in my life I have ever had such a sense of beauty and spiritual validity running together in one aesthetic illumination.[36]

> I began to feel the strange effect which the Buddha's face produced on me. All the gloom that rose from the depths of my soul seemed to clear up. It was as if the Buddha's face communicated its calm to me. Everything that up to now had troubled me and appeared so serious and important, now became so small, insignificant and unworthy of notice, that I only wondered how it could ever have affected me.[37]

PERSONAL WHOLENESS AND SOCIAL COHERENCE

Now to the question: How does Theravada Buddhism contribute to the conditions for achieving personal wholeness and social coherence? If the ancillary strategies are doing their job effectively, then the meanings of the myth of self-emancipation will come to dominate goal hierarchy mechanisms with the result that individuals will appraise their daily encounters relative to the values inherent in the myth. The question is whether these appraisals might be conducive to overlaps of self-interest.

In my view the effectiveness of Buddhism centers on mindfulness and compassion. To be mindful is to attend closely to whatever the mind is doing while the mind is doing it. Compassion, meanwhile, is a complex of emotional virtues, including love, sympathetic suffering, sympathetic joy, and evenmindedness—the Four Sublime States.

It seems to me that when mindfulness and compassion are brought together, they produce what psychologists call the coping process. Coping amounts to doing or thinking something that will affect the final outcome of a particular appraisal event. It involves attending to and appraising the meaning of what the mind is doing while the mind is doing it. It is a new appraisal of an ongoing appraisal in light of one's standards of emotional virtue. If I have disciplined myself to practice mindfulness, then I will take notice whenever an emotional response is swelling up in me, and I will immediately turn my attention toward the emergent emotion as an item for appraisal. If the emergent emotion violates my sense of emotional virtue (an element in my goal hierarchy), then I will appraise its meaning as a threat to my own interests, and will thus take measures to alter the outcome of the initial appraisal. For example, if I feel anger swelling up and if I disapprove of anger, then it is within my power as a mindful person to abort the anger response and come to a somewhat different view of the business at hand that I had initially appraised in an anger-provoking way.

The ancillary strategies of Buddhism are designed as means for effecting goal hierarchy transformations, such that various emotional virtues—the Four Sublime States in particular—will become biased into memory systems, making it likely that they will insinuate themselves into the processes of appraisal and coping. Therefore, to the extent that I practice the Four Sublime States, two things should follow. First, I will be less likely to make negative initial appraisals in my daily encounters, and second, when negative appraisals begin to unfold, I will be more likely to alter my emotional reactions midstream by introducing coping strategies.

The contribution of these dynamics to the twin teloi of personal wholeness and social coherence should be plain to see. If I internalize the character traits advocated by Buddhist teachings, then my self-esteem will become linked to emotions and behaviors that reflect those traits. Thus I will be motivated to be a less offensive and more cooperative person—that is, I will genuinely want to be less offensive and more cooperative, and I will genuinely want to reduce suffering wherever I encounter it. And thus whenever my emotions and actions give me evidence that I am an inoffensive, cooperative, compassionate person, I will have reason to feel a sense of fulfillment, or wholeness. These are the conditions for producing broad overlaps of self-interest.

PART
III

ON
THE FUTURE
OF RELIGION

The first part of this book set out to sketch a picture of human nature from the perspective of consilient scientific materialism. The argument was that human nature may be discerned by reviewing the evolutionary story of how our species came to be, and by examining the uniquely human strategies for carrying on. The sketch presented human beings as star-born, earth-formed creatures endowed by evolutionary processes to pursue the twin goals of personal wholeness and social coherence under the guidance of a plurality of systems for mediating behavior.

It was further shown how these systems are open to manipulation by symbolic means. Such means have been systematically deployed by religious traditions as strategies enabling humans to form and sustain counterintuitively large social groups. Religious traditions succeed to the extent that their integrated visions of reality and value (their myths) create the cognitive and emotional conditions for broad and stable overlaps of self-interest. When mythic traditions fail to do this—for whatever reasons—humans will be likely to default to the intuitive patterns of sociality practiced for hundreds of thousands of years by our hunting-gathering ancestors.

Part two surveyed five major mythic traditions in light of the sketch of human nature and the theory of religion developed in part one. There it was shown how Judaism, Christianity, Islam, Hinduism, and Buddhism have designed the means for managing aspects of human nature for the sake of promoting both personal wholeness and social coherence, thereby to assure a confluence of self-interests.

In the final chapters ahead our attention turns to consider the future of religion. Speculations about the future of religion have always been notorious for the magnitude of their inaccuracies. For example, during the Enlightenment the demise of religion was considered in the most sophisticated circles to be a fait accompli. And throughout the twentieth century many observers were confident that major traditions would converge into one universal religious body. Given the way things have turned out, one has to wonder what sort of information these bold forecasters of the religious life had to go on. Very little more, I reckon, than we have: a certain understanding of the nature of religion, a certain understanding of human nature, and an assessment of contemporary challenges facing the human community. Change one of these variables, and you will get a different reading on prospects for religion in the future.

I intend to give my own reading on the future of religion in the course of the next two chapters, but I will do so without much confidence that things will go as I expect them to. Then why bother? I have two reasons. First, it seems to me that saying something about the future of religion is somehow mandated by what has been said in earlier parts of this book. I have presented religion as a set of strategies designed to help human beings to carry on in the face of challenges to a full life. But if we are curious about our carrying on—and who isn't?—then we can hardly resist the temptation to spin out scenarios of future possibilities despite the sad performance record of such projections.

The other reason is that guesswork about the future is often useful even when it turns out to be wrong. Sometimes dire consequences may be prevented precisely because they are predicted. I mention dire consequences because my own expectations about the future are rather pessimistic. I believe that we are presently faced with the most threatening circumstances of our entire history as a species—bar none! Unsustainable patterns of human population and material consumption are now stressing natural and social systems near the point of no return. We are not living in harmony with reality, and the clearest measure of our folly is that the life-support systems of the earth are already in a state of serious decline. Further, I believe that these problems come at a time when our religious traditions are losing their powers to command an adaptive response.

It is not unlikely, in the course of the present century, that we will overshoot the carrying capacity of our planet, and the consequences of doing so will be grim. There may be little chance that we shall go extinct in the process of collapse—as many species already have, and as many more will—but in the near term we may expect to see chaos, misery, and death on a scale we can scarcely imagine.

The long-term prospects for human viability look more promising. In the aftermath of global calamity the ultimate cause of our misfortune will be unmistakable: failure to live in harmony with the limits of nature. This realization will then find expression in new ventures of mythmaking that will be notable for the central importance they give to the natural order. Intimations of the long-term future of religion are already evident in the emerging perspective of religious naturalism.

This reading of the future of religion will be presented in two chapters focused on the principal challenges faced by religious traditions today. Chapter 10 will examine the crisis of influence by exploring the impacts of modern science, religious diversity, and the rise of a consumer culture. Chapter 11 will examine the crisis of moral relevance by considering the urgencies of environmental degradation. The concerns underlying these two chapters are whether established mythic traditions can effectively address these crises, and what we might expect if they fail.

10

THE CRISIS OF INFLUENCE

I have claimed repeatedly that religious traditions are, essentially, mythic traditions. At the core of these traditions is an integrated narrative account of cosmology and morality. In the world of myth, reality and value are fused: Facts are pregnant with value implications, and values are imbued with the objective warrant of facts. The coalescence of cosmology and morality is achieved by means of a root metaphor. As we have seen, the metaphor of God as person achieves the fusion in Abrahamic traditions, while the concept of Dharma does it for Indian traditions.

I have further claimed that an attitude of realism regarding the root metaphor is critical to the effectiveness of a myth. If my realism about the root metaphor of a myth is compromised, then the fusion of reality and value is compromised; if my sense of the objective reality of certain prosocial values is compromised, then the linkage of those values to my self-esteem will be compromised; if my self-esteem is de-linked from certain values, then the power of these values to command a hearing in working memory will be compromised; and if the values in question fail to gain a hearing in working memory, then they cannot influence the manner of my appraising and coping with any business at hand. And further on, if the prosocial values of a mythic tradition fail to influence my appraisals—if they fail to educate my emotions—then my goal hierarchy will increasingly reflect the campsite values of intuitive morality, in which case overlaps of self-interest with others will be fewer and conflicts of interest will increase.

We see, then, that wherever the specter of nonrealism appears, it will constitute a major threat to the effectiveness of mythic traditions. The argument of the first two parts of this chapter will be that the specter of nonrealism already haunts the great religions of the world and will continue to do so. The discussions will focus on two principal sources of creeping nonrealism: the rise of modern science and the awareness of religious diversity. The final part will explore the rise of consumerism as a powerful source of competition for traditional myths. The cumulative impact of these phenomena has severely compromised the influence of religious traditions.

THE RISE OF MODERN SCIENCE

There is no point in fashioning arguments to show that the rise of modern science has interfered with the effectiveness of religious traditions. This has been the consistent lesson of countless surveys on the issue: Where scientific literacy is up, mythic realism is down. The question is not whether, but why.

Some defenders of mythic realism blame the fallacy of scientism.[1] Science, the argument goes, should not be confused with scientism. Science is the disciplined enterprise of finding out facts and explaining their occurrence by reference to general principles of nature. Science forms hypothetical beliefs about entities, events, properties, and relations in nature and seeks to confirm or refute them by the evidence of reason and observation. Scientism, meanwhile, is a set of philosophical beliefs about science, not a set of tested beliefs about nature. Scientism says that science is the sole authority on all claims about the natural order; that the limits of science are the true limits to what can be said about how things really are; that scientific claims are the only ones that warrant realist attitudes.

It should be clear that scientism is predisposed toward conflicts with mythic traditions in ways that science itself is not. For example, both science and scientism would reject on the strength of evidence the claim that the universe was created in six twenty-four-hour days. There is much reliable evidence to refute it and much reliable evidence to support an alternative claim. But science would *not* (as scientism would) reject claims like "God exists" or "God created the universe." These claims can be neither confirmed nor disconfirmed by the accepted principles and methods of science. Anyone wishing to affirm such claims (as theists do) or to deny them (as advocates of scientism do) must be aware that they are stepping well beyond the scope of science itself. Science qua science presents no obstacle to theistic belief. The implication is that anyone who believes that science is in conflict with religion must be guilty of confusing science with scientism. Nor is scientism itself a significant threat to theism, since it is merely a set of philosophical attitudes having no better justification than theistic attitudes: "[scientism] . . . is a kind of faith-commitment not entirely unlike the kind we find in religion."[2]

Well, I'm not so sure. The comparison of scientism with a faith commitment may sound promising to defenders of theological realism, but it overlooks important differences. Scientism declares confidence in a method, not in a set of doctrines. To be fair, the comparison would have to judge the merits of scientific methods against those of theological methods, and when this is done the similarities tend to vanish. For one thing, it is clear that the harshest critics of scientific findings are scientists themselves. Skepticism about scientific claims is a central virtue of the scientific method, but it is hardly a virtue of any faith commitment.

But never mind all that. One need not espouse scientism to see how familiarity with science might interfere with realism about root metaphors. It is enough simply to recognize how the onset of science displaced the explanatory power of the root metaphor of God as person. Animistic traditions attribute personal agency to many natural phenomena, including living things like animals and trees, and also to inanimate things like storms and wildfires.[3] Given the ability of our species to read minds, and thereby to understand and predict behavior, it is not surprising that we would use the mind-reading strategy to understand and predict the forces of nature. The metaphor of God as person is a vestige of this animistic mentality.

Metaphors (literally: "carry over") help us to achieve understanding by transferring the meaning of one thing to another. Thus, some lesser-known thing is apprehended metaphorically in terms of some better-known thing. For most of human history the domain of nature was the lesser-known reality and persons were the better-known reality. It therefore made perfect sense for us to come to terms with the mysteries of nature by reducing them metaphorically to the terms of personal agency. We personalized nature, thereby to demystify it. Nature came to be known to us as God's handiwork designed to express God's purposes, and these purposes gave us the ultimate explanation for all natural facts.

The explanatory power of the personal metaphor has, however, been severely compromised since the rise of modern science. The principal reason for this has been a reversal in the direction of explanation: Whereas we once explained nature in personal terms, we now explain personal reality in natural terms. The explainer has become the explained; personal agency has been naturalized; nature is now the better known reality by which we apprehend the lesser known reality of persons. Personality is hardly a plausible metaphor for giving us an ultimate explanation for all natural facts. Persons are clearly effects of nature, not its cause. In the process of naturalizing personal agency, we have naturalized the personal metaphor as well. God may once have been our best ultimate explanation for all natural facts, but we now have a fairly complete and satisfying naturalistic understanding of how mind-reading humans came up with the idea of God and came to use the metaphor for therapeutic and social purposes. These dynamics, not the fallacy of scientism, are responsible for a decline in realist attitudes toward the root metaphor of theistic traditions.

There are other dynamics as well. For example, we should not underestimate the extent to which standards of scientific inquiry have spilled over to become general habits of mind. In particular, the central norm of scientific rationality—that belief must be apportioned to the evidence—has become an intellectual virtue in all societies where science is well established, and it is reasonable to assume that this habit of mind has exercised an inhibitory effect with respect to mythic realism.

In my view, the many contentious issues between science and religion boil down to a single question: the perceived plausibility of a myth's root metaphor. If the root metaphor, the integrating concept, is deemed implausible, then realism flags. And in such cases the myth is unable to educate the emotions. Let this be clear: It is *not* the case that anything in science actually falsifies the root metaphor of a myth. As far as science is concerned, it is entirely possible that a personal God exists and it is possible that this personal God created the universe with a moral purpose. But possibility is not plausibility, and it is the plausibility of root metaphors that counts for mythic realism.

Consider some proposition, X. We say X is possible if we cannot decisively rule it out—that is, if we can see that X might be the case. But the threshold for plausibility is a bit higher. We say X is plausible only if we can see *how* it might be the case. And the demand to see how X might be the case is a demand that it cohere with our understandings of many other things. More than that, it is a demand that X enable a more complete and satisfying understanding of many other things. But determining the plausibility of X is sure to be a fairly sloppy process because "many other things" is likely to be a list of beliefs and attitudes that will vary considerably from one person to the next. This variable list of items is what some philosophers mean by the term "background beliefs."

My point here is to insist that in societies where science is well established, the list of background beliefs will include much of the scientific worldview, and such beliefs will make it more difficult to see how certain religious claims might be the case. For example, given what you know about human beings, it may be difficult for you to see how it might be the case that you will come back in some future life as a frog, regardless of how you behave in this life. It is possible, yes, but is it plausible? And if it is not plausible, then you are not likely to be motivated by fear of frogdom to change your goal hierarchy. Matters might just as well go the other way around. If you live in a society where a mythic vision dominates, and where faith (in the sense of "belief without evidence") is a stronger intellectual virtue than apportioning beliefs to evidence, then your background beliefs might tend to make the scientific worldview appear implausible.

When the scientific worldview makes inroads into traditional societies, as it will continue to do, it stirs up confusion and conflict. Inevitably, the arrival of science will trigger challenges to any vision that integrates cosmology and morality. And challenges to the central meanings of a myth will, in turn, provoke defensive responses. It is expected, therefore, that any society struggling to assimilate science will eventually concentrate its intellectual resources on the encounter of science and religion.

The encounter between science and religion has dominated the agenda of theological (especially Christian) discourse since the rise of modern science in the seventeenth century. According to Ian Barbour, four major options for thinking about the encounter have emerged in the course of debates: conflict, independence, dialogue, and integration.[4] I will summarize these four options in terms of the central question of the plausibility of a root metaphor.

CONFLICT

Mythic traditions make knowledge claims about how things are in the world—that is, they have an investment in cosmology. Science, too, is invested in formulating knowledge about the world. To the extent that science and religion work independently and with different methods, it is inevitable that their cosmologies will make conflicting claims. When Galileo presented evidence supporting the Copernican model of a heliocentric solar system, he was directly contradicting the Ptolemaic geocentric model that had been assimilated into the Christian myth.

When conflicts like this become evident, the most obvious intellectual strategy is to defend one option and to reject the other. The problem, however, is that conflicts between scientific and mythic knowledge claims can rarely be confined to specific issues. Galileo's defense of the Copernican system was based on the evidence of reason and observations of natural phenomena, while the church's defense was based on the evidence of reason and the testimony of scripture. It is not difficult to see how a particular conflict of this sort would quickly become a general conflict over authority, especially the authority of scripture. If the Bible is wrong about the shape of the solar system, then it is probably wrong about many other things as well.

The general challenge to defenders of the mythic vision became clear: how to relinquish authority on particular cosmological claims without relinquishing authority on the central claims where realist attitudes are essential to the effectiveness of the myth. The intellectual strategies of the Christian tradition have been preoccupied with answering this challenge since the seventeenth century. Much of the attention has been focused on reinterpreting the myth in ways that would accommodate scientific cosmology without compromising the plausibility of the root metaphor. The results, of course, have been mixed. New interpretations have enabled many individuals to continue seeing both *that* and *how* the personal God might be the case. But many others have become skeptical nonrealists about the myth, insisting that to see *how* the personal metaphor might be the case requires more reliable evidence than the new interpretations have been able to provide.

INDEPENDENCE

Another intellectual strategy for resolving tensions between science and religion is to divide the territory. Recall that the root metaphor of a myth provides the ultimate explanation for all facts as well as the ultimate justification for all values. The independence option attempts to resolve tension by separating these functions—that is, by giving to science the province of explaining natural facts and giving to religion the province of justifying moral values. Galileo himself took this approach.

More recently, Stephen J. Gould has advocated the independence strategy in his book *Rocks of Ages*.[5] Science and religion, he asserts, are "non-overlapping magisteria," with science in charge of cosmology and religion in charge of morality. Science is the domain of dispassionate apportionment of belief to objective evidence, while religion is the domain of passionate commitment to subjective values. Facts here, values there, and ne'er the twain shall meet.

This strategy appears to be a friendly settlement for religious traditions because it gives them a sphere of spiritual authority where science poses no threat. The strategy also safeguards the autonomy of scientific inquiry from the meddling of religious busybodies. The independence model claims that science and religion ask fundamentally different questions, requiring different kinds of answers. Science asks *how,* and religion asks *why.* They are, therefore, incommensurate spheres of meaning. The reality of God is sought and found in the collective experience of a historical community, not in discoveries by the scientific community. And further, the religious life is essentially about subjective and emotional responses to God's gracious acts in history, not about cognitive assent to knowledge claims about the natural world.

The independence model has been attractive to many mythic realists because it offers a convenient way to dismiss putative conflicts between science and religion as mere category mistakes. Others, however, would claim that the separation of categories in the independence model creates a false dichotomy. For one thing, it reflects a fundamental misunderstanding of the integrative structure of mythic traditions. And for another, it ignores the manner in which human existence and historical events are deeply embedded in the details of natural systems. The independence model would appear to be the cultural equivalent of surgically severing the brain's cognitive systems from its emotional systems, leaving the subject in a trout-like state, unable to apprehend a cosmos infused with meaning. At the end of the day, any theology that vacates the domain of cosmology also emasculates the root metaphor of theism.

DIALOGUE

The dialogue strategy for resolving tension between science and religion operates on a more general level and has a more analytical tone than either the conflict model or the independence model, although it shares important features with both. Like the conflict model, the dialogue approach agrees that mythic traditions are deeply invested in cosmological questions, but it stresses that the religious interest in cosmology is at the level of metaphysical truth, well beyond the scope of science. And like the independence model, the dialogue approach agrees that science and religion have ultimately different goals, but it stresses that there exist important points of contact and similarities between the two.

The dialogue approach is highly conversant with philosophical perspectives on both science and religion. It tries to clarify the boundaries between them and also to show how they may be of benefit to one another. Further, dialogue helps to identify and address moral issues that are raised by science but cannot be addressed by science. The dialogue approach also suggests ways in which mythic traditions can address certain "boundary questions" that escape the limits of science. For example, science assumes, but cannot explain, the rational intelligibility of nature. Here a mythic tradition can provide insight into the ultimate "ground" of the intelligible order that is formulated into laws by scientific inquiry. And more: Mythic traditions can foster attitudes toward the natural world in ways that are beneficial to the advancement of science.

The dialogue approach has also been instrumental in clarifying various similarities and differences among the methods of science and religion. For example, it points out that scientific data are often theory-laden, in a manner that parallels the way religious experiences are conditioned by mythic traditions. It also points out that both science and religion rely on the imagination; both make use of models, metaphors, and paradigms; both seek to test their beliefs by returning to data; and both science and theology are rational disciplines.

It is difficult to resist the impression that the dialogue approach has been worked out by liberal (Christian) theologians *for* liberal theologians. It is also difficult to resist the impression that they have depersonalized the root metaphor in a rush to accommodate the principles and methods of science. Traditional Christians must blanch at the appeal of vagaries like "the ground of rationality." How does one muster love or sympathy for that? The dialogue approach leaves one asking whether its theological abstractions can be reconciled with mythic images as easily as they can be reconciled with scientific principles.

Skeptics might find reason to be skeptical about the parallels dialogue thinkers draw between science and theology. What is the point, for example, of comparing theory-laden data in science with myth-laden experiences in religion? Does this parallel intend somehow to legitimate religious experience as data? But how odd it seems to legitimate something by comparing it to a flaw! Data certainly may be influenced by theory, but this is not exactly a desirable thing in science, where everything possible is done to minimize the effect. In the religious life, however, everything possible is done to enhance it.

A skeptic might also wonder whether a discussion of imagination, models, and metaphors in religious thought might actually contribute to the crisis of plausibility. To see God-as-person as a metaphor constructed by an act of imagination is to be well on the way to nonrealism.

Despite such caveats, the dialogue approach has had the overall effect of enhancing the plausibility of root metaphors by convincing many individuals that mythic traditions are compatible with, and even complementary to, the scientific enterprise.

INTEGRATION

The final strategy in Barbour's typology of responses to the crisis of plausibility is the integration approach. These strategists appreciate the importance of unifying the content of both science and religion within an integrated vision of how things are and which things matter. The integration approach recognizes that serious conflicts have appeared in the past (Galileo and Darwin are examples), but that such conflicts might vanish when the contents of science and religion are properly construed within a larger perspective. Integration strategists reject the independence model because it is willing to settle for a myth-busting bifurcation of cosmology and morality, and they go well beyond the dialogue model by attempting to unify science and religion at the level of content.

Some attempts to integrate science and religion begin with scientific content and then extrapolate the findings to enhance the plausibility of the root metaphor. In theistic traditions this approach is known as "natural theology." Richard Swinburne exemplifies this approach by arguing that science gives evidence of order in the world that is most plausibly accounted for by the intentionality of a personal creator.[6] He also maintains that consciousness cannot be naturalized, which adds to the plausibility of positing a supernatural creator. Similar arguments have been put forward by Michael Behe and William Dembski, advocates of the "intelligent design" movement.[7] The intelligent design thesis is that nature is loaded with highly intricate mechanisms that are far too complex to be plausibly explained by anything less than an intelligent designer.

Other attempts at integration use the content of science as a resource for reconstructing the content of existing mythic traditions. Arthur Peacocke, Holmes Rolston, John Polkinghorne, and Philip Hefner, among many others, have shown how modern science offers suggestive insights for qualifying the root metaphor of theism (as well as the central doctrines of creation, sin, and redemption) in ways that make the Abrahamic myths fully consonant with the details of evolutionary cosmology and biology.[8]

A third approach to integration is what Barbour calls "systematic synthesis." Here the strategy has been to work out a comprehensive metaphysical system, a single conceptual framework that makes both scientific cosmology and human values intelligible. The process philosophy of Alfred North Whitehead is a prime example of this approach.[9] According to Whitehead, all entities and events in nature (even quantum events) have experiential and evaluative properties. The fundamental indeterminacy of the quantum world cannot be resolved—thus nothing can happen—without evaluation and choice. There can be no facts without values. The process of reality is thus made intelligible by the concept of personal agency. This metaphysical vision is no less reminiscent of the Indian doctrine of elementary dharmas than it is of the theistic doctrine of divine providence.

The various attempts to integrate science and religion at the level of content or at the level of metaphysical principles have restored the plausibility of mythic visions for many contemporary men and women. But these efforts have not escaped serious criticism. Some critics argue that extrapolations from science to God are implausible because they either distort science or introduce too many ad hoc hypotheses. Others remain convinced that the efforts of integrationists have not repaired the damage done by David Hume's devastating critiques of natural theology and metaphysics in the eighteenth century.[10]

This is not the place to undertake a full-scale exposition and critique of the various strategies for reconciling the tensions between science and religion. For our purposes it has been enough to indicate that serious efforts are being made to address the crisis of plausibility. These strategists are doing precisely what they are called to do by their mythic traditions: They are attempting to mitigate the erosive effects that the scientific worldview has had on mythic realism.

The important question is whether such efforts will amount to anything more than fingers in the dike against relentless waves of doubt. There is little reason to think that the worst of the crisis has passed. Scientific literacy is still on the rise, and so is our awareness that science is not myth. That is, science can do much to broaden and deepen our understanding of how things are, but it cannot help us to determine which things matter. It can provide much in the way of explanation, but contributes nothing to our apprehension of meaning. This is a happy fact for some, because it draws attention to the need for a faith response. But for others it merely redraws attention to the specter of nonrealism by introducing the challenge of relativism.

AWARENESS OF RELIGIOUS DIVERSITY

One of the major themes of this book has been that cultural traditions require a broad consensus on matters of ultimate reality and value. Indeed, common adherence to a mythic vision is our best short definition of culture. When fundamental disagreements arise about cosmology and morality, we may expect to see declines in social solidarity and cooperation, together with increases in uncertainty and anxiety at the personal level. It follows that a multicultural society is something of an oxymoron. A diversity of myths competing for the goal hierarchies of individuals may have a certain democratic appeal to it, but in the long run religious pluralism constitutes a destabilizing factor, both socially and psychologically.

For diverse myths to flourish side by side within a single culture, it would be necessary for some sort of meta-myth to arise with sufficient resources to unify the particular myths. Hindu culture, as we have seen, manages to unify a plurality of religious orientations within the myth of many paths. The Roman Empire had its meta-myth of imperial religion, which enabled local myths to flourish. And American culture has had the meta-myth of civil religion to compensate for its constitutional tolerance of religious diversity.[11] As long as the meta-myths of diverse cultural traditions remain plausible and intact, there will be resources for solidarity and cooperation.

I intend to return to the topic of meta-myths later in the chapter, but for the moment our inquiry will be focused on whether an awareness of other traditions will contribute to the crisis of influence by compromising realism about one's own mythic vision and, if so, why. The short answers go like this: Yes, it is probable (though not necessary) that mythic realism will decline as one becomes more familiar with alternative religious orientations. And the reason is that recognizing the fact of religious diversity leads one to formulate a personal theory of religious diversity, and most of the options for such a theory tend to relativize the religious life.

The awareness of alternative mythic visions is not a new thing. Even a brief encounter with an alien culture will reveal fundamental differences in cosmology and morality. In the distant past, when cultural traditions lived in splendid isolation from one another, religious diversity could be accounted for easily. One simple and obvious inference would be that different groups have different gods. Just like parents: We have ours, they have theirs. No big deal. Another option might be simple exclusivism—that is, we have the only real God and other groups are obviously deluded in thinking that their "gods" are real. That's their problem.

These simplistic ways of dealing with the fact of religious diversity involve very little reflective thought about the relative contents of traditions, their effectiveness, the dynamics by which people come to have different myths, or the reasons that might be given for preferring one over another. Such questions grew more difficult to avoid as contacts between cultural traditions became more common and more extensive. In the contemporary world of globalized transportation and communication, they are virtually impossible to avoid.

What can be made of the fact that different cultural traditions have different mythic visions? This is not a very troublesome issue for those who take a secular scientific perspective on cultural diversity. It is obvious to the secular mind that different groups will be faced with accidentally different adaptive challenges and will come up with different responses to them, and these are sure to be reflected in different stories about how things are and which things matter. Again, no big deal. But if a person in a state of commitment to a particular religion encounters religious diversity, it is likely to be a very big deal.

Imagine a Christian student arriving at college from a provincial community and finding herself assigned to a Buddhist roommate. If there is any mutual curiosity, then an exchange of ideas will ensue, revealing fundamental incompatibilities between the two traditions. What then? Will the roommates attempt to convert one another? This may result in presenting reasoned arguments for one myth and against the other. But what if these fail? And what if there is no reason to believe that either roommate is happier, or more moral, than the other? What is entailed if traditions having radically different content are equally defensible (or objectionable) and equally satisfying? How can either roommate doubt that if she were born into the other's culture, she would hold the other's views? Given all this, is it possible for either student to remain unchanged in her attitudes about their differences?

It turns out that the range of available options for thinking about religious diversity is fairly limited. The literature generated by these questions is extensive, but may be simplified to three general perspectives: exclusivism, inclusivism, and pluralism.[12]

EXCLUSIVISM

Exclusivism is the view that only one religious tradition is true, and all others are false. For those mythic traditions that make exclusive claims (as most have done), this is the most immediate and clearly the most desirable initial response. But it is also the most offensive response from the perspective of other traditions. If others declare absolutely that your religion is wrong because theirs is right, you are likely to regard them as arrogant bigots.

But this does not mean that exclusivism is an untenable position. In his defense of exclusivism, Alvin Plantinga argues that there is nothing inherently immoral or irrational about the exclusivist position.[13] No one can be charged with arrogance or oppression for declaring resolutely that the Copernican system is superior to the Ptolemaic system or firmly believing that it is wrong to tell lies. Disagreement per se is neither arrogant nor oppressive, so taking offense at the religious exclusivist is unfounded. Plantinga further insists that religious exclusivism is intellectually defensible. It is never irrational to believe X when there is good evidence for X, even when the evidence is internal to the believer. The subjective evidence for X might be as strong as my conviction that I had breakfast this morning, and this evidence might have been planted in me by "the Internal Testimony of the Holy Spirit." [14] Whatever the details happen to be, asserting one's beliefs on good evidence—and excluding everything incompatible with them—is not arbitrary or irrational.

If our sole criterion is logic, then we must agree that Plantinga's defense of religious exclusivism is a coherent one. But it misses the point. It takes little reflection to see that equally coherent defenses of racism and sexism might be offered on the same grounds. Religious exclusivism may be a defensible position, but it is typically not a sustainable one. I have watched hundreds of college students work their way through the challenges of religious diversity, and the common pattern seems to be that they begin with the exclusivist position, but then move beyond it to accept some—no less coherent—form of inclusivism or pluralism. It is not clear why this pattern holds, but I suspect it involves the process of reconsidering whatever background beliefs and internal evidence it takes to sustain exclusivism. And to the extent that individuals move beyond exclusivism, they compromise their mythic realism and weaken the potential of mythic images to influence their emotional responses.

INCLUSIVISM

Inclusivism is a general category covering a range of views. One variation on inclusivism argues that all religious traditions grasp the essential core of the religious life, and all are therefore essentially true, even though they may disagree on specific doctrines. For example, one might argue that the Golden Rule is the essence of religion, and therefore all religions expressing this principle (as all do) are equally and essentially true. Or one might contend that the essence of religion is to achieve mystical insight, and therefore all religions enabling mysticism (as all do) are equally and essentially true. The essentialist view neutralizes the power of specific doctrines and images by making them relative to universal principles or to subjective experiences.

Another variation, the "approximation of truth" view, suggests that all mythic traditions may be inclusively ranked by the degree to which they express the absolute truth about reality and value. A Muslim might take this view by privileging Islam as absolutely true, but recognizing Judaism and Christianity as partially true. Non-Abrahamic monotheistic traditions would rank behind Judaism and Christianity, and polytheistic traditions would be ranked as least proximate to the absolute truth.

Approximationism has all the marks of veiled exclusivism, leaving intact realist attitudes toward specific doctrines and images. But the other forms of inclusivism compromise realism by relativizing the specific elements of religious traditions. Inclusivism appears designed for dodging the difficult questions and leaving everything where it was to start with. These views allow everyone to proceed religiously "as if" they have the whole truth, an unsatisfying conclusion for many thinkers.

PLURALISM

Some pluralists maintain that religious traditions represent different responses to the same transcendent reality. The sacred is revealed alike to all traditions, but the divine self-disclosure must of necessity be received and appraised according to socially constructed (and therefore plural) conventions of meaning. One divine reality, but many ways to experience and respond. Each mythic tradition manages to apprehend one limited aspect of the sacred. Like the legendary mountain climbers, each has a reliable impression of one narrow passage toward the summit, but none apprehends the whole reality.

But having come this far, the truncated realism of the pluralist is a mere half step from the full-fledged nonrealism of social constructivism and subjective relativism, both of which reduce all the phenomena of religious experience and expression to the side of human nature and historical accident.

I have argued that the rise of modern science and the growing awareness of religious diversity are challenging mythic realism to the point of generating a complex crisis of influence among the world's religious traditions. But there may be even more to the story than creeping nonrealism. The power of traditional myths to educate our emotions and to influence our goal hierarchies might be further compromised by competition from the intensely aggressive myth of consumerism.

THE MYTH OF CONSUMERISM

I am not entirely convinced that consumerism qualifies as a genuine myth in the full sense of the term, but I am convinced that its influence in shaping the attitudes and values of contemporary men, women, and children in Western culture has in large measure displaced the influence of traditional myths. In what follows I will present the consumerist myth (if it is one) according to the outline used in previous chapters. This will help us to see how consumerism has further compromised the influence of traditional mythic visions, and will open the way to the closing arguments of the next chapter.

HISTORICAL CONTEXT

Material consumption is anything but a new phenomenon, and we may even say that the potential for greedy behavior has always been a significant element in human nature. But mass consumption and the ideology of consumerism are relatively recent developments, historically contingent on the rise of capitalist economies in sixteenth century Europe. Karl Polanyi has characterized the rise of capitalism as a "great transformation" that resulted in "disembedding" economic dynamics from their social, political, and religious context.[15]

Prior to the rise of capitalism the activities of production and consumption were supervised within comprehensive feudal institutions, which were themselves regulated by values inherent in the Judeo-Christian myth. Economic activity was steered by the custodians of communal needs, not driven by individual anxiety and ambition. Usury laws, market licensure, and "just price" constraints are examples of the embedded character of precapitalist economics.[16]

The rise of capitalism occurred as economic mechanisms were disembedded from the constraints of feudal institutions. Decisive for this process was the emergence of labor markets. In the aftermath of the Black Plague and the Hundred Years' War, there was a shortage of labor, which meant it was possible for workers to leave their feudal bondage to the land and sell their labor in towns. Meanwhile, gold discoveries and enclosure laws resulted in surplus capital in the hands of merchants. Thus, "Merchants with money to spend and landless peasants with labor to sell hooked up, and capitalism began."[17] The result was a surge in both population and economic growth.

Doug Brown, following Polanyi, argues that the rise of capitalism transformed Europe from a culture of security to a culture of insatiable freedom. The creation of labor markets meant that both merchants and workers were free agents, disembedded in the economic sense. Workers could decide for themselves where to sell their labor and what to purchase with their wages, and merchants were free to invent new ways to use the labor force to produce goods for sale.

But the freedom of the new system was counterbalanced by a loss of security. Capitalism meant that everyone was out there on their own, to sink or swim. Workers might have been free agents, but they were also at risk. If you had marketable skills and if you worked hard, then there were personal rewards. But if you were lazy or unskilled, or if you squandered your earnings, then you would be in trouble. The same was true for merchants: If you could find workers to produce and buyers to consume, then you might do well, but if you couldn't, you would sink.

Players in the new game were utterly dependent on markets for their security, but they were also rewarded by markets for their skill, ingenuity, and hard work. So the whole system came to depend on the self-interested motives of the individual players: fear of being left behind and desire to get ahead. The new economy was disembedded and driven.

But it would be a long time before the capitalist system would morph into a consumerist culture. Per capita consumption in the new system was anything but lavish. Most people consumed only what they needed, partly because that's all they could afford, and partly because anything more was generally considered wasteful. The wealthy elite consumed with extravagance, as they had since ancient times, but the masses continued to live frugally. They could hardly have done otherwise, for resources were in such scarce supply that production could barely keep up with demands for basic necessities. But this was all to change as the Enlightenment unfolded in the eighteenth century.

The Enlightenment brought with it a deepened understanding of nature as well as corresponding breakthroughs in engineering and technology, and these conspired to produce an enthusiastic ideology of unfettered material and social progress. With the secrets of nature unlocked, the ambitions of humanity could be unleashed. In no time at all, the industrial revolution was under way, with the result that capacities for production soon exceeded demands for goods. In fact, by the late nineteenth century excessive production had resulted in economically stagnating levels of surplus goods.

Mass production was in desperate need of mass consumption. Now that the system was geared up to give everyone whatever they wanted, the trick was to get them to want more. In a system already driven by fears of falling behind and desires for getting ahead, the solution was obvious: find ways to play on these strings of human nature. Business leaders responded to the crisis with an astonishing blitz of aggressive strategies designed to maximize consumption. These included the creation of enticing department stores, mail-order retailing, a revolution in advertising (from 30 million dollars spent in 1880 to 600 million spent in 1910), media campaigns to hype fash-

ion trends and brand names, and fundamental changes in social institutions. "Educational and cultural institutions, governmental agencies, financial institutions, and even the family itself changed their
meaning and function to promote the consumption of commodi-
ties."[18] The radical transformation taking place between 1890 and
1930 was not unlike being conquered by an alien culture. Consumerism had arrived.

THE MYTH OF MARKET PROVIDENCE

Capitalism emerged in the process of disembedding economic activities from an authoritarian mythic tradition that insisted on steering
the economy toward its vision of the good life. It isn't that capitalism
abhors mythic traditions or conceptions of the good life. Quite the
contrary, it eagerly tolerates all worldviews and all values, so long as
they come to bear upon economic activity exclusively at the level of
individual choices. If individuals want to express their Christian piety
in their economic activity, then fine. The same goes for Buddhists,
Muslims, Sierra Club members, and all the rest. Individuals are
free to bring their myths into the marketplace, and the marketplace
couldn't care less.

But to ask that the marketplace be embedded within a mythic
tradition cannot be tolerated. Disembedding the mechanisms of the
economy from the custodial directives of a particular understanding
of reality and value creates a free and open field of competition for
the hearts and minds (read: goal hierarchies and appraisals) of individual consumers. So far said, this is straightforward capitalist doctrine. Capitalism becomes consumerism only when it attempts to em-
bed itself in a mythic vision of its own making—that is, the Myth of
Market Providence:

> The root metaphor of the consumerist myth is the providential market,
> the idea that the free market has a self-regulating and prudential sovereignty
> about it, and if we can manage to keep the market full of activity, it will pro-
> vide for everyone's needs. The market is our savior from want, our path to
> fulfillment. It is in the nature of human beings to realize their full potential for
> happiness. Happiness will be maximized, and human nature fulfilled, by con-
> suming material goods and services in the marketplace. The more consump-
> tion the better, for increased consumption means economic growth, and eco-
> nomic growth means that more people can more efficiently maximize their
> happiness. If everyone realizes their potential for production and consump-
> tion, then the infinite bounty of the earth's resources and the providence of
> the market will do the rest.

Evangelists for consumerism could not care less whether any-one explicitly professes this myth, or even hears it. They care only that people act as if they believe it. Consumerism is advanced, not by preaching doctrines, but by selling products. Tacit realism about market providence is sufficient.

Before moving on to the emotional appeals and the ancillary strat-egies of consumerism, I will attempt to clarify its status as a myth. Consumerism may be fairly characterized as a meta-myth, compara-ble to Roman imperial religion, Shinto in Japan, or American civil re-ligion. A brief look at the phenomenon of civil religion will help to sharpen the thesis. Robert Bellah opens his classic article "Civil Reli-gion in America" with these words:

> While some have argued that Christianity is the national faith, and others that church and synagogue celebrate only the gener-alized religion of "the American Way of Life," few have realized that there actually exists alongside of and rather clearly differen-tiated from the churches an elaborate and well-institutionalized civil religion in America.[19]

Bellah's argument is that the separation of church and state has not denied the political realm a religious dimension. In many ways American civil religion draws on the substance of Judeo-Christian tradition, but it has an identity and integrity of its own, replete with sacred documents (Declaration of Independence, Constitution), saints (Washington, Jefferson), holy days (Memorial Day, Thanksgiv-ing), creed (Bill of Rights), sacred places (Washington, D.C., Gettys-burg), symbols (flag, Liberty Bell, bald eagle), rituals (Pledge of Alle-giance, parades), monuments (Statue of Liberty, Mount Rushmore), and a concept of salvation (freedom and prosperity). At critical mo-ments in American history (Revolutionary War, Civil War, world wars, post-election transfers of political power), political leaders invoke these elements of civil religion to galvanize a spirit of national piety, thereby inducing individuals to sacrifice for the good of the Republic.

There exists an ambiguous relationship between American civil piety and Judeo-Christian piety. Normally they are mutually support-ive, but in times of national crisis, when civil religion comes out into the open, there may be deep conflicts. In wartime, for example, the tension between civic duty and religious virtue is unmistakable.

I submit that it is not unreasonable to regard consumerism as an emergent meta-myth—with saints, symbols, rituals, and a way to sal-vation—asserting itself alongside, and in occasional conflict with, tra-ditional and civil religion. Just as the separation of church and state has not denied the political realm a religious dimension (civil religion), we may say that disembedding economic activities from authoritarian myths has not denied the economic realm a mythic dimension (consumerism).

The initial response of public officials to the crisis of 9/11 was revealing. When we might have expected civil religion to rise to new heights, we were treated instead to consumerist sermons by Mayor Giuliani and George W. Bush, imploring us to wipe our tears and get back to the serious business of shopping. And who can forget the hoopla of the Wall Street reopening, when media celebs were assembled for the highly symbolic bell-ringing ceremony? Nothing can harm us if we honor our duty to keep the market active.

EMOTIONAL APPEALS

Today's readers need no help in appreciating the extent to which consumerism plays on the strings of fear and desire to goad us into buying things we don't need. The arousal and exploitation of emotional systems is precisely what marketing is all about. In a recent marketing text bearing the title *Why People Buy Things They Don't Need,* Pamela Danziger gets right to the point on page one: "In today's consumer-driven society, satisfying consumer needs has less to do with the practical meeting of physical needs and everything to do with gratifying desires based on emotions."[20] The point is that when marketing strategies arouse emotions, then people *do need* the things they buy.

The Yale lock company tried to make people need their products by running an ad showing a woman lying alone in bed, while cast against her bedroom wall was the ominous shadow of an approaching man. The caption read: "Night loneliness . . . the sound of stealthy tampering at the door . . . a moment of helpless terror . . ." Elsewhere, a baby food ad pictured an abandoned pair of baby shoes together with alarming statistics about infant death rates.[21] Such fear tactics in advertising are easily justified: "Since time began, fear has been a regulatory part of humanity—our primitive religion taught the vengeance of the gods, our modern revivalists, like Billy Sunday, frightened people with damnation. So what's a little fear in advertising?"[22] Fear in advertising can even improve the world: "We've a better world with a bit of the proper kind of fear in advertising . . . fear in women of being frumps, fear in men of being duds."[23]

Appeals to desire are even more common. Advertisers promise consumers that they will become more attractive, more productive, more envied, sexier, healthier, and generally more happy and fulfilled if only they will buy the right products. Consider what the Ronco

Showtime Grill can do for a person: "By spending only $99, in five easy payments, you suddenly become a better person, a better home-maker and wife, admired by your friends and family, and more ful-filled in all aspects of your life."[24]

The hot topic in advertising these days is "branding," that is, building product loyalty by appealing to emotional systems. In a book entitled *Emotional Branding*—featuring chapters like "Tastes that Tantalize" and "Scents that Seduce"—Marc Gobé offers up "The Ten Commandments of Emotional Branding" and "The Four Pillars of Emotional Branding." Here we discover that companies want to "reach the hearts of consumers" and remain "profoundly in touch" with them so that "relationships of mutual respect" can develop into "sound and permanent connections" and "rewarding partnerships." We also discover that products will "mean something on an emo-tional level" to consumers if the products are given a "personality, character and charisma" that "provokes an emotional response."[25]

The original purpose of advertising was to give consumers in-formation about products themselves, such as material components, specifications, craftsmanship, price, and availability. But contempo-rary advertising is more about the alleged benefits and experiences that products will bring to consumers. The point of much advertising is to create associations between products and ideal outcomes, asso-ciations that are intended to arouse self-esteem deficits and create self-esteem linkages to the purchase of a product. An advertisement, don't forget, constitutes an item of business at hand that must be ap-praised for its meaning, and whenever the appraisal results in an emotional response, it produces exactly what the consumerist myth needs to function: a motivated consumer. And such items of business are constantly at hand. American youth, for example, endure more than 20,000 such encounters each year in television ads alone.

ANCILLARY STRATEGIES

The myth of consumerism is supported and advanced by the full range of ancillary strategies: intellectual, institutional, ritual, aesthetic, and experiential. A thorough discussion of each of these strategies would take us too far off point, so this will have to be brief and cursory.

INTELLECTUAL STRATEGIES. Consumerism manages to per-sist under constant pressure from numerous critics who would like to see economic activity re-embedded in, and regulated by, a larger vi-sion of reality and value. Such critics represent the most worrisome challenge of all to the consumerist myth: creeping nonrealism about market providence. The intellectual strategists of consumerism, concentrated in university business schools, independent think tanks, and government agencies, make it their business to defend the plausibility of the free market and economic growth against the critics.

On the cosmological front, consumerism has been accused of presupposing a naive and unscientific understanding of the natural world, especially its delicate ecosystems and material cycles. In short, critics charge that consumerism is a driving force behind global environmental degradation and cannot possibly be sustained. The apologists for consumerism typically respond by saying that environmentalist critiques are grossly exaggerated, and to the extent that problems arise, we may safely expect them to be solved by new technologies and economic growth.

Other critics weigh in on the moral side, insisting that consumerism presupposes a shallow and ignoble view of human reality and fulfillment. Consumerist responses vary. Some defenses offer reasoned arguments on utilitarian or pragmatic principles, while others resort to drawing down resources from Judeo-Christian tradition. Here, for example, is the circuit preacher Russell Conwell in his famous sermon: "I say, then, you ought to have money. If you can honestly attain to riches in [this city], it is your Christian and Godly duty to do so." [26] He then says that we are called into the marketplace by "that divine inspiration toward better things." [27] Conwell delivered this sermon more than six thousand times around the turn of the twentieth century.

INSTITUTIONAL STRATEGIES. The institutional strategies for supporting and advancing consumerism are virtually everywhere. Every corporation, every franchise, every neighborhood convenience store amounts to an organized institution for selling "more things to more people more often for more money." [28] But there are many institutions committed to the cause that have nothing to sell but consumerism itself. Of these, the most important are government agencies, business alliances, and financial institutions. According to Bush administration Secretary of Commerce Donald Evans, "It is government's job to create the right conditions for America's workers and businesses to flourish." [29] During the 1920s the role of the U.S. government in promoting consumption changed dramatically, as shown by the mammoth 8,000 percent budget increase for the Bureau of Foreign and Domestic Commerce. [30]

In the formative years of consumerism (1890–1930), many business alliances appeared that had the same mission as the Commerce

Department, to accelerate consumption. In response to a White House initiative, the U.S. Chamber of Commerce was established in 1912 to become "a rallying point for promoting and defending free enterprise and individual opportunity."[31] This national alliance defines itself by its "aggressive representation [of business interests] before Congress, government agencies and the courts to curb overzealous regulators, reduce taxes, and open markets to U.S. products and services around the world."[32] It is also committed to "an unprecedented campaign to elect a pro-business Congress" that will actively support growth and prosperity by encouraging consumption. Virtually every city and town in America has its own local incarnation of this consumerist institution.

Financial institutions gave consumerism a tremendous boost as they helped to transform the American ethos concerning debt. Benjamin Franklin considered debt a moral failing, as did most Americans at the beginning of the twentieth century. By 1960, however, nearly a half million Americans carried credit cards. By 1997, 100 million Americans carried 614 million credit cards in their wallets and purses, with 73 percent of their spending involving debt.[33] Credit cards play an enormous role in the consumerist culture because they enable emotionally aroused shoppers to buy on impulse things they wouldn't buy if they had to go home and save money.

RITUAL STRATEGIES. Consumerism is also advanced by a range of individual and collective ritual strategies. The activity of shopping itself is individually ritualized by setting aside special shopping days, perusing newspaper and magazine ads, compiling lists of wanted items, selecting appropriate clothing to wear, making a procession to the mall, advancing methodically from shop to shop, stopping in for a snack, examining displays, interacting with sales personnel, trying out various items, and, of course, sacrificing money. Shopping is considerably more than buying. "Buying is an activity understood by economists. Shopping is a phenomenon of interest to anthropologists and sociologists. Shopping provides opportunities for dreaming and playing—it's an escape and, ultimately, an art."[34]

Many of the public ritual events of consumerism originated by commercializing traditional religious festivals. For example, if you consider Christmas in terms of how people invest their time and attention (and their money), then there is no question that the holiday belongs more to commercial interests than to Christian piety. It is not simply that consumerism has exploited Christmas, Hanukkah, Easter, Thanksgiving, Halloween, the Fourth of July, and other holidays—it has gone one step further, to transform their meaning. If you compare the pre-1920 character of these holidays with our present day commercial extravaganzas, you can't miss the point.

Consumerism has even invented ritual events of its own. Cities and towns all over consumerland have their own community festivals, where the pretext might be "Gold Rush Days," "Frontier Days," "Watermelon Days," "River Fest," or whatever, but the subtext is always "Shop!"

AESTHETIC STRATEGIES. The powerful new genre of commercial art was born in 1902 when some of the most creative minds in America were assembled to design product displays for the opening of the Marshall Field department store in Chicago. Ever since, aesthetic strategies have played a central role in the development of the consumerist culture. In 1917 Walter Dill wrote: "To substitute the standard of the artist for the standard of the capitalist would be impossible in business." Nonetheless, he added, "A harmonious working of the two is possible."[35]

We now find art and commerce harmonized everywhere, in advertising, in shopping venues, and in products themselves. Advertisers routinely co-opt the fine arts by parodying or alluding to masterworks. How often have we seen da Vinci's *Mona Lisa* and Grant Wood's *American Gothic* featured in ads? But the advertising industry (which continues to attract some of the best talent around) routinely manages to create masterworks of its own. Anyone who thinks advertising is not art should spend an afternoon paging through magazines like *Vanity Fair* or *Vogue*. The artful presentation of products attracts attention and creates memorable images that invite consumers into an ideal world of the artist's making.

Shopping venues harmonize art and commerce, too. The Bon Marché in Paris, Harrod's in London, Stewart's Astor Palace in New York, and Marshall Field in Chicago (among many others) are important architectural works as well as commercial outlets. The interiors of many department stores and shopping malls artfully orchestrate space, lighting, sounds, and odors to create the effect that one has actually entered the ideal world.

And who can deny that many products themselves are works of art? The iMac computer has been described as sculpture with an operating system, and a trip to Fortnam and Mason in London is enough to convince anyone that some candies are just too beautiful to eat. Jewelry, clothing, kitchenware, light fixtures, furniture, and even packaging are often much more than mere goods. A university professor once remarked (and many would agree) that the difference between an art museum and Crate and Barrel is that "one can afford and buy the art presented in the store!"[36]

EXPERIENTIAL STRATEGIES. The experiential strategies of consumerism follow a simple principle: Find out what consumers want (or arouse them to want something), and then give them exactly that. We may wish to argue that consumer aspirations are vulgar, trivial, or selfish, but that would only miss the point. People want neither more nor less than what their goal hierarchies dictate. If the market gives consumers what they want, they will experience happiness, and their faith in market providence will be reinforced. Here, for example, are two pious consumers describing their shopping experiences:

> You get satisfaction, and you're thrilled about where you've been, and you plan the next one. Greater satisfaction builds more anticipation for the next time.[37]

> There is anticipation in the search for something wonderful, then stress, then ultimate satisfaction when the right purchase is made. Then you want to do it all over again.[38]

Market research has determined that most consumers (89 percent) go shopping for the purpose of improving their quality of life, which they define in terms of achieving pleasure, beautifying their homes, relaxing, being entertained, achieving emotional satisfaction, relieving stress, and elevating their social status.[39] As long as these are the terms of personal fulfillment reflected in goal hierarchies, and as long as the marketplace can manage to satisfy them, the Myth of Market Providence will flourish.

PERSONAL WHOLENESS AND SOCIAL COHERENCE

Myths play on the strings of human nature for the sake of enhancing the conditions for personal wholeness and social coherence. It is clear that the consumerist myth appeals to self-interest by satisfying personal aspirations, but how can it be shown that consumerism offers anything more than the self-centered ethic of hedonism? In other words, how does consumerism enhance the conditions for overlaps of self-interest?

The consumerist response is that overlapping self-interest is the very essence of a market transaction. Let's say you have a piece of kitsch and I have four dollars; and let's further suppose that I would rather have the kitsch than the four dollars, and you would rather have four dollars than the kitsch. If we ever got together on these particulars, then we would probably make an even trade, to our mutual happiness and self-interest. Everybody wins because we both satisfy our desires without making any net sacrifice.

This is precisely the promise of the Myth of Market Providence: Come to the market with your money and your kitsch and you will go away happy and fulfilled. The important thing is that we all accept our share of responsibility for keeping the market alive with activity. This should not be burdensome, however, since every market transaction a person makes will result in personal satisfaction. And every transaction will result in overlapping self-interests as well. So why not go shopping? It's fun, it makes us happy, it's fulfilling, and it creates jobs for other people. Personal wholeness and social coherence are both maximized by a growing market, with no discernible sacrifices for anyone.

Earlier in this chapter I argued that the rise of modern science and the awareness of religious diversity have raised the level of religious skepticism and mythic nonrealism, and have thereby compromised the effectiveness of traditional myths to influence the thinking, feeling, and acting of contemporary men and women. The influence of traditional myths has been compromised by the rise of consumerism as well, but by a slightly different process. Consumerism does not undermine the meanings of traditional myths by advancing skepticism and nonrealism; instead, it simply blows away these meanings in the competition for mind space. Do North Americans and Europeans spend as much time in synagogues, churches, and mosques as they do in shopping malls? Do they spend as much time reading scriptures as they do paging through advertisements? Can they list as many biblical characters as they can brand names? Do they pray as often as they watch TV? In each case, a resounding No!

Who can seriously doubt that consumerism has vastly more influence on our goal hierarchies, self-esteem links, memory systems, and appraisals than religious traditions do? Less and less do we bring our values and attitudes *to* the marketplace, while more and more we take them *from* the marketplace. The ancillary strategies of consumerism amount to a full-court press to this end.

The complicated historical process of disembedding economic activity from the constraints imposed by moral traditions has resulted in a freewheeling global economic system that refuses to answer to anything but its own internal imperative to grow. What was not fully realized along the way, however, was that disembedding the economy from the moral systems of a culture did not disembed it from the natural systems of the earth. This leaves us with a huge problem because it is now clear that consumerism is a major force driving us toward an environmental crisis of global proportions. The other major force is overpopulation. If material consumption and human reproduction are not carefully regulated in the future, then we will eventually exceed the capacity of the earth to sustain human life.

The future of religion is bound up with these problems. The critical question is whether religious traditions can inspire a process of demolishing the Myth of Market Providence and re-embedding economic (and reproductive) activity within an ecologically savvy moral vision before it is too late. But if it is true that the influence of religious traditions has been seriously compromised, then there is some reason to expect the worst.

DOOMSDAY AND BEYOND

The argument ahead is simple. The life-support systems of the earth, upon which the survival of our species depends absolutely, are in a state of serious decline on a global scale, and we now have good reason to believe that the global environmental crisis has been induced by excessive human impact on natural systems. This leaves us with a critical choice between unpleasant options. One option is to respond to the crisis by engineering a massive reduction of our collective impact. The other option is not to respond, in which case increased impact will eventually trigger environmental disruptions, resulting in a massive dieback of the human population. In other words, either we make a correction or nature will make one for us.

The first option requires significant changes in values and attitudes, changes that might be inspired only by a mythic tradition, or perhaps by a coalition of mythic traditions. The second option, I fear, is the more likely of the two, because there is reason to doubt that our religious traditions possess either the will or the influence to nurture us toward an adaptive and timely response.

HOW DENSE CAN PEOPLE BE?

No one is foolish enough to believe that human population and consumption can grow without ceasing. Even the most passionate advocates for growth recognize that we occupy a finite planet with finite resources. There must be limits, but who can say with confidence what those limits are? Presently, the human population exceeds 6.4 billion people. Some are prodigious consumers, but most are not. Could the earth sustain 6.4 billion prodigious consumers? Could it sustain 12 billion modest consumers? Both of these are doubtful, but nobody knows for sure what the earth's carrying capacity is.

Carrying capacity has been defined in several ways, but the basic idea is this: The carrying capacity of an ecosystem is the maximum population of a given species that can be sustained indefinitely without degrading the biological productivity of the ecosystem. The only certain way to determine the earth's carrying capacity for humans is to exceed the limits beyond which natural systems begin to collapse. Here's a thought experiment: Place a pair of guppies into a fish tank and let them go on reproducing toward the limits. Suppose that the guppy population doubles every twenty-four hours. On day two, then, the population will soar to four guppies; on day three we have eight; sixteen on day four, and so on. By day nine there will be more than 500 guppies in the fish tank and the system may appear as stable as ever. But suppose on day ten we discover that the surface of the water is thick with dead fish. The carrying capacity of the fish tank, we may deduce, is somewhere between 500 and 1,000 guppies. When the carrying capacity was exceeded, a system-wide collapse was triggered, endangering all the guppies.

It is disconcerting to discover that we are actually running a similar experiment with our own planetary ecosystem and our own species. Humans are already responsible for consuming 40 percent of the net primary productivity of terrestrial ecosystems, and this percentage continues to rise.[1] We don't know for sure what the earth's tolerance for human life is, yet we persist in growing toward the breaking point.

Ecologists and environmental scientists are eager to find constructive ways to think about the earth's carrying capacity, so that if we choose to deal with the problem, we might have a decent chance at avoiding our own Doomsday. But humans are not guppies, and the earth is not a fish tank. Any attempt to specify the earth's carrying capacity for humans is complicated by a range of variables introduced by our technology.[2] The IPAT formula proposed by Paul Ehrlich and John Holdren suggests that carrying capacity should be expressed in terms of maximum sustainable impact. The formula $I = P \times A \times T$ stipulates that human impact (I) is a function of population size (P), per capita affluence (A) as measured by consumption, and technological damage (T) done to natural systems in the production of items consumed.[3] As population, consumption, or technology change, so will human impact change.

A more promising attempt to determine carrying capacity is the ecological footprint idea, developed by Mathis Wackernagel and William Rees.[4] An individual's ecological footprint is an estimate of the total area of biologically productive land and sea that would be necessary to produce everything consumed by the individual and to absorb all the wastes generated by the individual. In other words, the ecological footprint estimates how many acres it would take to support an individual's lifestyle. To calculate ecological footprint, we would need to measure levels of consumption and waste and then convert these figures into area equivalents. This has been done for sixty categories of resources and wastes. What we find is that the *actual global per capita ecological footprint* is 5.6 acres (2.3 hectares). By contrast, the *maximum sustainable* footprint for our current population is estimated at 4.7 acres (1.9 hectares). This means that our current demands on the ecological capacity of the earth exceed the maximum sustainable demands by more than 20 percent (using 1999 data).[5] According to the ecological footprint analysis, human beings are already exceeding the carrying capacity of the earth, and have been doing so for more than a generation.

If the ecological footprint analysis is a reliable indicator of carrying capacity overshoot, then it should be possible by now to detect undeniable signs of degradation to nature's life-support systems. Indeed it is possible, as the following survey shows.

THE AIR CRISIS

Air pollution is a natural phenomenon. Natural systems themselves release harmful chemicals into the atmosphere independently of human activity. Human activities, however, have dramatically increased the levels of airborne pollutants to the point of overloading the material recycling systems. For example, about a third of atmospheric carbon dioxide can be traced to human sources. Humans release into the atmosphere twice the level of sulfur and three times the natural levels of nitrogen oxides and ammonia.[6]

Natural systems are unable to recycle these materials as fast as we produce them, resulting in harmful levels of air pollutants mixing vertically and horizontally in the atmosphere where they react chemically with each other, often returning to the earth's surface as solid particles or droplets, or as dirty rain. Most of the serious air pollutants come from cars, trucks, ships, airplanes, power plants, factories, and waste incinerators, but rural communities also contribute to the overload by burning wood, dung, and crop residues. Fires associated with deforestation projects are also implicated.

The effects of overstressing nature's gaseous recycling systems are already intolerable and are certain to get worse. The threats to human health are serious. The World Health Organization estimates that 20 percent of humans now live in areas where the air is unfit to breathe. Half the population of the United States is continuously exposed to contaminated air, leading to 120,000 deaths each year.[7] In some Brazilian cities conditions are so bad that breathing the air induces vomiting. Air pollution contributes to cancer, bronchitis, asthma, emphysema, and birth defects. Children, the elderly, pregnant women, and heart patients are especially vulnerable to air pollution.

Excessive air pollution is also devastating to plant life. Air pollutants break down leaf coatings that help to retain moisture and to prevent damage from diseases, pests, and frost. They interfere with photosynthesis, reduce nutrient uptakes, and kill decomposers that are essential to plants. Air pollution is responsible for killing off 35 percent of the forests in twenty-eight European countries.[8] Early stages of this process, called "forest death," are now evident in the United States.

The life-support systems of freshwater lakes have been severely impacted by air pollution as well. An estimated 16,000 lakes in Norway and Sweden are no longer able to support fish because of excessive acid deposits, and another 52,000 have lost most of their acid-neutralizing capacities.[9] In the United States 80 percent of northeastern and midwestern lakes are threatened by excessive levels of acid due to air pollution.

By far the most troubling aspect of the air crisis is the prospect of dramatic disruptions in global climate systems. By the end of the present century the mean surface temperature of the earth is expected to increase anywhere between 2°C and 6°C.[10] What makes this an alarming prediction is that the change will be too rapid for many life forms and ecosystems to adjust. Climate changes have occurred naturally throughout the history of the planet, but none so quickly as the present one.

The principal cause of global warming is by now familiar to everyone: the greenhouse effect. Concentrations of carbon dioxide and other greenhouse gases have increased in the earth's atmosphere, preventing solar heat from escaping into space. Human activities since the industrial revolution—especially fossil fuel combustion and deforestation—are responsible for these increases. Rapid global warming is expected to result in elevated sea levels, heat waves, floods, droughts, fires, expansion of deserts, alteration in ocean currents, rapid shifts in growing seasons and vegetation zones, and increased pestilence. More frequent, intense, and variable weather events are expected as well.

The consequences of these changes for delicate ecosystems and the human population are expected to be devastating. Patterns of bird migration are already changing and will continue to do so. Plants, animals, and insects will migrate toward the poles and toward higher elevations. Many plant species, unable to migrate quickly enough, will go extinct, together with some animal species that depend on them. Water supplies will become overabundant in some regions and scarce in others. Water-borne and insect-borne diseases will increase. Crops will fail, resulting in unpredictable shortages in food supplies. Low-lying regions, including many important coastal cities, will be flooded and many freshwater ecosystems will be disrupted by rising oceans. As much as a third of humanity may be forced to migrate as oceans rise and crop zones shift, causing profound and unpredictable social and economic disruptions.[11]

THE LAND CRISIS

During the past fifty years gains in agricultural productivity have resulted in surplus supplies of food, creating widespread confidence that agriculture might continue to satisfy the world's nutritional needs despite projected increases in population. The dramatic increases in productivity were due to three factors: improved hybrid plants, substantial increases in irrigation, and more extensive use of fertilizers. Present indications, however, suggest that no further gains can be expected in any of these areas.[12] The agricultural boom of the last century is likely to be followed by an agricultural bust in the present century, as both the amount and the productivity of cropland are expected to decline.

Soil erosion is already a global problem, and it will continue to worsen as the century unfolds. Soil erosion is a natural phenomenon, but prior to the advent of intensive agriculture, the process of soil renewal kept pace with soil losses to erosion. The natural process of soil formation is slow, taking on average 500 years to replace one inch of topsoil. Presently, the global rate of soil erosion is 20 to 100 times the rate of soil renewal.[13] About one-third of the original topsoil in the United States has been lost to erosion, and the rate of topsoil loss in agricultural states like Iowa and Missouri is about 35 times the natural renewal rate. The earth as a whole is losing nearly one percent of its topsoil each year. Soil erosion ultimately results in desertification, a decline and eventual loss of agricultural productivity. Each year an area the size of West Virginia goes barren, while another the size of Kansas becomes marginally productive.[14]

Since the 1950s the area of irrigated farmland has grown

dramatically, with the result that about 40 percent of the world's food is now grown on irrigated soil. Irrigation has been an important factor in the boom agriculture of the past half-century, but there are now clear signs of diminishing returns in the future. For one thing, the supply of water for irrigation will decrease sharply in the present century. But equally worrisome is the problem of salinization. Irrigation water contains various quantities of salts that accumulate in topsoil, eventually rendering the soil unfit for farming. Salinization is already reducing yields on a quarter of the world's irrigated cropland, and this percentage is certain to increase steadily in the years ahead.[15]

In addition to the significant amount of cropland that will be lost to erosion and desertification, we can expect the trend of converting cropland to nonagricultural uses to continue. Presently the United States loses more than 5,000 acres of cropland and 1,500 acres of pastureland per day to nonfarming uses. European countries are losing about one percent of their farmland to nonfarming uses every four or five years.[16] The chief culprits are urban sprawl and industrialization, but the transportation industry takes a share as well. In addition to these nonfarming uses, we may expect to see an increase in energy farming—growing crops for the production of ethanol instead of food.

THE WATER CRISIS

Most of the earth's surface is covered by water, yet a miniscule 0.003 percent is readily available for human use, and this amount is very unevenly distributed.[17] Fresh water is available to humans in lakes, rivers, rainfall, and underground aquifers. In nature, it is collected, purified, and distributed through the hydrological cycle, a slow and continuous process of recycling. The hydrological cycle is capable of providing ample amounts of fresh, clean water for human consumption, so long as the system is not overstressed.

Unfortunately, we are depleting and degrading the earth's fresh water supply much faster than it can be replenished by natural systems. Presently, a quarter of the earth's human population lacks access to safe supplies of drinking water. Many people in developing countries are forced to walk several miles each day to secure small amounts of (often contaminated) water. Each year five million people die from preventable water-borne diseases.[18] By the year 2020, a third of the world's population could be faced with acute water shortages.[19]

The depletion of underground aquifers has resulted in rapidly falling water tables on every continent. The areas most heavily stressed by water shortages are northern Africa, most of India, central and northern China, the Middle East, Mexico, and much of the southwestern United States. Water tables in the breadbasket of India are dropping more than three feet per year. Forty percent of China's grain crop—the world's largest—grows in the north China plain, where water tables are dropping more than five feet per year.[20] Libya is drawing down fossil aquifers (which cannot be replenished) at a rate that is likely to dry them up within fifty years.[21] Texas, New Mexico, Oklahoma, and Colorado are drawing down the level of the vast Ogallala aquifer at one hundred times the natural recharge rate. Some of the wells in these states are now more than a mile deep. By 2020, a quarter of the Ogallala's original supply of water will be gone.[22]

The world's rivers are being drawn down at an alarming rate as well. The Colorado River is so heavily siphoned by Arizona, Nevada, and California that it dries up before reaching the sea. The same is true of the Amu Dar'ya in central Asia. The Yellow River in China now runs dry for part of each year. The mighty Nile is reduced to a trickle by the time it reaches the Mediterranean Sea, and the Ganges barely reaches the Bay of Bengal during the dry season.[23] Worldwide water shortages are already critical as a result of excessive drawdowns, and the problems are certain to be intensified in the current century by the effects of global warming and increasing population.

The water crisis might disappear quickly if we could only figure out an effective method of desalinating seawater. There are presently about 7,500 desalination plants in operation, mostly in the Middle East and North Africa, but these together supply less than 0.1 percent of global water consumption.[24] The problems with desalination are that it requires huge amounts of energy and generates harmful wastes. To the extent that desalination might relieve the water crisis, it would greatly intensify other environmental problems.

Globally, the major consumers of water are agriculture (70 percent), industry (20 percent), and municipalities (10 percent). As populations grow, especially in cities, competition between these sectors for dwindling water supplies will become intense. Urbanization in China is already pulling supplies away from the agricultural sector. The industrial sector in Indonesia illegally draws water directly from irrigation canals, and some factories have purchased (then abandoned) rice fields just for the access to water. In the southwestern United States, cities are buying water, water rights, and water-endowed land as supplies grow scarce. "As supplies tighten," writes Sandra Postel, "water will shift to where it is more highly valued."[25] The global distribution of water use is rapidly shifting to the

American pattern, where industrial and municipal uses exceed agricultural use. The potential fallout of this trend for future food supplies is not encouraging.

Competition for water supplies will also intensify between nations, and is likely to become a frequent cause of hostilities in this century. More than 200 river basins are now shared by two or more countries. Argentina and Brazil contest each other's claims to the La Plata River; India and Pakistan lock horns over the Indus River; Jordan, Israel, and Syria battle over access to the Jordan River; Egypt, Ethiopia, and Sudan are about to come to blows over access to the Nile; India and Bangladesh dispute over the Ganges; and Turkey's dam projects on the Euphrates River threaten the vital interests of Syria and Iraq. Water is already an important factor in shaping the foreign and domestic policies of many nations, and there can be no doubt that its role will increase as the global water crisis worsens.

THE BIODIVERSITY CRISIS

The extinction of biological species is a natural phenomenon, just as the emergence of new species is. Indeed, extinction and speciation are the ultimate themes of biological evolution. Evolutionary biologists estimate that about 99 percent of the earth's historic life forms have gone extinct. Nobody knows for sure how many species presently inhabit the planet—we have identified about 1.7 million—but estimates range from 3 to 30 million. Most biologists are comfortable with a range of 5 to 10 million extant species.

In some periods of natural history the rate of speciation has exceeded the rate of extinction, and in other periods extinction has exceeded speciation. For the past 600 million years conditions have favored a general increase in the total number of species. During this same period, however, five massive extinction events have occurred, the most recent being 65 million years ago when an asteroid struck the planet, killing off the great dinosaurs.

Biologists use the concept of background extinction to refer to the vaguely normal rate of species demise that results from the ongoing dynamics of natural selection. In other words, when you factor out the effects of global mass extinction events, you get the natural, or background, rate of extinction. The background extinction rate recognized by most biologists is about one extinction per million species per year.[26] Thus, if there are five million extant species, the background rate of extinction would be five species per year. We will appreciate the magnitude of the crisis of biodiversity when we learn that the current rate of species extinction is somewhere between ten and one hundred extinctions *per day*.[27] Even if we work with the highest background extinction estimate (10 per year) and the lowest current extinction estimate (3,650 per year), we are forced to conclude that the current rate of extinction is at least 365 times the natural rate, and very probably much more. By anyone's definition we are presently witnessing a global mass extinction event.

The consensus is that the current mass extinction is ultimately attributable to excessive human impact on natural systems. The proximate causes are destruction and fragmentation of natural habitat (especially due to deforestation), displacement by introduced species, overharvesting, and pollution of air, soil, and water. By these means, humans may be responsible for extinguishing as many as one-half of all species of plants, insects, fish, birds, and mammals by the end of the present century.[28]

This is an ancient problem. The archaeological record shows that wherever humans arrived in new territories, they posed an immediate threat to many nonhuman species. For example, soon after humans arrived in North America 12,000 years ago, more than fifty species of large animals became locally extinct, including the giant ground sloth, camels, elephants, lions, and the saber-toothed tiger. When humans arrived on Madagascar, they promptly extinguished fourteen species of lemurs and many bird species. Upon their arrival in New Zealand humans wiped out reptiles and more than twenty species of birds, including the giant flightless moa, the largest bird ever to evolve.[29] Yet the carnage of our ancestors came nowhere near the devastation to biodiversity caused by humans in the twentieth century. It is not that contemporary humans are more vicious and thoughtless than our ancestors were, we are merely more numerous and efficient.

There are many ways to demonstrate the value of biodiversity, and thus the folly of our past and present interactions with the non-human world. For example, we eat other species; they perform important ecological functions; they provide medicines for us; they entertain us; and so on. But in the end it all comes down to one general point: The biological diversity of the planet constitutes an essential life-support system for human beings. We need lots of other species in order to carry on. The general rule is: The larger and more complex the species (that's us), the more ways there are to show that its survival depends on oodles of smaller and less complex ones. If we recklessly eliminate half the nonhuman species on the planet, then it will be unlikely that very many places on earth could manage to sustain human populations.

THE ENERGY CRISIS

For the vast majority of history humans have relied on renewable energy resources—plants, animals, sun, wind, water—to satisfy their energy needs. It is only in the past century that a majority of energy needs have been met by nonrenewable fossil fuels. The world's first oil well was drilled in 1859, when there were yet few uses for oil. By 1900 the petroleum economy was beginning to take hold, and oil was being extracted at a rate of a few thousand barrels per day. At that time only a few thousand automobiles were in use worldwide. Today, at the peak of the petroleum economy, there are more than a half-billion automobiles in service and we are pumping more than seventy million barrels of oil per day.

Fossil fuel combustion—oil, natural gas, coal—now accounts for 75 percent of global energy use. Industrialized nations rely on fossil fuels for 90 percent of their energy needs. The petroleum economy, however, is doomed because the supply of fossil fuels, especially oil, is diminishing rapidly. If present rates of oil consumption continue, the world's reserves of crude oil will be gone by mid-century, just in time for the bicentennial of the age of petroleum. The United States has already exhausted more than half of its reserves. Even if the wildest dreams for new discoveries come true, the crude oil of the planet will be depleted well before the end of the century.

In one way of thinking, there is no energy crisis. We have at our disposal more energy than we could possibly use, even if demand continues to rise. I mean, of course, the sun. Except for radioactive materials, the sun is the ultimate source of all the earth's energy. The oil, gas, and coal we presently use are themselves derived from fossilized life forms that trapped solar energy eons ago. But these energy forms represent a mere fraction of what is on offer. If we added up the total amount of energy stored in the remaining reserves of fossil fuels, it would amount to less than two weeks' supply of the earth's solar energy.[30] If there is an energy crisis, it has far less to do with the supply of energy than with the form in which we use it. Fossil fuels are cheap, versatile, convenient, and produce a high yield of net useful energy. That's why we use them almost exclusively.

Exhausting the world's remaining supplies of fossil fuels would constitute no serious threat to nature's life-support systems. Quite the opposite—the combustion of these materials is the principal cause of damage to air, climate, soil, water, and biodiversity. If all the fossil fuels disappeared overnight, it would come as a great relief to natural systems.

The same, however, is not true of social systems. The social consequences resulting from a sudden disappearance of fossil fuels would be devastating. The global economy would collapse, the majority of humanity would be rendered jobless, transportation systems would stop dead, food distribution would halt, and the billions who didn't starve or freeze to death would be at each other's throats over diminishing scraps of food and firewood. This scenario sounds far-fetched until one recalls the disruptions and hostilities associated with the minor fuel shortage of 1973.

The real energy crisis is a social and economic one, and it turns on a single question: Can the global economic system manage to reconstruct itself on the foundations of renewable energy resources quickly enough to avert widespread social chaos? A smooth and socially un-disruptive transition from our present petroleum economy to a solar-hydrogen and labor-intensive economy would be costly and time consuming. It would require an enormous amount of investment and planning, and would entail fundamental changes in virtually every industry. And perhaps most important of all, the prospect of such a transition runs directly against the short-term interests of every sector of the economy we can name. The science for the transition may be mostly in place, and the technology for it may already be feasible, but the requisite courage is nowhere in sight.

The foregoing glimpse of global environmental problems is no substitute for serious reading on these issues. It is intended merely to underscore what is already a growing consensus of informed opinion: If deep measures are not taken very soon to reduce human impact on natural systems, then there will be hell to pay. What kind of hell, when and where, we can only guess at. Population and consumption are already overstressing the natural and social systems of the planet, and they continue to increase. Resources are dwindling as competition for them grows more intense. Overlaps of national interests are shifting and thinning. Something, as the bumper sticker says, has to give. We are now—ecologically, socially, politically, economically—at a point where almost anything could happen. Something surely will. In any event, we may fairly ask whether we are in any position to take courageous steps toward solving the problem during the critical first half of this century.

It seems clear that reducing human impact on natural systems is a global concern that calls for a concerted global response. We certainly cannot expect to make any progress unless a majority of governments in the world come to some sort of agreement about what progress would look like. A reduced ecological footprint, for sure, but how do we achieve that? We might attempt to reduce our collective footprint by cutting way back on human consumption and stabilizing global population at something like ten billion by the end of the century. Or we might take measures to reduce the world's population substantially—say, down to two or three billion—and improve material conditions among the most impoverished nations. Or perhaps we should aim for something in between.

Assuming that agreeable target levels for population and consumption can be achieved, how likely is it that nations will comply with them? The record so far is not very impressive. Bearing in mind the fate of the Kyoto Protocol, isn't it likely that nations will pull out of agreements as soon as they become too specific, or whenever the political breezes change direction? How far have we come in establishing the global institutional conditions necessary to enforce mandatory compliance? What good are international agreements when local officials everywhere proceed as if their particular communities were exempt from global problems? The evidence so far is that we will continue to stutter and stall until nature does the dirty work of imposing a correction. Meanwhile, we are left to suffer discouragement when elected officials lack courage and when courageous candidates cannot win votes. If ever there was a time when humanity needed the ministrations of mythic traditions to inspire sacrifices of immediate self-interest, this must be it. But we may fairly ask whether our received traditions possess the will to pursue the adaptive option.

THE MORAL RELEVANCE OF RELIGIOUS TRADITIONS

We know the problem: exceeding the earth's carrying capacity. We know the urgency: red alert, no time to lose. We know the causes of the problem: excessive human population and excessive material consumption. We know the general solution: reduce human impact to sustainable levels of population and consumption. We know on what the general solution depends: changes in social, political, and economic goals and policies. We know the fundamental forces that can drive such changes: values, attitudes, goal hierarchies, self-esteem links. And finally, we know how to manipulate these fundamental forces: a full-court press on behavior mediation systems, utilizing the ancillary strategies of a morally relevant mythic tradition.

And now our question is whether we can expect the world's received traditions to respond to the challenge in a manner that is both timely and effective. Doing so would require that these traditions possess both the influence and the will to effect significant change. In the previous chapter I raised some doubts about the influence factor. Influence, however, can change where there is sufficient commitment to particular outcomes. Where there's a will there's a way. Our question therefore comes down to whether the world's received traditions possess the will to lead a morally relevant response to the crisis.

No one can give a decisive answer to a question like this. The question itself is simple, but the variables involved are too complex to allow anything more than a hunch. My hunch is that an effective and timely response will not be forthcoming, and that we will continue on our course of self-destruction and impending doom. This is a hunch, remember, and not a carefully reasoned forecast based on reliable indicators. Still, anyone who puts hunches out in print should be prepared to support them, or at least to clarify the hunch-formation process. My hunch that the received mythic traditions will fail is based largely on the disappointing record so far.

We have been aware of serious environmental problems for decades, and yet the engagement of religious traditions with these issues has been tardy and tentative. There are promising developments here and there, but these are offset by signs of indifference and outright hostility toward environmental concerns. Given the urgency of the challenge, and given the lack of unity of purpose within and among mythic traditions, there can be little hope that they will make a meaningful difference.

Alarms about environmental degradation were sounded throughout the 1960s, and it began to appear that a vigorous and persistent response was taking shape. When the first Earth Day was celebrated in 1970, environmentalists were confident that lasting systematic changes would be forthcoming, and indeed the environmental legislation of the 1970s appeared to confirm the optimistic notion that human beings were prepared for the sacrifices of living in harmony with nature's limits. Religious voices were neither the earliest nor the loudest in calling for change, but they did start to chime in after Earth Day.

Religion had been put on the defensive by a 1967 article appearing in *Science.* The author, Lynn White Jr., argued that the Judeo-Christian tradition has been uniquely responsible for the ecological crisis because it promoted the view that nature's purpose was to serve humans. He further argued that the environmental crisis would continue to worsen until the Judeo-Christian tradition were either rejected or transformed.[31] Theological responses to White's thesis tended to agree that Judeo-Christian tradition contributed to the crisis, but only because it had been wrongly understood and practiced. Thus, from the early 1970s onward, theologians busied themselves with working out the biblical and theological resources to underpin a starchy environmental ethic.

After decades of reflection the general opinion seems to be that whereas the great mythic traditions of the world have never placed ecological responsibility at the core of piety, they do nevertheless possess sufficient resources to inspire a forceful response to the ecological crisis. This was the overwhelming impression left by a remarkable series of conferences held at Harvard's Center for the Study of World Religions between 1996 and 1998. This series assembled some of the best minds to represent all the major mythic traditions of the world. Conference organizers challenged participants to address the difficult questions:

> It is important to ask where the religions have been on these issues and why they themselves have been so late in their involvement. Have issues of personal salvation superseded all others? Have divine-human relations been primary? Have anthropocentric ethics been all-consuming? Has the material world of nature been devalued by religion? Does the search for otherworldly rewards override commitment to this world? [32]

I had the pleasure of attending several of these conferences and was deeply impressed with the presenters, most of whom were knowledgeable about the crisis at hand, disappointed with the contributions of their own traditions to the crisis, and fiercely committed to transforming their traditions in ways that would enable a timely and effective response to the crisis. The tone of these proceedings was one of cautious yet passionate optimism that mythic traditions can and will rise to the challenge. The Harvard series of conferences on religion and ecology convinced me that the intellectual strategies of our received traditions are virtually in place for an adaptive response. Whether the other ancillary strategies are consilient with the intellectual dimension is not so clear.

If we could be confident that the Harvard conferences told the whole story, then no one could doubt that the received traditions possess a concerted will to lead a radical global transformation. But unfortunately there is much more to the story. While liberal theologians were busy working out the foundations for an environmental ethic, religious conservatives were busy forging alliances with right-wing politics.

During the 1970s a broad coalition of conservative religious organizations began to assert itself in the political domain, determined to restore traditional religious values that many felt were being un-

dermined by the liberal agenda.[33] The new religious right was opposed to abortion rights, gay rights, the Equal Rights Amendment,
pornography, and instruction about sex and evolution in the schools.
Most of all, they feared secular humanism and communism. Environ-
mentalism was lumped in with these no-no's because it was perceived
that the environmental movement was dominated by left-wing liber-
als. Many conservative religious groups therefore lobbied vigorously
against environmental interests, insisting that the ecological crisis
was a fabrication of left-wing ideology.[34]

We can begin to appreciate the odds against achieving a unity
of purpose within just one mythic tradition when we consider how
polarized contemporary Christians are on the fundamental issues of
population and consumption. The Roman Catholic Church, for example, appears to be suffering from a disconnect between its intellectual and institutional dimensions. The Vatican position on birth
control still insists that the purpose of sexual intercourse is the generation of offspring, and that defeating this purpose by artificial means
of birth control amounts to a mortal sin. And this despite the fact that
many Catholic nations in the developing world (where people still listen to the Vatican) are stressed by overpopulation. Notwithstanding
the valiant efforts of many committed Roman Catholic environmentalists (well represented at the Harvard conference on Christianity
and ecology), there are no signs that the Vatican might abandon its
pro-natalist position to espouse zero population growth.

The odds for transforming conservative protestant attitudes toward population are no less discouraging. The pro-family position
adopted almost universally by conservative Christian groups has consistently expressed itself in aggressive opposition to family planning,
sex education, and reproductive choice. Many conservative groups
actively lobby Congress for increased tax incentives for large families.
Environmentalist appeals for population control are likely to be seen
by many conservative Christians as a phony pretext for advocating
abortion rights. In was on these grounds that conservative groups
persuaded the Reagan and Bush administrations to withdraw U.S.
support for United Nations population programs.[35]

Nor should we expect to see a united Christian front to dismantle consumerism and move toward an ecologically sustainable global
economy. There are many well-informed Christian activists and theo-
reticians who abhor consumerism on both spiritual and ecological
grounds.[36] In particular, the National Conference of Catholic Bishops
has consistently held U.S. economic policy to a firm moral standard,
often appearing on the verge of calling for re-embedding economics
within a religious vision.

But against these voices are millions of conservative Christians who agree with Jerry Falwell that God favors capitalism and free enterprise. Christians have never been of one mind about money, some associating wealth with a sinful nature and others associating it with God's blessings. These polar attitudes came to loggerheads in America at the end of the nineteenth century, with the "Gospel of Wealth" advocates espousing the idea that God's blessings can be measured in dollars, and the "Social Gospel" group contrasting wealth and privilege to the Kingdom of God. A similar juxtaposition has emerged in contemporary American Christianity. At one extreme are Christian critics of consumerism and the disembedded economy, and at the other extreme are the glitzy televangelists, especially the phenomenally successful "prosperity gospel" movement.[37]

The prosperity gospel movement has deep roots in the mind-cure groups of the nineteenth century. These groups espoused a cluster of ideas that tended to blur the distinction between entrepreneurial and religious enthusiasm: ". . . it was wish-oriented, optimistic, sunny, the epitome of cheer and self-confidence, and completely lacking in anything resembling a tragic view of life."[38] The mind-cure movement was very instrumental for advancing consumerism in the late nineteenth century (it also inspired feel-good books like *Pollyanna* and *The Wonderful Wizard of Oz*).

The mind-cure movement's contemporary manifestation, the prosperity gospel, has millions of followers and has made significant inroads into mainstream Christian churches. The largest Methodist congregation in America, for example, is led by prosperity gospel preacher Kirbyjon Caldwell (who is also a spiritual advisor to George W. Bush). As a measure of the success of this movement, consider that *The Prayer of Jabez,* a prosperity gospel book, sold one million copies during February 2001 alone![39]

Basically, the prosperity gospel movement teaches that verbal confessions of faith possess the metaphysical power to compel God's blessings. When you are in a state of genuine faith, therefore, "whatever comes out of your mouth shall be produced in your life."[40] In other words, if you ask for riches while in a state of true faith, then God cannot deny your wish. Not that God would want to deny it, for God wants you to be wealthy. As one prosperity gospel preacher assured his congregation, "I have heard God speak, and I can tell you, I have heard the sound of abundance."[41]

One may argue that the prosperity gospel is simply a crude form of magic that appeals to the base material aspirations of mostly poor and credulous people. It is not genuine Christianity at all. But that would miss the point, which is that the prosperity gospel goes hand-in-glove with consumerism and is now a major force in American culture. It is also a serious contender for religious followers in South America and Africa. I mention this movement because it indicates the kind of support that consumerism is able to derive from religious traditions. It also represents a massive and well-funded source of resistance against any attempts to dismantle consumerism and to reconstruct the economy around sustainable principles, goals, and policies.

It is difficult to say why the Judeo-Christian tradition—and other traditions as well—has been so late in addressing environmental concerns, and why, once doing so, it has failed to achieve a unified commitment to respond. One plausible explanation is that the natural world has not been sufficiently valued in the received traditions. Nature has been featured neither as an object of salvation or liberation nor as the ultimate creative condition for human existence. The Harvard conferences made a compelling case that the major traditions are all consistent with a robust environmental ethic, but the stronger claim—that such an ethic is firmly implicit in the myths—has not been persuasive. It is not even persuasive for a majority of devotees to these traditions, which accounts for the lack of consensus regarding environmental concerns.

Our question has been whether the world's received traditions possess the influence and the will to lead a morally relevant response to the global environmental crisis. The potential is clearly present, and events like the Harvard series on religion and ecology offer some encouragement. But in my judgment a timely and effective response is not likely. The requisite changes are too radical, the sacrifices are too deep, the time is too brief, the value of nature is too peripheral, and the forces of resistance internal to these traditions are too great. My remarks to this point have been drawn mostly from circumstances prevailing in the Judeo-Christian tradition, but these circumstances have parallels in other traditions. In any event, it is fairly certain that any global response would fail without cooperation and leadership from the Judeo-Christian world.

So what does this mean? It means, I am afraid, that the process of correcting our ecological footprint deficit will not be one of our choosing. There are no good reasons to believe that we might be moved in sufficient numbers to make the necessary sacrifices before they are forced upon us. Sacrifices are difficult for us, and we need lots of inspiration to make them willingly. So it is not surprising that the pattern of ecological overshoot and collapse has been repeated many times in human history.[42] We are well into the pattern again—this time on a global scale—because we have too long ignored the proverbial warning that unless we change directions, we will end up where we are heading. And where is that? Extinction? Probably not, although the possibility is real enough.

DESCENDING INTO HELL

For anyone who likes to dwell on Doomsday scenarios, there is plenty of material to work with these days. Imagine the prospect of genetically engineered organisms replicating with such violent efficiency that they smother the entire planet with inedible scum within a matter of days. Or imagine that some physicist succeeds in producing "strange-quark matter" that attracts, then consumes, ordinary matter. Or what about a high-energy laboratory experiment that triggers a vacuum capable of expanding at nearly the speed of light, destroying everything in its path? These scenarios are theoretically possible, but fortunately they are too improbable to provoke much more than amusement.[43]

Unfortunately, however, there are several Doomsday scenarios falling well within the realm of plausibility, many of them envisioned by speculators in the new field of environmental surprise. These investigators fret about sudden and devastating effects arising from ecological discontinuities, synergisms, unnoticed trends, and unpredictable feedback loops.[44] Ecological discontinuities are abrupt systematic changes, often triggered when a threshold is reached. The melting of Greenland's ice, for example, would dilute salt water to a threshold point where its buoyancy might effect a sudden change in the circulation pattern of the Gulf Stream. If such a discontinuity in Gulf Stream flow were to occur, then northern Europe could be plunged into a devastating ice age within a few years.[45]

Synergisms involve the interactions of several phenomena to produce new system properties capable of magnifying effects. Two problems, for example, might interact to produce a megaproblem. Or a half-dozen chemical compounds might interact to produce persistent toxins or bioactive agents. Virtually every human body on the planet contains traces of roughly 500 synthetic chemical compounds that were nonexistent before 1920, some of which are hormonally active and capable of disrupting the body's endocrine and immune systems.[46] The ever-increasing number of new compounds released into the environment heightens the potential for unpredictable chemical synergisms.

Positive feedback loops (where effects contribute further to the causes) can magnify and accelerate problems. Some experts fear that positive feedback loops may generate a runaway greenhouse disaster.

As global warming continues, there is a risk that it will increase the carbon dioxide output and decrease the carbon dioxide uptake of natural systems, thereby accelerating the release of carbon dioxide and methane gas into the atmosphere.[47] A runaway greenhouse event could push temperatures on earth to near the boiling point of water.

The fundamental concern of environmental surprise assessors is that as natural systems are brought under increasing stress, it becomes far more likely that sudden and unpredictable events will occur, and that some of these will seriously threaten human survival. The growing risk of infectious diseases is especially worrisome. New diseases are being identified at a rate of about one per year. Some of the newcomers are Lyme disease, the Ebola virus, Legionnaires' disease, the Hanta virus, the SARS virus, and HIV. Also, new drug-resistant strains of tuberculosis and malaria have appeared in recent years. Given the concentration of about half the world's population in big cities, and given the way people and foodstuffs travel around the globe, it is possible that some new disease, or a mutant form of a known one, could wipe out billions of people within a few months. HIV may yet turn out to be the great scourge many thought it would be. There are now about forty million HIV-positive victims worldwide. This virus mutates quickly, and some believe it is merely a matter of time before airborne or mosquito-borne strains appear.

Doomsday might well arrive on the heels of persistent worldwide crop failures that could destabilize the global economy and result in massive starvation. We already witness the effects of sporadic and regional crop failures, but if there occurs a coincidence of failures, or if droughts continue for three or four years running, or if some virulent plant disease whirls out of control, the knock-on effects would be catastrophic for human populations.

But perhaps after all the day of reckoning will be socially induced. The events of September 11, 2001 may have been just the beginning of a terminal Age of Terror. Consider that the United States, with just 6 percent of the world's population, commands a third of the world's energy resources. How long will such discrepancies be tolerated? Say what we will, deep resentment of the industrialized world's exploitative policies continues to spread among the world's poor and dispossessed. For most Americans the question is not whether there will be more terrorist attacks, but when and by what means. Bioterrorism is perhaps the most insidious threat. A polio virus was artificially created in 2002, using a genetic blueprint downloaded from the Internet.[48] How long before amateur biotechnicians, armed with a political agenda, are able to create deadly artificial viruses and release them randomly among unsuspecting victims?

Enough of this! There are no limits on the number of possible Doomsday scenarios one might generate, and there is no point in piling them up here. The important thing is to realize that human life on the planet is highly contingent and vulnerable, and it becomes much more precarious as we exceed the limits of sustainability. It becomes even more precarious as we create extensive webs of economic interdependence among human groups. A globalized economy has the effect of placing too many eggs in one basket—so if one large sector of the global system were to be struck by a major disaster, there would be severe consequences elsewhere in very short order. Human life has always been contingent and vulnerable, of course, but the potential for global disaster has never been greater than it is right now. And this potential will grow. If a major calamity struck today, we might be in pretty good shape to recover before a cascade of devastation let loose. But fifty years from now—when the global population exceeds eight billion, when many countries are hanging by a thread, and when the International Monetary Fund can no longer afford bailouts—then the story might be very different.

The specific details of a Doomsday event are anyone's guess, but it is likely that the general consequences would tend to be similar in most scenarios. Most generally, the underlying conditions for personal wholeness and social coherence would begin to collapse. Supplies of vital resources diminish, and competition for them grows increasingly ugly and dangerous; important industries begin to falter and fold; unemployment skyrockets; homelessness increases; individuals become progressively more fearful, anxious, suspicious, uncooperative, devious, and desperate; inflation soars; gangs coalesce; crime becomes rampant; raids on hoarders are commonplace; vigilante groups organize; public services decline; cities grow unmanageable and squalid; utilities become undependable; riots, looting, and fires ravage whole cities; schools close; nothing gets repaired; water and food become increasingly scarce and putrid; diseases spread; healthcare systems buckle; sickness and death at every hand; armed conflicts flare up in city streets; refugees, scavengers, and shanty towns everywhere; border incidents escalate into minor wars; and the lamps of civilization go dark.

THE PHOENIX WILL RISE

The aftermath of a global collapse would not be a pleasant sight anywhere, but we may expect that the death, destruction, and decivilization will be somewhat uneven. By fifty years from now there will be more than twenty megacities with populations exceeding twenty million, and hundreds of cities above five million. Large urban centers will be the hardest hit. Some areas might come to resemble a moonscape, where humans are locally extinct. In many areas the remnant will be reduced to living hand-to-mouth, with warlords on the prowl.

Other areas will be relatively fortunate, having suffered only moderate damage to life and property. In such places the process of restoring civilization will be much faster. The best places to be in the aftermath of the holocaust will be those communities that had learned beforehand how to live sustainably. The place to look for would be an off-the-grid community where mechanisms for local food production and distribution are well established. From such communities the wisdom and the will for a new order of bioregional sustainability will radiate.

No place, however, would be completely unscathed, and every person left alive would be profoundly affected. The collapse would be the most harrowing, momentous, and central event of everyone's life. Ultimately, such an event would have to be explained, for humans cannot endure without apprehending the meaning of their suffering.

In 586 B.C.E. Jerusalem was destroyed by the Babylonians and a small remnant of survivors were carried off into exile. The Jews were perplexed by this event because they had believed that God would never allow such a humiliating defeat to befall his chosen people. In their grief and bewilderment the Jews received an explanation for their suffering: God was punishing them because they had not been living wisely. Similarly, in the coming dark age the remnant will clamor for ways to make their experience intelligible and to restore the conditions for personal wholeness and social coherence. Explanations offered by the received traditions will not be convincing, for against them will stand the plain fact that the old myths failed to prevent the most horrific event in human history.

The most compelling stories in the aftermath will be those featuring the overshoot and collapse phenomenon—stories that explain our suffering as a consequence of excessive and unsustainable demands on finite natural systems. These explanations will eventually expand to mythic proportions—that is, they will develop a mythic core of ideas about how things ultimately are and which things ultimately matter. These stories will explain that the holocaust happened because human beings had failed to acknowledge and embrace their true status as natural beings. Instead, they had the presumption to believe that Nature's laws and limits were ultimately irrelevant to human destiny. In other words, the holocaust was a direct consequence of devaluing the order of Nature. But Nature, they will explain, is the ultimate source of truth and value and the ultimate context for human fulfillment. To be wise is to live in harmony with Nature.

In the aftermath of global disaster various Nature myths will begin to emerge independently in different parts of the world. These myths will reflect diverse experiences and may to varying degrees incorporate material from received traditions, but they will tend to share a common piety of reverence for Nature, and they will be marked by a deepening sense of how important it is to understand the workings and the limits of natural systems. These two fundamental elements—knowing about Nature and valuing the integrity of natural systems—represent essential first steps leading toward explicit cosmological and moral conceptions.

Mythmakers will be drawn to what remains of science and natural history as principal resources for their explanations of the collapse as well as their claims about how to live wisely in the future. The various Nature myths of the aftermath will therefore have two very important things in common. They will share the holocaust itself as a common point of departure (though it will be variously experienced), and they will be informed in some measure by the common resources of science, especially ecology and natural history. In other words, aftermath mythmakers will be struggling to appraise the meaning of the same global event, and their appraisals will tend to be influenced by the same general notions about cosmology and morality.

To my mind, these would be extremely significant commonalities having potential for generating a relatively unified global wisdom tradition. In the short term we may expect to endure much death and devastation resulting from our failure to respond to the environmental challenges at hand. But there is reason to hope that in the aftermath we will assemble the conditions for an optimal long-term future for humankind. From the ashes of global collapse we may expect to see a phoenix arise in the form of a new Nature-centered meta-myth: the Myth of Religious Naturalism.

Prophets of the myth of religious naturalism have already started to appear, but it is still too early to identify a coherent and energized movement, let alone an established mythic tradition. And it is unfortunately too late in the day to hope that the myth of religious naturalism will take shape and substance in time to prevent the coming dark age. Nevertheless, if there were to be an eleventh-hour flourishing of religious naturalism—that is, if the story is widely told, and ancillary strategies begin to develop in the coming decades—then so much the better for the remnant.

Naturalism is the view that nothing transcends Nature—the real is natural and the natural is real. The phenomena of Nature—entities, events, properties, relations—can be rendered intelligible only by reference to the states of natural systems. References to extra-natural realities have no currency to explain the phenomena arising from natural systems. Naturalists seek to justify their stance on grounds that humans are wholly embedded within Nature and therefore have no access to extra-natural realities, which naturalism regards as oxymoronic monstrosities. One violates the principle of naturalism by positing transcendent realities that stand juxtaposed to Nature. Familiar dualisms, such as nature-culture, nature-history, nature-spirit, or nature-God, are therefore rejected by naturalists. Post-holocaust myths are likely to condemn such dualistic thinking as the fatal flaw that allowed humans to abstract themselves from Nature and ignore the signs of environmental degradation.

If this is naturalism, then what is *religious* naturalism? Presumably, religious naturalism is centered on an integrated vision of cosmology and morality. The cosmology of religious naturalism is the evolutionary cosmology of contemporary science, a story about the natural history of the cosmos, the creation and emergence of the universe from the big bang to the environmental crisis and beyond.

A post-holocaust mythic vision would seek to integrate this cosmology with an eco-centric morality, the imperative to sustain human life on the planet by addressing needs for personal wholeness and social coherence within the limits of natural systems. An eco-centric morality judges policies and actions relative to their environmental impact. It treats the integrity of natural systems as an absolute value, implied by the principle that any vision of the good life presupposes life, and that life presupposes the integrity of natural systems. Personal wholeness and social coherence, by which we judge human fulfillment, are themselves ultimately contingent on the integrity of natural systems. The integrity of natural systems may appear to be a minimalist moral standard, but in fact its implications are extensive. It insists, for example, that economic activity must be rigorously held to account by a moral vision of sustainable ecology. Religious naturalism cannot tolerate a disembedded economy.

All of this sounds plausible enough until one asks about the root metaphor that would make evolutionary cosmology and eco-centric morality commensurate. As we have seen repeatedly, mythic visions derive their power from a root metaphor, an integrating concept (such as God as person, or Dharma) that provides ultimate explanations for all facts as well as ultimate justifications for all values. But it is far from clear what metaphoric device, if any, might explain the facts of evolutionary cosmology and justify the values of eco-centric morality without violating the principle of naturalism. The personal metaphor of God would both explain and justify, of course, but it would compromise the principle of naturalism by introducing an extra-natural reality.

The problem of specifying a root metaphor for religious naturalism has to do with specifying a goal or purpose (telos) for the natural order. Does the universe have an agenda? Is there some good for Nature, such that we might know whether we are living in harmony with it? Eco-centric morality appears to assume such a good, namely, unperturbed natural systems. But where is the warrant for such an assumption? Indeed, one may ask whether it is ever possible to live out of harmony with Nature. What makes eco-centric morality more in keeping with Nature than egocentric morality? If we could answer this question, then we might have enough insight for a root metaphor.

Perhaps we could argue that eco-centric morality is more conducive to sustaining human life in the long term. But the assumption fueling this argument is that sustaining human life is in keeping with Nature. Again, what warrants this assumption? Can we seriously claim that human extinction would be out of keeping with Nature? Such a claim might be difficult to maintain in light of the fact that 99 percent of all historic species are already extinct. Extinction (even ours) is surely in keeping with Nature.

It appears that the meta-myth of religious naturalism suffers from the problem of a missing metaphor. In order to make evolutionary cosmology fully commensurate with eco-centric morality, there must be a root metaphor to close the gap between facts and values, some device to warrant the assertion of natural values without introducing extra-natural realities. Without this core integration of cosmology and morality, religious naturalism will itself bear the marks of an oxymoronic monstrosity.

I am not convinced, however, that the missing metaphor constitutes a serious problem for a meta-myth. Myths without root metaphors are ultimately incoherent, but a meta-myth may be significantly different on just this point. I am inclined to think that the missing metaphor is not a problem for the reason that nobody ever really practices a meta-myth. It may be enough that particular versions of religious naturalism are prepared to install their own conceptual integrations of reality and value. Perhaps the only critical point for religious naturalism is that evolutionary cosmology becomes integrated with eco-centric morality by means of some conceptual device. Whatever metaphors do the trick are fine, so long as they don't compromise the principles of naturalism. Religious naturalists must remain open to a range of options for resolving the problem of the missing metaphor.

Resources are already at hand for warranting the assertion of natural values without introducing transcendent realities. Within the received traditions themselves there are intimations of religious naturalism, especially among people who are well versed in the complexities of contemporary science. One of the profound lessons of contemporary science has been that matter is astonishingly creative stuff. There was a time in the early history of the universe when the only possible scientific discipline was physics. But as the universe expanded, genuinely new realities—molecules, life, mind—emerged spontaneously. The more we learn about the details of natural processes, the more evident it becomes that these processes are themselves creative. Nothing transcends Nature like Nature itself.

This important insight is reflected in various contemporary theologies that emphasize the immanence of God. Gordon Kaufman, for example, leaves behind the personal metaphor, preferring to speak of a deeply mysterious and serendipitous creativity manifesting itself in cosmic-historical trajectories.[49] Similarly, Holmes Rolston speaks of the cosmic evolutionary process in terms of struggle, endurance, and achievement. Nature appears to be informed by internal, systemic values as it struggles through to something ever higher.[50]

Some strains of process theology assert that all natural phenomena arise from elementary moments of experience and valuation. Thus, without natural values there can be no Nature. God is viewed as an essential part of the natural process, and to this extent God is a natural entity, not a supernatural one.[51] Another theologian, Sallie McFague, speaks of the earth as "God's body."[52] And another, Rosemary Ruether, speaks of Gaia, the living and sacred earth, and God, the transcendent deity, as deeply interrelated realities.[53] The most interesting feature of these theologies is that they narrow the distinction between Nature and God. Talk about nature and talk about God are very close to being the same thing. God is becoming naturalized.

The gap between Nature and God is being narrowed from the other direction as well—that is, Nature is becoming divinized. Brian Swimme and Thomas Berry insist that the earth is holy, and that Nature is the primary revelation of the divine.[54] Ursula Goodenough, a prominent cell biologist, writes with passion about the sacred depths of Nature.[55] Margaret Atwood reinforces the point beautifully: "God is not the voice in the whirlwind. God *is* the whirlwind."[56] And after centuries of obscurity under the pall of heresy, pantheism is staging a comeback:

> Its central tenet is that the Universe is the ultimate reality and the highest object of human reverence, while nature and each one of us is a sacred part of the totality. Our naturalistic approach accepts and reveres the universe and nature just as they are, and promotes an ethic of respect for animal and human rights and for lifestyles that sustain rather than destroy the environment.[57]

These intimations of religious naturalism make the integration of evolutionary cosmology and eco-centric morality appear less problematic. As the gap between the natural and the sacred narrows—as God is naturalized and Nature is divinized—the problem of the missing metaphor begins to fade away and the central core of religious naturalism becomes clear: *Nature is the sacred object of humanity's ultimate concern.* Nature is the ultimate ground of natural facts, and eco-centric values are justified by the claim that Nature is sacred.

The values inherent in Nature are obscure (as are the laws of Nature), and our apprehension of them will always be heavily conditioned by our biology and culture. It will always be too much to claim that we know Nature's agenda, just as it has always been too much to claim knowledge of God's will. But if we attend carefully to the feedback, we will see clearly enough that Nature does not tolerate everything. Some human values are not natural values, some historical trajectories are not in harmony with Nature's tolerance for human life. It is for Nature to determine which human values will be consistent with our carrying on, just as Nature determined which values were consistent with our coming to be. The fact of human survival presupposes a narrow range of values, and it is these values, these natural commandments, this Dharma, that will be the ultimate concern of religious naturalists.

Religious naturalism is already in the air, but it is not yet a robust mythic tradition because the ancillary strategies are not in place to exert a full-court press on behavior mediation systems. We may see some movement in this direction during the coming decades, but it is unlikely that religious naturalism will become a dominant influence until the events of history render alternative mythic visions irrelevant and unpalatable.

One becomes a religious naturalist in the same way that one becomes a pious theist: by the process of goal hierarchy transformation.

Theists are known by their acceptance of the reality of God, but *pious* theists are known by whether their goal hierarchies conform to God's will. Likewise, naturalists universally accept that the real is natural and the natural is real, but *religious* naturalists will be known by their personal responses to Nature. It will be the work of ancillary strategies to instill a pattern of eco-centric piety by shaping attitudes and educating the emotions. Religious naturalists will then be known by their reverence and awe before Nature, their love for Nature and natural forms, their sympathy for all living things, their guilt for enlarging ecological footprints, their pride in reducing them, their sense of gratitude directed toward the matrix of life, their contempt for those who abstract themselves from natural values, and their solidarity with those who link their self-esteem to sustainable living.

IN THE END, IRONY

Occasionally, in reflective moments, humans find themselves rising above the nitty-gritty of everyday life to ask what it's all about—life, the universe, the whole shebang. Suddenly, everything there is becomes the business at hand that cries out to be appraised for its meaning. There will always be a significant fraction among us who will come away from their reflections agreeing with Steven Weinberg's famous comment that the more deeply we understand Nature, the more clear it becomes that Nature is pointless. There will also be among us those for whom Nature remains unintelligible apart from the designs of a transcendent God. Nihilists will always be among us because there are no objective means by which to discern natural values, and theists will always be among us because naturalism can never explain Nature.

Religious naturalists will come down at various points between the extremes of hardcore naturalism and theistic supernaturalism, but they will reject the extremes strenuously because they will recognize that nihilism and theism are both consistent with devaluing Nature. This will leave religious naturalists open to critics from both sides—they will be seen as too dogmatic and credulous for the nihilists and too skeptical and reductionistic for the theists. But this will not be a serious problem, for in the aftermath of global disaster they will have the ultimate defense that both nihilism and supernaturalism are potentially, as well as historically, maladaptive.

In the end, however, there may be a fair amount of irony all around. Nihilists may reject the certitude of values, yet they cannot possibly live a value-free existence. Theists will insist that religion is about a transcendent God, yet the God worshipped and served is always incarnate in natural forms. And religious naturalists may affirm the sacredness of Nature and practice eco-centric piety sincerely, yet deep down they must know that religion is no more about Nature than it is about God.

NOTES

Introduction

1. Daniel Pals, *Seven Theories of Religion* (New York: Oxford University Press, 1996), 278.

2. Pals (esp. chapters 7 and 8) has an excellent discussion of the particularist position, which he associates with the influential anthropologist Clifford Geertz and his followers.

3. Mircea Eliade, *The Sacred and the Profane* (New York: Harcourt, Brace and World, 1957).

4. Edward O. Wilson, *Consilience: The Unity of Knowledge* (New York: Alfred A. Knopf, 1998).

Chapter One: The Epic of Evolution

1. Eric Schulman, "The History of the Universe in 200 Words or Less," http://www.cv.nrao.edu/~eschulma/histcom.html [March 4, 2004].

Chapter Two: The Evolution of Behavior

1. Jean-Pierre Changeux, *Neuronal Man* (New York: Oxford University Press, 1985), 28.

2. Leda Cosmides and John Tooby, "From Evolution to Adaptations to Behavior," in *Biological Perspectives on Motivated Activities,* ed. Roderick Wong (Norwood, N.J.: Ablex Publishing, 1995), 46.

3. Steven Pinker, *How the Mind Works* (New York: W.W. Norton & Co., 1997), 30.

4. Daniel Schacter, *Searching For Memory* (New York: Basic Books, 1996), 57.

5. Richard F. Thompson, *The Brain: A Neuroscience Primer,* 2d ed. (New York: W. H. Freeman & Co., 2000), 384.

6. Jerome Kagan, *The Nature of the Child* (New York: Basic Books, 1984), 185ff.

7. Pinker, *How,* 143.

8. Wilson, *Consilience,* 164.

9. Much of what I will have to say about the emotions in this chapter and the next is based on the work of Richard Lazarus. See especially *Emotions and Adaptation* (New York: Oxford University Press, 1991).

10. Joseph LeDoux, *The Emotional Brain* (New York: Simon and Schuster, 1996), 172.

11. Ibid.

12. Antonio Damasio, *Descartes' Error* (New York: Grosset/Putnam, 1994), 70.

13. Kagan, *Nature,* 172–184.

14. Paul Churchland, *Matter and Consciousness* (Cambridge, Mass.: MIT Press, 1988), 122.

15. See Frans deWaal, *Chimpanzee Politics* (New York: Harper and Row, 1982) for a detailed study of the emotionally influenced social system of chimpanzees.

16. Loyal Rue, *By the Grace of Guile* (New York: Oxford University Press, 1994), 125.

17. LeDoux, *Emotional,* 301.

18. Damasio, *Descartes,* 197.

19. Changeux, *Neuronal,* 133.

20. Schacter, *Searching,* 71.

21. Changeux, *Neuronal,* 133.

22. Pinker, *How,* 301.

23. Ibid., 144.

24. LeDoux, *Emotional,* 274.

25. Ibid., 277.

26. For a review of current models of working memory, see Karl Haberlandt, *Cognitive Psychology,* 2d ed. (Boston: Allyn and Bacon, 1997), chapter 8.

27. Richard A. Carlson, *Experienced Cognition* (Mahwah, N.J.: Lawrence Erlbaum Associates, 1997), 154.

28. Howard Kaplan, *Self Attitudes and Deviant Behavior* (Pacific Palisades, Calif.: Goodyear, 1975), 10.

29. See Terrence Deacon, *The Symbolic Species* (New York: W. W. Norton & Co., 1997) for a defense of early dating, and Jared Diamond, *The Third Chimpanzee* (New York: Harper Perennial, 1992) for a defense of later dating.

30. Deacon, *Symbolic,* 251.

31. Robert Seyfarth, Dorothy Cheney, and Peter Marler, "Monkey Responses to Three Different Alarm Calls," *Science* 210 (1980): 801–803.

32. Richard Dawkins, *The Selfish Gene* (New York: Oxford University Press, 1976).

Chapter Three: The Education of Emotion

1. LeDoux, *Emotional,* 24.

2. Ibid., 25.

3. Richard Lazarus, *Emotion and Adaptation* (New York: Oxford University Press, 1991), 7.

4. See the following for discussion of the history and status of the debate: R. W. Levenson, "Emotion in the Autonomic Nervous System: A Prospectus for Research on Autonomic Specificity," in *Social Psychophysiology and Emotion,* ed. H. Wagner (London: Wiley, 1988).

5. Paul Ekman, et al., "Universals and Cultural Differences in the Judgments of Facial Expressions of Emotion," *Journal of Personality and Social Psychology* 53 (1987): 712–717.

6. R. Plutchik, *Emotions: A General Psychoevolutionary Theory* (New York: Harper and Row, 1984), 202.

7. N. H. Frijda, *The Emotions* (Cambridge: Cambridge University Press, 1986), 88.

8. Catherine Lutz, *Unnatural Emotions* (Chicago: University of Chicago Press, 1988).

9. James Averill, "A Constructivist View of Emotion," in *Emotion: Theory, Research and Experience,* vol. 2, ed. R. Plutchik and H. Kellerman (New York: Academic Press, 1980), 326.

10. Lazarus, *Emotion,* 56.

11. Antonio Damasio, *The Feeling of What Happens* (New York: Harcourt Brace & Co., 1999), 71–78.

12. Ibid., 285–287.

13. Lazarus, *Emotion,* 47.

14. Jerome Kagan, *Galen's Prophesy* (New York: Basic Books, 1994), xvii.

15. Ibid., 131–133.

16. Ibid., 76.

17. Richard Petty, Duane Wegener, and Leandre Fabrigar, "Attitudes and Attitude Change," in *Annual Review of Psychology,* vol. 48 (Palo Alto, Calif: Annual Reviews, 1997), 610.

18. Barbara Rogoff, *Apprenticeship in Thinking: Cognitive Development in Social Context* (New York: Oxford University Press, 1990).

19. Lauren B. Adamson and Duncan McArthur, "Joint Attention, Affect, and Culture," in *Joint Attention: Its Origins and Role in Development,* ed. C. Moore and P. Duhman (Hillsdale, N.J.: Erlbaum, 1995).

20. Damasio, *Feeling,* 136.

21. Paul Taylor, *Respect for Nature* (Princeton, N.J.: Princeton University Press, 1986), 246.

22. Damasio, *Feeling,* 194.

23. I find myself in agreement with Richard Lazarus that happiness cannot be fruitfully pursued as a goal in its own right, but results from achieving other goals. Emotions are goal relevant, but are not themselves goals. Whenever we hear someone profess to be in pursuit of happiness, it is well understood that they are focused on specific goals offering the means to this elusive and abstract end.

24. Carroll E. Izard, "Organizational and Motivational Functions of Discrete Emotions," in *Handbook of Emotions,* ed. Michael Lewis and Jeannette M. Haviland (New York: Guildford Press, 1993), 634.

25. LeDoux, *Emotional,* 21.

26. Damasio, *Descartes,* 149–150.

27. Lazarus, *Emotional,* 159.

28. Randolph Cornelius, *The Science of Emotion* (Upper Saddle River, N.J.: Prentice Hall, 1996), 51.

29. Lazarus, *Emotional,* 112.

30. See Milalyi Csikszentmilalyi, *Flow: The Psychology of Optimal Experience* (New York: HarperCollins, 1991) for a discussion of moods and mood altering activities.

31. Kagan, *Galen,* 271.

32. William James, *The Principles of Psychology,* vol. 2 (New York: Holt, 1890), 103.

33. Cited in Stanley Coren, Clare Proac, and Lawrence Ward, *Sensation and Perception* (New York: Academic Press, 1979), 416.

34. Colin Turnbull, "Some Observations Regarding the Experiences and Behavior of the Bambuti Pigmies," *American Journal of Psychology* 74 (1961): 304–308.

35. Coren, et al., *Sensation,* 400.

36. Ibid., 418.

37. Abraham Maslow, *Motivation and Personality* (New York: Harper and Row, 1970).

38. Lazarus, *Emotion,* 359.

39. My support for this view is tentative, for I suspect that differences in temperament and lifestage might be shown to weaken claims for the predominance of either a constitutive or a regulative role. Those who theorize about moral and emotional development should seriously consider the possibility that cultural influences are primarily regulative but become more constitutive with maturity. We are still a long way from a deep understanding of these matters.

40. Romans 8:19. All quotations from the Bible are taken from *The New Oxford Annotated Bible* (New York: Oxford University Press, 1991).

41. Carol Stearns, "Sadness," in *Handbook of Emotions,* ed. Michael Lewis and Jeannette Haviland (New York: Guildford Press, 1993), 547–561.

42. Robert C. Solomon, quoted in Keith Oatley, "Social Construction in Emotions," in *Handbook of Emotions,* ed. Michael Lewis and Jeannette Haviland (New York: Guildford Press, 1993), 342.

43. Stearns, "Sadness," 550.

44. Richard Shweder, "The Cultural Psychology of the Emotions," in *Handbook of Emotions,* ed. Michael Lewis and Jeannette Haviland (New York: Guildford Press, 1993), 426–427.

Chapter Four: The Nature of Religion

1. J. L. Saver and J. Rabin, "Neural Substrates of Religious Experience," *Journal of Psychiatry* 9, no. 3 (Summer 1997): 498–510.

2. Isaiah 44:9–20; Jeremiah 10:3–16.

3. John Calvin, *Institutes of the Christian Religion,* trans. Ford Lewis Battles (London: SCM, 1960), 554.

4. John Hick, *God and the Universe of Faiths* (London: Macmillan, 1973).

5. Steven Mithen, *The Prehistory of the Mind* (London: Thames and Hudson, 1996), 174ff.

6. Stewert Guthrie, *Faces in the Clouds* (New York: Oxford University Press, 1993), 21.

7. Sigmund Freud, *The Future of an Illusion* (Garden City, N.J.: Anchor Books, 1964).

8. Emile Durkheim, *The Elementary Forms of the Religious Life* (New York: The Free Press, 1965).

9. Guthrie, *Faces,* 52–53.

10. Mithen, *Prehistory,* 178.

11. Gerhard Lenski and Jean Lenski, *Human Societies,* 4th ed. (New York: McGraw-Hill, 1982), 113–114.

12. H. V. Vallois, "The Social Life of Early Man: The Evidence From Skeletons," in *Social Life of Early Man,* ed. Sherwood Washburn (Chicago: Aldine, 1960), 214–235.

13. Peter J. Wilson, *The Domestication of the Human Species* (New Haven: Yale University Press, 1988), 40.

14. There has been much interest lately in the topic of evolution and commitment. A person holding a subjective commitment to some value X may be counted on to make personal sacrifices in the service of X. The capacities for making and as-

sessing such commitments are prosocial traits having strong adaptive potential. These capacities may well be the mechanisms underlying the phenomenon of self-esteem linkage. See *Evolution and the Capacity for Commitment,* ed. Randolph Nesse (New York: Russell Sage Press, 2001).

Chapter Five: Judaism

1. Hershel Shanks, ed., *Ancient Israel* (Englewood Cliffs, N.J.: Prentice-Hall, 1988), 73–74.

2. Albert Camus, *The Plague* (New York: Random House, 1946), 66.

3. Jacob Neusner, *The Way of Torah: An Introduction to Judaism* (Belmont, Calif.: Dickenson Publishing Company, 1970), 32.

4. Moses Maimonides, *Guide to the Perplexed,* trans. Shlomo Pines (Chicago: University of Chicago Press, 1963), 1.53.

5. From the "ethical will" of Eleazar of Mainz; quoted in Neusner, *The Way,* 51.

6. See Raphael Patai, *The Jewish Mind* (New York: Charles Scribner's Sons, 1977), pages 134–151, for an account of tantalizing parallels between Kabbalah and the mystical traditions of Hinduism and Islam.

7. Quoted in Philip Birnbaum, *Encyclopedia of Jewish Concepts* (New York: Hebrew Publishing Company, 1979), 234.

8. It is well-known that memory systems are biased according to moods (for example, depressed subjects are biased to recall negative information). See Susan Mineka and Kathleen Nugent, "Mood-Congruent Memory Biases in Anxiety and Depression," in *Memory Distortion,* ed. Daniel Schacter (Cambridge, Mass.: Harvard University Press, 1995), 178.

9. Patai, *Jewish Mind,* 363.

10. Peregrine Horden, *Music as Medicine: The History of Music Therapy Since Antiquity* (Burlington, Vt.: Ashgate, 2000).

11. Kathleen Higgins, "Musical Idiosyncrasy and Perspectival Listening," in *Music and Meaning,* ed. Jenefer Robinson (Ithaca, N.Y.: Cornell University Press, 1997), 99.

12. Deryck Cooke, *The Language of Music* (New York: Oxford University Press, 1959).

13. Irene Heskes, *Passport to Jewish Music* (Westport, Conn.: Greenwood Press, 1994), 37.

14. Ibid., 58–59.

15. Ibid., 18–19.

16. Quoted in Gila Flam, *Singing for Survival* (Urbana: University of Illinois Press, 1992), 53.

17. Ibid., 154–155.

18. Neusner, *The Way,* 32.

19. Abraham Heschel, *The Sabbath: Its Meaning for Modern Man* (New York: Farrar, Strauss and Young, 1951), 10.

20. Maimonides, quoted in Birnbaum, *Encyclopedia,* 560.

21. From the Jewish Prayerbook, quoted in Neusner, *The Way,* 31.

22. I speak from experience on this matter, for I have a comparable document from my own father's hand. Nothing I have read has ever moved me so deeply as his expressions of hope for his children's moral and spiritual development.

23. From the *Testaments of the Twelve Patriarchs,* quoted in Birnbaum, *Encyclopedia,* 523–524.

24. Daniel Maguire, *The Moral Core of Judaism and Christianity* (Minneapolis: Fortress Press, 1993), chapter 7.

Chapter Six: Christianity

1. Norman Perrin, *The New Testament: An Introduction* (New York: Harcourt Brace Javanovich, 1974), 66.

2. Henry Bettenson, ed., *Documents of the Christian Church* (New York: Oxford University Press, 1967), 26.

3. Rom. 5–7.

4. Rodney Stark, *The Rise of Christianity* (San Francisco: Harper, 1996), 6–7.

5. Acts 14:15–17; 17:22–31; Romans 1:18–32.

6. William James, *The Varieties of Religious Experience* (New York: Longmans, 1902).

7. Quoted in John P. Kildahl, "The Personalities of Sudden Religious Converts," in *Current Perspectives in the Psychology of Religion*, ed. H. Newton Malony (Grand Rapids, Mich.: Wm. B. Eerdmans, 1977), 242.

8. Ibid.

9. Ibid., 238.

10. Margaret Miles, *Image As Insight* (Boston: Beacon Press, 1985), 59–62.

11. G. Chastain, P. S. Seibert, and F. R. Ferraro, "Mood and Lexical Access of Positive, Negative, and Neutral Words," *Journal of General Psychology* 22 (1995): 137–157. Also A. L. Bouhuys, G. M. Bloem, and T.G.G. Groothuis, "Induction of Depressed and Elated Mood by Music Influences the Perception of Facial Emotional Expressions in Healthy Subjects," *Journal of Affective Disorders* 33 (1995): 215–226.

12. Douglass Seaton, *Ideas and Styles in the Western Musical Tradition* (Mountain View, Calif.: Mayfield, 1991), 7.

13. From "Homily on the First Psalm" by Basil the Great. Excerpted in Oliver Strunk, *Source Readings in Music History*, vol. 2 (New York: W. W. Norton & Co., 1998), 12.

14. Hans Belting, *Likeness and Presence: A History of the Image before the Era of Art* (Chicago: University of Chicago Press, 1994).

15. Ibid., 208–211.

16. Miles, *Image*, 53.

Chapter Seven: Islam

1. Alfred Guillaume, *Islam* (Baltimore: Penguin Books, 1956), 4.

2. Karen Armstrong, *Muhammad: A Biography of the Prophet* (New York: HarperCollins, 1992), 58.

3. F. E. Peters, *Muhammad and the Origins of Islam* (Albany, N.Y.: SUNY Press, 1994), 110–114.

4. Caesar Farah, *Islam: Beliefs and Observances*, 6th ed. (Hauppauge, N.Y.: Barron's Educational Series, 2000), 33.

5. Armstrong, *Muhammad*, 55–57.

6. All quotations from the Qur'an are taken from Abdullah Yusuf Ali, *The Holy Qur'an: Text, Translation and Commentary* (Brentwood, Md.: Amana Corporation, 1989).

7. John Esposito, *Islam: The Straight Path* (New York: Oxford University Press, 1998), 35.

8. Sachiko Murata and William Chittick, *The Vision of Islam* (New York: Paragon House, 1994), 41.

9. Ibid., 42.

10. Ibid., 41.

11. See the Qur'an, 2:43; 5:55; 9:71.

12. Murata and Chittick, *Vision,* 21.

13. Ibid., xxv.

14. Quoted in Murata and Chittick, *Vision,* 178.

15. Bernard Lewis, *The Arabs in History* (New York: Harper & Row, 1966), 131.

16. Quoted in Esposito, *Islam,* 102.

17. Murata and Chittick, *Vision,* 305.

18. See part three of Murata and Chittick, *Vision,* for an extended discussion of the centrality of beauty in Islamic tradition.

19. Alexandre Papadopoulo, *Islam and Muslim Art* (New York: Harry N. Abrams, 1979), 194.

20. Richard Ettinghausen, "The Man-Made Setting," in *Islam and the Arab World,* ed. Bernard Lewis (New York: Alfred A. Knopf, 1976), 68.

Chapter Eight: Hinduism

1. John Keay, *India: A History* (New York: Grove Press, 2000), 16–17.

2. Thomas Hopkins, *The Hindu Religious Tradition* (Belmont, Calif.: Dickenson Publishing, 1971), 28.

3. Ramakrishna Puligandla, *Fundamentals of Indian Philosophy* (Nashville: Abingdon Press, 1975), 205.

4. Sarvepalli Radhakrishnan and Charles Moore, eds., *A Sourcebook in Indian Philosophy* (Princeton N.J.: Princeton University Press, 1957), 178.

5. *The Bhagavad-Gita* 12:24–25, in Radhakrishnan and Moore, *Sourcebook,* 150.

6. *The Bhagavad-Gita* 18:64–66, in Radhakrishnan and Moore, *Sourcebook,* 162.

7. From the *Chandogya Upanishad,* in Robert Hume, trans., *The Thirteen Principal Upanishads* (New York: Oxford University Press, 1971), 233.

8. Surdas, "Sur's Lord," in Kenneth Bryant, *Poems to the Child-God* (Berkeley: University of California Press, 1978), 208.

9. Surdas, "The Butter Thief," in Bryant, *Poems,* 177.

10. In Edward C. Dimock, ed., *In Praise of Krishna: Songs From the Bengali* (Garden City, N.Y.: Anchor Books, 1967), 29. The series of poems presented here were composed by unknown poets of the bhakti movement, which flourished in Bengal in the 16th and 17th centuries.

11. Ibid., 4.

12. Ibid., 58.

13. Edward Dimock, "Doctrine and Practice Among Vaisnavas of Bengal," in *Krishna: Myths, Rites and Attitudes,* ed. Milton Singer (Honolulu: University of Hawaii Press, 1966), 43.

14. Ibid., 42.

15. Ibid., 48.

16. Ibid., 52.

17. Milton Singer, "The Radha-Krishna *Bhajanas* of Madras City," in Singer, *Krishna,* 90–138.

18. S. N. Dasgupta, *Hindu Mysticism* (New York: Frederick Ungar, 1959), 124–126.

19. See A. C. Bhaktivedanta, *The Nectar of Devotion* (New York: The Bhaktivedanta Book Trust, 1970). This lengthy text is a commentary on Rupa Gosvami's lengthy book *Bhakti-Rasamrta-Sindhu,* which is a commentary on the even lengthier *Srimad-Bhagavatam.*

20. Quoted in Bhaktivedanta, *Nectar,* 109.

21. David R. Kinsley, *Hinduism: A Cultural Perspective* (Englewood Cliffs, N.J.: Prentice-Hall, 1982), 64.

22. M. S. Randhawa, *Kangra Paintings of the Gita Govinda* (New Delhi: National Museum, 1963), 54.

23. Ibid., 58.

24. Phillip Zarrilli, *Kathkali Dance-Drama: Where Gods and Demons Come to Play* (New York: Routledge, 2000), 74–78.

25. Balwant Gargi, *Folk Theater of India* (Seattle: University of Washington Press, 1966), 123.

26. Bhaktivedanta, *Nectar,* 139.

27. From the *Srimad-Bhagavatam,* quoted in Bhaktivedanta, *Nectar,* 103.

28. Bhaktivedanta, *Nectar,* xiv.

29. Ibid., 103.

Chapter Nine: Buddhism

1. Richard Gombrich, *Theravada Buddhism* (New York: Routledge, 1988), 50.

2. Stephan Beyer, "The Background to Buddhism," in *Buddhism: A Modern Perspective,* ed. Charles Prebish (University Park, Pa.: The Pennsylvania State University Press, 1975), 8.

3. Gombrich, *Theravada,* 55–59.

4. From *Majjhima-Nikaya,* in *A Buddhist Bible,* ed. Dwight Goddard (Boston: Beacon Press, 1966), 22.

5. Bhikshu Sangharakshita, *The Three Jewels* (Garden City, N.Y.: Anchor Books, 1970), 56.

6. From *Samyutty-Nikaya,* in Henry Clark Warren, trans., *Buddhism in Translations* (New York: Atheneum, 1972), 151.

7. Edward Conze, *Buddhist Thought in India* (Ann Arbor, Mich.: University of Michigan Press, 1970), 96–99.

8. Walpola Rahula, *What the Buddha Taught* (New York: Grove Press, 1974), 54.

9. The Buddha is said to have dwelt in a state of nirvana "with bases" (that is, with sense experiences) for the last forty-five years of his life, and after death he continued in a state of nirvana "without bases." See Richard Robinson and Willard Johnson, *The Buddhist Religion* (Belmont, Calif.: Wadsworth, 1982), 50.

10. Conze, *Buddhist Thought,* 56.

11. Donald Swearer, *The Buddhist World of Southeast Asia* (Albany, N.Y.: SUNY Press, 1995), 6.

12. Richard Gombrich comes close to this assertion with his distinction between "soteriological religion" and "communal religion." *Theravada,* 24–26.

13. From *Milindapanha,* in Warren, *Buddhism,* 215.

14. Charles Hallisey and Ann Hansen, "Narrative, Sub-Ethics, and the Moral Life," *Journal of Religious Ethics* 24, no. 2 (1996): 130.

15. E. B. Cowell, ed., *The Jataka* (New Delhi: Munshiram Manoharlal Publishers, 1990), 60–61.

16. Ibid., 63.

17. Rafe Martin, "Thoughts on the *Jatakas*," in *The Path of Compassion*, ed. Fred Eppsteiner (Berkeley: Parallax Press, 1988), 99.

18. Stephan Beyer, "The Doctrine of Meditation in the Hinayana," in Prebish, *Buddhism*, 138.

19. From *Anguttara-Nikaya*, in Goddard, *Buddhist Bible*, 59.

20. Eugene d'Aquili and Andrew Newberg, "Religious and Mystical States: A Neuropsychological Model," *Zygon: Journal of Religion and Science* 28, no. 2 (1993): 177–200. Also, Andrew Newberg, Eugene d'Aquili, and Vince Rause, *Why Gods Won't Go Away* (New York: Ballantine Books, 2001).

21. d'Aquili and Newberg, "Religious and Mystical States," 189.

22. For a discussion of major doctrinal disputes, see Conze, *Buddhist Thought*, 121–158.

23. Sangharakshita, *Three Jewels*, 143.

24. Edmund Perry and Shanta Ratnayaka, "The Sangha as Refuge," in *The Threefold Refuge in the Theravada Buddhist Tradition*, ed. John Ross Carter (Chambersburg, Pa.: Anima Books, 1982), 55.

25. From *Sigalaka Sutta*, in Swearer, *Buddhist World*, 165.

26. Michael Carrithers, *The Forest Monks of Sri Lanka* (Delhi: Oxford University Press, 1983), 292.

27. Sangharakshita, *Three Jewels*, 232.

28. Ibid.

29. Swearer, *Buddhist World*, 49–51.

30. Mohan Wijayaratna, *Buddhist Monastic Life* (Cambridge: Cambridge University Press, 1990), 126.

31. Swearer, *Buddhist World*, 24.

32. Robert Elinor, *Buddha and Christ: Images of Wholeness* (New York: Weatherhill, 2000), 84.

33. Rajeshwari Ghose, *In the Footsteps of the Buddha* (Hong Kong: University of Hong Kong, 1998), 18.

34. Ibid.

35. Swearer, *Buddhist World*, 26.

36. From Thomas Merton, quoted in Nancy Wilson Ross, *Buddhism: A Way of Life and Thought*, (New York: Vintage Books, 1981), 98.

37. From P. D. Ouspensky, quoted in Ross, *Buddhism*, 100.

Chapter Ten: The Crisis of Influence

1. John Haught, *Science and Religion: From Conflict to Conversation* (New York: Paulist Press, 1995). See especially 16–17; 31–34.

2. Ibid., 16.

3. Stewart Guthrie, *Faces in the Clouds* (New York: Oxford University Press, 1993), 16.

4. Ian Barbour, *Religion and Science: Historical and Contemporary Issues* (San Francisco: HarperSanFrancisco, 1997), chapter 4.

5. Stephen J. Gould, *Rocks of Ages* (New York: Ballantine Publishing Group, 1999).

6. Richard Swinburne, *The Existence of God* (Oxford: Clarendon Press, 1979).

7. Michael Behe, *Darwin's Black Box* (New York: The Free Press, 1996) and Willian Dembski, *The Design Inference* (New York: Cambridge University Press, 1999).

8. Arthur Peacocke, *Theology for a Scientific Age* (Cambridge, Mass.: Basil Blackwell, 1990); Holmes Rolston, *Science and Religion* (New York: Random House, 1987); John Polkinghorne, *The Faith of a Physicist* (Princeton: Princeton University Press, 1994); Philip Hefner, *The Human Factor* (Minneapolis: Fortress Press, 1993).

9. Alfred North Whitehead, *Process and Reality* (New York: Macmillan, 1929).

10. David Hume, *Dialogues Concerning Natural Religion* (New York: Social Science Publishers, 1948).

11. Robert Bellah, "Civil Religion in America," *Daedalus* 96, no. 1 (Winter 1967): 1–21.

12. Philip Quinn and Kevin Meeker, eds., *The Philosophical Challenge of Religious Diversity* (New York: Oxford University Press, 2000), 3.

13. Alvin Plantinga, "Pluralism: A Defense of Religious Exclusivism," in Quinn and Meeker, *Religious Diversity,* 172–192.

14. Plantinga, "Pluralism," 181.

15. Karl Polanyi, *The Great Transformation* (Boston: Beacon Press, 1957).

16. The medieval church imposed "just prices," that is, prices that would not alter the financial status of either party in an exchange of goods. See Doug Brown, *Insatiable is Not Sustainable* (Westport, Conn.: Praeger, 2002), 67.

17. Brown, *Insatiable,* 66.

18. Richard Robbins, *Global Problems and the Culture of Capitalism* (Boston: Allyn and Bacon, 1999), 17.

19. Bellah, "Civil Religion," 4.

20. Pamela Danziger, *Why People Buy Things They Don't Need* (Ithaca N.Y.: Paramount Market Publishing, 2002), 1.

21. Stuart Ewen, *Captains of Consciousness* (New York: McGraw-Hill, 1976), 98.

22. Quoted in Ewen, *Captains,* 98.

23. Ibid., 99.

24. Danziger, *Why People,* 32.

25. Marc Gobé, *Emotional Branding* (New York: Allworth Press, 2001), xxviii–xxxii.

26. Russell Conwell, *Acres of Diamonds* (Old Tappan, N.J.: Spire Books, 1960), 23.

27. Ibid., 33.

28. Sergio Zyman, quoted in Danziger, *Why People,* 13.

29. www.commerce.gov/bios/evans_bio.html [March 25, 2004].

30. Robbins, *Global Problems,* 17.

31. www.uschamber.com [March 3, 2004].

32. Ibid.

33. Ann Satterthwaite, *Going Shopping: Consumer Choices and Community Consequences* (New Haven: Yale University Press, 2001), 142.

34. Gobé, *Emotional Branding,* 162.

35. Ewen, *Captains,* 61.

36. Gobé, *Emotional Branding,* 118.

37. Danziger, *Why People,* 63.

38. Ibid.

39. Ibid., 60.

Chapter Eleven: Doomsday and Beyond

1. Robert Ricklefs, *The Economy of Nature,* 5th ed. (New York: W. H. Freeman and Co., 2001), 125.

2. Bill McKibben, *Maybe One* (New York: Simon and Schuster, 1998), 73.

3. Paul Ehrlich and John Holdren, "Impact of Population Growth," *Science* 171 (March 26, 1971): 1212–1217.

4. Mathis Wackernagel and William Rees, *Our Ecological Footprint: Reducing Human Impact on the Earth* (Gabriola Island, B.C.: New Society Publishers, 1996).

5. "Ecological Footprint Accounts," www.RedefiningProgress.org [March 3, 2004].

6. Tyler G. Miller, *Environmental Science: Sustaining the Earth,* 4th ed. (Belmont, Calif.: Wadsworth, 1993), 183.

7. Hilary French, "You Are What You Breathe," in *The Worldwatch Reader,* ed. Lester R. Brown (New York: W. W. Norton and Co., 1991).

8. Miller, *Environmental Science,* 195.

9. Ibid., 196.

10. Ricklefs, *The Economy,* 511.

11. J. F. Rischard, *High Noon: Twenty Global Problems, Twenty Years to Solve Them* (New York: Basic Books, 2002), 70–74.

12. Lester Brown, "Feeding Nine Billion," in *State of the World 1999,* ed. Lester Brown (New York: W. W. Norton and Co., 1999), 127.

13. Miller, *Environmental Science,* 264.

14. Ibid., 266.

15. Ibid., 273.

16. Daniel Chiras, *Environmental Science,* 3rd ed. (Redwood City, Calif.: Benjamin Cummings, 1993), 150.

17. Miller, *Environmental Science,* 224.

18. Ibid., 231.

19. Rischard, *High Noon,* 83.

20. Brown, "Feeding," 24.

21. Sandra Postel, "Redesigning Irrigated Agriculture," in *State of the World 2000,* ed. Lester Brown (New York: W. W. Norton and Co., 2000), 41.

22. Miller, *Environmental Science,* 236.

23. Brown, "Feeding," 124–125.

24. Miller, *Environmental Science,* 236.

25. Postel, "Redesigning," 45.

26. Edward O. Wilson, *The Diversity of Life* (New York: W. W. Norton and Co., 1992), 280.

27. Some ecologists put the current extinction rate as high as 100,000 per year. See Richard Leakey and Roger Lewin, *The Sixth Extinction* (New York: Doubleday, 1995), 240–241.

28. Rischard, *High Noon,* 77.

29. Ricklefs, *The Economy,* 492.

30. Christopher Flavin and Seth Dunn, "Reinventing the Energy System," in Brown, *State of the World 1999,* 23.

31. Lynn White Jr., "The Historical Roots of Our Ecologic Crisis," *Science* 155 (1967): 1203–1207.

32. Mary Evelyn Tucker and John Grim, *Hinduism and Ecology* (Cambridge, Mass.: Harvard University Press, 2000), xix.

33. Glenn H. Utter and John W. Storey, *The Religious Right: A Reference Handbook* (Santa Barbara, Calif.: ABC-CLIO, Inc., 1995). This book is a useful guide to the personalities and organizations that compose the new religious right.

34. *Environmental Agenda,* John Hagee Ministries, San Antonio, Tex., 1992, videocassette.

35. Ian Barbour, "Scientific and Religious Perspectives on Sustainability," in *Christianity and Ecology,* ed. Dieter T. Hessel and Rosemary R. Ruether (Cambridge, Mass.: Harvard University Press, 2000), 393.

36. See especially John B. Cobb, *Sustainability: Economics, Ecology and Justice* (Maryknoll, N.Y.: Orbis Books, 1992).

37. This movement has associations with the "power of positive thinking" tradition of Norman Vincent Peale and Robert Schuller, and also with the Positive Confession and Word-Faith groups. See http://www.rapidnet.com/~jbeard/bdm/Psychology/posit.htm [March 3, 2004].

38. William Leach, *Land of Desire* (New York: Pantheon Books, 1993), 225.

39. http://www.letusreason.org/BookR5.htm [March 25, 2004].

40. Robert Tilton, quoted at: http://www.rapidnet.com/~jbeard/bdm/Psychology/posit.htm [March 3, 2004].

41. http://www.chron.com/cs/CDA/printstory.hts/religion/2177559 [October 15, 2004].

42. Clive Ponting, *A Green History of the World: The Environment and the Collapse of Great Civilizations* (New York: Penguin, 1991).

43. See Martin Rees, *Our Final Hour* (New York: Basic Books, 2003), and John Leslie, *The End of the World* (New York: Routledge, 1996).

44. Chris Bright, "Anticipating Environmental 'Surprise'," in Brown, *State of the World 2000,* 24–25.

45. Rees, *Our Final Hour,* 111.

46. Anne Platt McGinn, "Phasing Out Persistent Organic Compounds," in Brown, *State of the World 2000,* 80.

47. Leslie, *The End,* 61.

48. Rees, *Our Final Hour,* 55.

49. Gordon Kaufman, *In Face of Mystery: A Constructive Theology* (Cambridge, Mass.: Harvard University Press, 1993).

50. Holmes Rolston, "Does Nature Need To Be Redeemed?" *Zygon* 29, no. 2 (June 1994): 205–29.

51. David Ray Griffin, *Religion and Scientific Naturalism* (Albany, N.Y.: SUNY Press, 2000).

52. Sallie McFague, *The Body of God: An Ecological Theology* (Philadelphia: Fortress Press, 1993).

53. Rosemary Ruether, *Gaia and God* (San Francisco: HarperSanFrancisco, 1992).

54. Thomas Berry and Brian Swimme, *Universe Story* (San Francisco: HarperSanFrancisco, 1994).

55. Ursula Goodenough, *The Sacred Depths of Nature* (New York: Oxford University Press, 1998).

56. Margaret Atwood, quoted at: http://www.pantheism.net/ [March 3, 2004].

57. World Pantheist Movement, quoted at http://www.pantheism.net/ [March 3, 2004].

INDEX

Abgar, (King of Edessa), 218
Abraham, 171, 228–229, 237
action tendencies, 104–109, 174, 219
Adam and Eve, 171, 176, 219
adaptation, 1, 30–33, 39
Advent, 220
advertising, 330–331, 333–334, 337
aesthetic strategies: in Buddhism, 306–308; in consumerism, 337–338; in Christianity, 216–220; in Hinduism, 274–277; in Islam, 246–248; in Judaism, 183–186
affection, 44. *See also* emotions
agape (love), 108–109, 120, 206. *See also* emotions
agriculture, 345–347
Al-Ghazali, 243–245, 247
algorithms: in neural systems, 8, 41, 48, 58, 61
Allah, 225, 228–232, 234–235, 237. *See also* God
al-Lat, 225
Allport, Gordon, 84
al-Uzza, 225
American Gothic (Wood), 337
amygdala, 42, 299
ancillary strategies: in Buddhism, 300–308; in Christianity, 208–222; in consumerism, 334–338; in Hinduism, 267–277; in Islam, 234–248; in Judaism, 179–191. *See also* myth; religion
anger, 81, 95–96, 101–102, 105, 119. *See also* emotions
Anguttara-Nikaya, 298–299, 377n19
animism, 146, 150, 225, 317
antinomialism, 182
anti-reductionist thesis, 13
apocalypticism, 195–198, 200–201, 208
apologetics, 208–210
appraisal process, 47–48, 80–81, 86–

88, 94, 97, 106, 113–117, 162, 171, 282; components of, 87–88; examples of, 100–104
apprenticeship in thinking, 85–86
Arabia, 224
Arabic, 227, 232, 242
Aristotle, 13, 49, 119
Arjuna, 261, 263
Armstrong, Karen, 225n2
art: function of, 137–141; and theology, 139–141, 216. *See also* aesthetic strategies
artisans-producers, 254. *See also* caste system
Aryanization, 253
Aryans, 253–255
asceticism, 280
Asoka (King of India), 261, 280, 305
Atharva Veda, 255
atman (self), 256, 258
atomism, 13
atonement, 200
attitudes, 84–86, 111, 162, 178, 206–207
Atwood, Margaret, 366
autobiographical self. *See* self
Averill, James, 81
awe, 44, 219, 229, 231–232. *See also* emotions; humility
Axial Period, 145

Baalism, 135, 168
background feelings, 83, 109–111, 206
Balarama, 264
baptism, 199, 221
Barbour, Ian, 319–324
Bar Mitzvah, 189
Basil (the Great), 218
Bat Mitzvah, 188–189
Beautiful Names, 231, 246
Bedouin, 118–119, 224, 227

self-concept, 93, 98, 100, 157

self-emancipation, 281. *See also* myth of

self-esteem, 63–66, 114; links, 64–65, 114–115, 162, 191, 223, 249, 309, 315, 334

self-interest: overlaps of, 114–115, 161–162, 165–166, 178, 191, 223, 249–250, 255, 278, 304, 309–311, 315, 338–339

self-monitoring, 63–64, 98, 100, 157

self-understanding: crisis of, 158–159

self-worth, 99, 114, 223

Semon, Richard, 38

Separatism, 243

September 11 (2001), 333, 359

Sermon on the Mount, 205

shahada (confession), 235

shari'ah (law), 239–241

Shavuot, 187

Shaw, George Bernard, 222

Shindler's List, 185

Shroud of Turin, 218

shukr (gratitude), 232

Shulman, Eric, 21–22

Siddhartha Gautama. *See* Buddha

Sierra Club, 331

Sigalaka Sutta, 303

sin, 176, 188, 202–203, 207, 222, 238

Singer, Milton, 270

Singing for Survival (Flam), 185–186

skandas, 286–288, 290

Skeptics, 208

social coherence, 9–10, 75–77, 118, 160–164, 251, 255. *See also* personal wholeness

social constructivism, 81

Social Gospel, 356

social roles, 63; and performance standards, 98, 100, 157

sociality, 9, 158, 165, 311

soil erosion, 345–346

Solomon (King of Israel), 169

Solomon, Robert C., 372n42

spiritual traditions. *See* religious traditions

Stearns, Carol, 118

Stewart's Astor Palace, 337

Stoics, 208

suffering, 280, 282–283, 286, 288, 293. *See also* dukkha

Sufism, 245–246

Sukkot, 187

sunna, 240

supernatural, 12

Surdas, 375nn8–9 (chapter 8)

survival: threats to, 1, 359. *See also* environmental crisis

sustainability, 343, 350, 352, 357, 361, 367

Swimme, Brian, 365

Swinburne, Richard, 323

symbolic reference, 68–70, 72, 149–150

symbolic systems, 19, 66–74

symbols, 15, 66

sympathy, 44; in Buddhism, 295–297; in Christianity, 201–202, 218–219. *See also* emotions

synagogue, 190

synergisms (environmental), 358

taking refuge, 283, 304–305

Talmud, 185, 187

Taylor, Paul, 89

telos (teloi): in behavior mediation systems, 48; biological, 56–57, 60, 63, 75. *See also* personal wholeness; social coherence; goals

temperaments, 83–84, 110–111

temple deities, 271–273, 275

Ten Commandments, 191–192

Ten Precepts, 304

terrorism, 359

tertiary emotions. *See* emotions

theism, 316–317, 321, 367–368

theocracy, 215, 240

theology, 208–210, 216

Theravada Buddhism, 281, 291–295, 298, 301, 303–305, 308; esoteric piety in, 292–293, 297, 299–300, 307; laity in, 293, 302–305; meditation in, 298–300; merit-making in, 293, 297, 303, 305–306, 308; monks in, 293, 302–305; popular piety in, 292–293, 297, 300, 307; salvation-seeking in, 293, 297. *See also* Buddhism

Three Gems, 305

Tilton, Robert, 380n40

ABOUT THE AUTHOR

Loyal Rue is professor of philosophy and professor of religion at Luther College in Decorah, Iowa. In addition to numerous articles and reviews, he has authored several books, including *Amythia: Crisis in the Natural History of Western Culture* (1989), *Everybody's Story: Wising Up to the Epic of Evolution* (1999), and *By the Grace of Guile: The Role of Deception in Natural History and Human Affairs* (1994), which was a New York Times Notable Book of 1994.